Lecture Notes in Computer Science 14531

Founding Editors

Gerhard Goos
Juris Hartmanis

The series Lecture Notes in Computer Science (LNCS), including its subseries Lecture Notes in Artificial Intelligence (LNAI) and Lecture Notes in Bioinformatics (LNBI), has established itself as a medium for the publication of new developments in computer science and information technology research, teaching, and education.

LNCS enjoys close cooperation with the computer science R & D community, the series counts many renowned academics among its volume editors and paper authors, and collaborates with prestigious societies. Its mission is to serve this international community by providing an invaluable service, mainly focused on the publication of conference and workshop proceedings and postproceedings. LNCS commenced publication in 1973.

Bong Jun Choi · Dhananjay Singh ·
Uma Shanker Tiwary · Wan-Young Chung

Editors

Intelligent Human Computer Interaction

15th International Conference, IHCI 2023
Daegu, South Korea, November 8–10, 2023
Revised Selected Papers, Part I

 Springer

Editors
Bong Jun Choi 🅳
Soongsil University
Seoul, Korea (Republic of)

Dhananjay Singh 🅳
Saint Louis University
St. Louis, MO, USA

Uma Shanker Tiwary 🅳
Indian Institute of Information Technology
Allahabad, India

Wan-Young Chung 🅳
Pukyong National University
Busan, Korea (Republic of)

ISSN 0302-9743   ISSN 1611-3349 (electronic)
Lecture Notes in Computer Science
ISBN 978-3-031-53826-1   ISBN 978-3-031-53827-8 (eBook)
https://doi.org/10.1007/978-3-031-53827-8

This Springer imprint is published by the registered company Springer Nature Switzerland AG
The registered company address is: Gewerbestrasse 11, 6330 Cham, Switzerland

Paper in this product is recyclable.

Preface

IHCI is an annual international conference in the field of Human-Computer Interaction, dedicated to exploring research challenges arising from the intricate interaction between machine intelligence and human intelligence. The fifteenth edition of this event focuses on the theme of "AI-Powered IHCI," delving into the dynamic intersection of artificial intelligence and human-computer interaction.

We are pleased to introduce the proceedings of the 15th International Conference on Intelligent Human-Computer Interaction (IHCI 2023), hosted both on-site and online by EXCO Daegu, South Korea, from November 8 to 10, 2023. The event took place in Daegu, South Korea, and featured 4 special sessions, 2 workshops, 2 tutorial sessions, and 2 panel discussions, all aligned with the conference's central theme. Out of 139 submitted papers, the program committee accepted 71 for oral presentation and publication, following the recommendations of at least 2 expert reviewers in each case.

The 15th IHCI conference included keynote and invited talks facilitated by expert session chairs with substantial experience in both industry and academia. It drew more than 200 participants from over 25 countries.

IHCI has become the primary global gathering for academic researchers, graduate students, leading research think tanks, and industry technology developers in the field. We believe that participating in IHCI not only contributes to personal and professional growth but also fosters success in the business realm, ultimately benefiting society as a whole.

Our sincere gratitude goes to all the authors who submitted their work to IHCI 2023. The Microsoft CMT portal conference system played a crucial role during the submission, review, and editing stages, and we extend our thanks to the technical program committee (TPC) and organizing committee for their invaluable efforts in ensuring the conference's success. Finally, we express our appreciation to the speakers, authors, and participants whose contributions made IHCI 2023 a stimulating and productive conference. The continued success of the IHCI conference series relies on their unwavering support in the future.

November 2023

Bong Jun Choi
Dhananjay Singh
Uma Shanker Tiwary
Wan-Young Chung

Organization

General Chairs

Wan-Young Chung   Pukyong National University, South Korea
Uma Shanker Tiwary   Indian Institute of Information Technology,
     Allahabad, India

Advisory Chairs

Ajay Gupta   Western Michigan University, USA
Dae-Ki Kang   Dongseo University, South Korea
Jan Treur   Vrije Universiteit Amsterdam, The Netherlands
Venkatasubramanian Ganesan   National Institute of Mental Health &
     Neurosciences, India

Program Chairs

Madhusudan Singh   Oregon Institute of Technology, USA
Sang-Joong Jung   Dongseo University, South Korea
Jong-Hoon Kim   Kent State University, USA

Technical Program Chairs

David (Bong Jun) Choi   Soongsil University, South Korea
Dhananjay Singh   Saint Louis University, USA
Uma Shanker Tiwary   Indian Institute of Information Technology,
     Allahabad, India
Wan-Young Chung   Pukyong National University, South Korea

Tutorial Chairs

Hanumant Singh Shekhawat   Indian Institute of Technology, Guwahati, India
Ikechi Ukaegbu   Nazarbayev University, Kazakhstan
Annu Sible Prabhakar   University of Cincinnati, USA

Workshop Chairs

Nagamani Molakatala	University of Hyderabad, India
Ajit Kumar	Soongsil University, South Korea
Rajiv Singh	Banasthali Vidyapith, India

Session Chairs

Jong-Hoon Kim	Kent State University, USA
Ikechi Ukaegbu	Nazarbayev University, Kazakhstan
Irish Singh	Oregon Institute of Technology, USA
Annamaria Szakonyi	Saint Louis University, USA
Young Sil Lee	Dongseo University, South Korea
Sang-Joong Lee	Dongseo University, South Korea
Swati Nigam	Banasthali Vidyapith, India
Jan-Willem van't Klooster	University of Twente, The Netherlands
Ajit Kumar	Soongsil University, South Korea
Tatiana Cardona	Saint Louis University, USA
Bong Jun Choi	Soongsil University, South Korea
Manoj Kumar Singh	Banaras Hindu University, India
Aditi Singh	Cleveland State University, USA
Naagmani Molakatala	University of Hyderabad, India
Maria Weber	Saint Louis University, USA
Madhusudan Singh	Oregon Institute of Technology, USA
Saifuddin Mahmud	Bradley University, USA

Industrial Chairs

Mario José Diván	Intel Corporation, USA
Garima Bajpai	Canada DevOps Community of Practice, Canada
Sandeep Pandey	Samsung, India

Publicity Chairs

Zongyang Gong	Vanier College, Canada
Mohd Helmy Bin Abd Wahab	University Tun Hussein Onn Malaysia, Malaysia
Sanjay Singh	Motilal National Institute of Technology Allahabad, India

Local Organizing Chairs

Sang-Joong Jung	Dongseo University, Korea
David (Bong Jun) Choi	Soongsil University, Korea
Jong-Ha Lee	Keimyung University, Korea
Dhananjay Singh	Saint Louis University, USA

Technical Support Chairs

Swati Nigam	Banasthali Vidyapith, India
Irish Singh	Oregon Institute of Technology, USA
Aishvarya Garg	Banasthali Vidyapith, India

Plenary Speakers

Wan-Young Chung	Pukyong National University, South Korea
Uma Shanker Tiwary	Indian Institute of Information Technology, Allahabad, India
Dhananjay Singh	Saint Louis University, USA

Keynote Speakers

Uichin Lee	Korea Advanced Institute of Science and Technology, South Korea
Mary Czerwinski	Microsoft Research Lab – Redmond, USA
Henry Leung	University of Calgary, Canada
KC Santosh	University of South Dakota, USA
Shiho Kim	Yonsei University, South Korea

Invited Speakers

Yashbir Singh	Mayo Clinic, USA
Hyunggu Jung	University of Seoul, Korea
Hyungjoo Song	Soongsil University, Korea
Gaurav Tripathi	Bharath Electronics Limited, India
Jan-Willem van't Klooster	University of Twente, The Netherlands
K. S. Kuppusamy	Pondicherry University, India
Siba K. Udgata	University of Hyderabad, India

Jungyoon Kim   Kent State University, USA
Anshuman Shastri   Banasthali Vidyapith, India
Madhusudan Singh   Oregon Institute of Technology, USA

Tutorial Speakers

Urjaswala Vora   Pennsylvania State University, USA
Shael Brown   McGill University, Canada
Colleen M. Farrelly   Staticlysm LLC, USA
Yashbir Singh   Mayo Clinic, USA

Panel Discussion Speakers

Jong-Hoon Kim   Kent State University, USA
Maria Weber   Saint Louis University, USA
Tatiana Cardona   Saint Louis University, USA
Annamaria Szakonyi   Saint Louis University, USA
Urjaswala Vora   Pennsylvania State University, USA
Wan-Young Chung   Pukyong National University, South Korea
Uma Shanker Tiwary   Indian Institute of Information Technology,
                                  Allahabad, India
Dhananjay Singh   Saint Louis University, USA

Workshop Speakers

Shiho Kim   Yonsei University, South Korea
Ashutosh Mishra   Yonsei University, South Korea
Dimple Malhotra   Imatter Institute of Counselling and Behavioral
                                  Sciences, USA

Technical Program Committee

Abhay Kumar Rai   Central University of Rajasthan, India
Aditi Singh   Cleveland State University, USA
Ahmed Imteaj   Southern Illinois University, USA
Aishvarya Garg   Banasthali Vidyapith, India
Ajit Kumar   Soongsil University, South Korea
Andres Navarro- Newball   Pontificia Universidad Javeriana, Colombia

Ankit Agrawal	ADGITM, India
Anupam Agrawal	IIIT Allahabad, India
Arvind W. Kiwelekar	Dr. Babasaheb Ambedkar Technological University, India
Awadhesh Kumar	Banaras Hindu University, India
Bernardo Nugroho Yahya	HUFS, South Korea
Bhawana Tyagi	Banasthali Vidyapith, India
Wan-Young Chung	Pukyong National University, South Korea
David (Bong Jun) Choi	Soongsil University, South Korea
Dhananjay Singh	Saint Louis University, USA
Irina Dolzhikova	Nazarbayev University, Kazakhstan
Irish Singh	Oregon Institute of Technology, USA
Jeetashree Aparajeeta	VIT Chennai, India
Jong-Hoon Kim	Kent State University, USA
Jungyoon Kim	Kent State University, USA
Kanike Sreenivasulu	DRDO Hyderabad, India
Lei Xu	Kent State University, USA
Manju Khari	JNU, India
Manoj Kumar Singh	Banaras Hindu University, India
Manoj Mishra	Banaras Hindu University, India
Madhusudan Singh	Oregon Institute of Technology, USA
Mohammad Asif	Indian Institute of Information Technology, Allahabad, India
Nagamani Molakatala	University of Hyderabad, India
Nidhi Srivastav	SKIT Jaipur, India
Pooja Gupta	Banasthali Vidyapith, India
Pooja Khanna	Amity University, Lucknow, India
Rajiv Singh	Banasthali Vidyapith, India
Raveendra Babu Ponnuru	Cleveland State University, USA
Ravindra Hegadi	Central University of Karnataka, India
Renu Dalal	Guru Govind Singh Indraprastha University, India
Ritu Chauhan	Amity University, Noida, India
Ruchilekha	Banaras Hindu University, India
Sachchida Nanad Chaurasia	Banaras Hindu University, India
Saifuddin Mahmud	Bradley University, USA
Sakshi Indolia	Narsee Monjee Institute of Management Studies, India
Sarvesh Pandey	Banaras Hindu University, India
Shakti Sharma	Bennett University, India
Siddharth Singh	University of Lucknow, India
Sonam Seth	Banasthali Vidyapith, India
Sudha Morwal	Banasthali Vidyapith, India

Contents – Part I

Human Centred AI

Human-Robot Interaction and Intelligent Interfaces

User Centred Design

Contents – Part II

Intelligent Systems

Mobile Computing and Ubiquitous Interactions

Social Computing and Interactive Elements

Natural Language and Dialogue Systems

HGAN: Editable Visual Generation from Hindi Descriptions

Varsha Singh$^{(\boxtimes)}$ (ID), Shivam Gupta, and Uma Shanker Tiwary

Indian Institute of Information Technology, Allahabad, India
{rsi2018002,mit2020003,ust}@iiita.ac.in

Abstract. Visual content generation is an active area of research nowadays with considerable work. Still leaving possibilities for enhancement. However, in this area, the domain of the Hindi language has remained unexplored, largely due to the scarcity of linguistic resources. As Hindi is the fourth most spoken language in this world, such work will be a considerable contribution to the research. This paper introduces the HGAN (Hindi Generative Adversarial Network) for creating high-quality visual content from textual descriptions in Hindi and allows the user to change specific image attributes. To ensure diversity, word-level spatial attention and channel-wise attention mechanisms are employed to enable the model to offer control over the attributes of generated content. HGAN shows considerably good results, a 20.3% increase in IS and a 16.6% decrease in reconstruction error on the Hindi dataset than state-of-the-art methods.

Keywords: Visual Content · Textual Description · Generative Adversarial Network (GAN) · Hindi dataset · Attention Mechanism (AM)

1 Introduction

In natural language processing and computer vision, image captioning is among the most prevalent and challenging topics. Image captioning creates subtitles for an image, and visual content generation using text descriptions is the inverse of Image Captioning [1–3]. Image generation from a given text description is challenging in computer vision and language processing. Text-to-image synthesis technology may be employed in various industrial applications once it gets more efficient and precise. It's a difficult task to create realistic visuals that semantically match supplied language descriptions. Still, it has many potential applications in areas like image editing, video games, and computer-aided design [4]. Considerable progress has been made in text-to-image generation using generative adversarial networks(GANs) [5–7]. Conditional GANs [8–10] allow us to generate realistic images conditioned on given text descriptions [11]. The proposed text-to-image GANs might be able to minimise the time it takes to publish a comic by automatically producing plot sequences and supporting the creative process in the comic business.

B. J. Choi et al. (Eds.): IHCI 2023, LNCS 14531, pp. 3–14, 2024.
https://doi.org/10.1007/978-3-031-53827-8_1

This study aims to create a synthetic image from text descriptions and allow users to adjust it by changing text descriptions. The focus is on changing written descriptions to change specific visual features (e.g., colour, texture, and category) of the object in the generated images. With the change in input language other than English, the existing models show significant degradation in the model's performance. This work is done on the Hindi text descriptions. To do this, Hindi captions for the CUB dataset were built. All necessary hyperparameter modifications are performed, employing a hybrid encoding strategy and added dropout layers to prevent overfitting. The results show that the used approach performs better than the state-of-the-art and successfully modifies generated visual content using natural language descriptions.

The developed model generates different subregions independently and then merges them into a single image. It consists of 3 generators and follows a multi-stage design [12,13] that synthesizes images from noise. The proposed GAN model has three modules: 1) an Encoding module to do all preprocessing on Hindi text, 2) The word-focused channel attention module, a channel-wise attention-driven generator with word-level spatial for generating semantically identical visuals for each channel, and 3) an Editing module which uses perceptual loss to keep the unattended part of the image intact.

In the first module, a hybrid encoding strategy is used with hyperparameter modification to process the text in Hindi. This module gives the Hindi textual description as input to convert it into tokens, which are further encoded using a text encoder whose output is given as input to the Long-Short Term Memory (LSTM), which gives sentence vector and word vector as input to GAN and Word-Focused Channel Attention module.

The proposed model follows the divide and conquer technique, which divides the entire image into many sub-regions and constructs each sub-region independently, and later on, all sub-regions will be merged and form a complete image. Therefore, deciding which subregion will form first is important; channel-wise attention determines this. This helps generate the subregion connected with the most relevant word. The word-level discriminator helps to discriminate different subregions with respect to words given in the text description. The channel-wise attention with word-level discriminator combined form the second module, the Word-Focused Channel Attention module.

This is the most important component in the regeneration process. It helps determine which subregions must be modified for new text descriptions. The new text description is the old text with information about the change required in the sub-region, that is, which part needs modification and what. Editing, the third module, uses perceptual loss [14], which is computed to control the randomness in the image. Perceptual loss measures the difference between the high-level features of two images (old and new).

The proposed model allows us to modify specific attributes of a generated image, like its colour, shape, texture, etc. The perceptual loss function helps to modify a certain attribute of an image and keeps other attributes as it is. The dropout layer is added to prevent overfitting after the second module. The proposed model consists of 3 generators; the first generator generates a low-resolution image containing the basic outlines and texture of an image, and as that image passes through subsequent generators, its resolution improves. The last generator produces a high-quality image.

An exhaustive investigation has been done to validate the model, which shows that our model can generate high-quality images and allows modifying specific attributes of a generated image without compromising diversity. The self-created dataset is used for experiments, and results show that our model surpasses the current state of the art in terms of quality.

2   Background

2.1   Text-to-Image Generation

Text-to-image generation has become one of the most important problems in the research and development area. Researcher Mansimov et al. [15] developed the AlignDRAW model, which works on the divide and conquer technique. His model generates subregions of an image in different phases using the attention mechanism. The attention mechanism helps decide the order in which subregions must be generated. To synthesize visuals from text, Nguyen et al. [16] proposed an approximation Langevin technique. The cGAN was initially used by Reed et al. [10] to produce believable visuals which depend on the text description. Researcher Zhang et al. [12] have developed a completely new technique in which the entire generation process is divided into many steps. The starting step takes a latent point as input and generates a low-quality image as output, which is given as input to the next step. The last step produces the high-resolution images as output.

2.2   Attention

Building a model that can update certain parts of an image has potential applications in real-life scenarios. To achieve this, the attention mechanism has played a very important role and has proven effective in many ways. It has been used in many applications like Object detection [17,18], Image captioning [19], machine translation [20], and visual question answering [21]. The attention mechanism helps the model decide which portion of the image contains the most relevant information and which portion contains the least relevant information. This way, it can reduce the time and allow the model to update only a certain portion. Xu et al. [12] developed the AttnGAN model, which uses a word-level attention mechanism that helps generate the subregion connected with the most relevant word. Word-level attention alone cannot work efficiently. CNN has a variety of featured channels, and by default, all channels play an equally important role. The model needs to use different channels with different priorities for better results. A channel-wise attention mechanism is required to connect the word with the correct subregion, considering priorities using attention.

3   Material and Method

Given text description S, the aim is to generate a realistic image 'I' that is semantically as close as possible to sentence S. and also allows us to modify a certain portion of the generated image. If sentence S is modified into another sentence, K, the synthetic image should follow the changes made in 'S' while preserving irrelevant content in image 'I' when the sentences are given in Hindi. To achieve this:

- Captions were generated in Hindi for the CUB dataset, and extensive pre-processing is done.
- LSTM and GRU encoding techniques are explored for Hindi text.

Hindi captions are created for the Birds, whereas some of the sentences were not correctly translated, and those sentences were replaced by correct sentences manually. For the Hindi dataset, we first tokenized the sentence into a list of words, and each word was converted into a vector. After exploring encoding techniques, it is found that no encoding technique is explicitly required for the Hindi dataset. LSTM and GRU are used to extract context from Hindi sentences. To avoid overfitting the CNN model, a dropout layer is introduced for visual feature extraction, nullifying some of the neurons' effects and leaving others unmodified. If the dropout layer is absent, the learning is significantly affected by the first batch of training samples. As a result, traits that only show in later samples or batches would not be learned.

3.1   Architecture

The architecture used as the backbone of the model is Multi-stage AttnGAN [12]. The Word2Vec tokenization algorithm generates tokens, and generated tokens are encoded using UTF-8. Sentence S is given to the pretrained bidirectional RNN text encoder, which encodes sentence features $s \epsilon R^D$ in which D is the dimension which describes the whole sentence, whereas word features $w \epsilon R^{D x L}$ with length L (i.e. the number of words) and the dimension of each word is D. Conditioning augmentation is applied on s. A conditioning augmentation (CA) network selects random latent variables from a distribution denoted by the symbol. A random vector z is concatenated with the augmented sentence feature s as the input to the first step.

The image-generating process is divided into several parts. The network generates a hidden visual feature v_i at each level, which is fed into the associated generator G_i to create a synthetic image. As a result, the entire framework develops an image in stages, from coarse to fine size. The attentive word-context features are extracted using the Word-Focused Channel Attention module (Fig. 1).

3.2   Word-Focused Channel Attention

In Word-Focused Channel Attention, two attention mechanism, spatial attention and channel-wise attention mechanism, are used, which helps to modify a

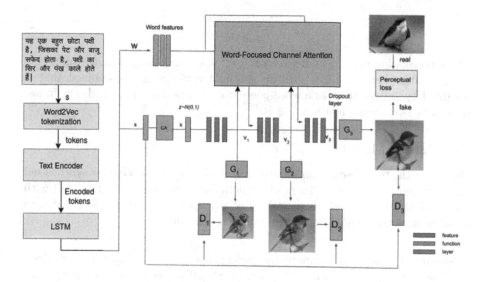

Fig. 1. The architecture of proposed HGAN model.

certain attribute of a generated image and keep other attributes as it is. The spatial attention [22] associates different subregions of an image with the most related word so that if a certain word is changed in a sentence, the region connected with that word will be changed. It basically helps to modify certain parts of a synthesized image. Whereas channel-wise attention helps to pick the most relevant word from the sentence, subregions related to that word will be constructed first. It basically helps to decide the order of construction of different subregions.

Channel-Wise Attention. In the channel-wise attention mechanism, at every stage, it generates correlation values between channel and words for all the spatial locations. Channel-wise attention module helps to decide the order of formation of different subregions. The mathematical formula for channel-wise attention is given below

$$D_{i,j}^k = \frac{\exp\left(m_{i,j}^k\right)}{\sum_{L=0}^{L-1}\exp\left(m_{i,l}^k\right)} \tag{1}$$

where $m_{i,j}^k$ is a correlation value between i^{th} channel and j^{th} word at stage k

Word-Level Discriminator. The word-level discriminator offers precise feedback, assisting the generator in creating the subregion associated with the most significant word and forcing the generator to adjust the image's area per the text. [14] inspired word-level discriminator to investigate the relationship between picture subregions and each word.

$$\beta_{i,j} = \frac{\exp(m_{i,j})}{\sum_{l=0} \exp(m_{i,l})} \tag{2}$$

where correlation coefficient between the ith word and the jth image subregion is $\beta_{i,j}$. After that, $b = \bar{n}\beta^T$, Entire spatial knowledge gets combined that is weighted by matrix know as a word-context correlation matrix, yields the image subregion-aware word features $b \in R^{D \times L}$.

3.3   Editing Module

Modifying the text description slightly may lead to a completely new synthesized image. Some attributes of the newly generated image may not align well with the attributes of the old image. We used perceptual loss [14], which helps to preserve irrelevant components and reduce randomness.

Perceptual Loss. To compute the perceptual loss, a 16-layered, pretrained CNN on the ImageNet dataset, VGG network [23] is used. Semantic properties are extracted from generated image I' and real image I, and the Euclidean distance between them is calculated.

$$\mathcal{L}_{per}(I', I) = \frac{1}{C_i H_i W_i} \|\phi_i(I') - \phi_i(I)\|_2^2 \tag{3}$$

where H_i and W_i represent the height and width of the feature map, respectively, and $\phi_i(I)$ indicates the activation of the ith layer of the VGG network.

3.4   Objective Functions

It is divided into two sub-models: the generator and discriminator models. The discriminator model is taught to determine if a sample is real or fake by using false samples created by the generator model. Both the generator model and discriminator model compete with each other. The generator model produces generator loss, and the discriminator model produces discriminator loss. Both model tries to minimize loss respectively.

3.5   Generator Objective

The generator loss L_G as Eq. 4 consist of an adversarial loss L_{G_k} and correlation loss between text and image L_{corre}. To reduce the randomness, we have perceptual loss L_{per} and a matching loss between text and image L_{DAMSM} [12], which determines the correctness of an image.

$$\mathcal{L}_G = \sum_{k=1}^{K} (\mathcal{L}_{G_k} + \lambda_2 \mathcal{L}_{per}(I'_k, I_k) + \lambda_3 \log(1 - \mathcal{L}_{corre}(I'_k, S))) + \lambda_4 \mathcal{L}_{DAMSM} \tag{4}$$

From true image distribution, I_k is taken as a sample picture, and K is the number of steps. P_{data} at stage k, generated picture I'_k, sampled from the model distribution P_{G_k} at the k_{th} step. λ_2, λ_3 and λ_4 are hyper-parameters that regulate various losses. Cosine similarity is used to compute the text-image matching score, and L_{DAMSM} [12] is used to access that matching score, and taking spatial information into account, L_{corre} represents the connection between the generated picture and the given text description

$$L_{G_k} = -\frac{1}{2}E_{k'} \sim PG_k \left[\log\left(D_k\left(I_k'\right)\right)\right] - \frac{1}{2}E_{I_{k'}} \sim PG_k \left[\log\left(D_k\left(I_{k'}'S\right)\right)\right]$$

The adversarial loss L_{G_k} comprises conditional and unconditional adversarial losses.

3.6   Discriminator Objective

The below loss function is used for training the discriminator

$$L_D = \sum_{k=1}^{K} \left(L_{D_k} + \lambda_1 \left(\log\left(1 - L_{corre}\left(I_{k'}S\right)\right) + \log L_{corre}\left(I_{k'}S'\right)\right)\right)$$

where L_{corre} is the correlation loss that determines whether the image contains word-related visual attributes, $S_{\prime}$ is a mismatched and unrelated text description that was randomly selected from the dataset, and λ_1 is a hyper-parameter that regulates the significance of subsequent losses. The conditional adversarial loss analyses if the presented image matches the textual description, whereas the unconditional adversarial loss checks whether the image is real. S:

$$\mathcal{L}_{D_k} = \underbrace{-\frac{1}{2}E_{I_k \sim P_{data}} \left[\log\left(D_k\left(I_k\right)\right)\right] - \frac{1}{2}E_{I_{k'} \sim PG_k} \left[\log\left(1 - D_k\left(I_k'\right)\right)\right]}_{\text{Unconditional Adversarial Loss}}$$
$$\underbrace{-\frac{1}{2}E_{I_k \sim P_{data}} \left[\log\left(D_k\left(I_k, S\right)\right)\right] - \frac{1}{2}E_{I_k} \sim PG_k \left[\log\left(1 - D_k\left(I_k', S\right)\right)\right]}_{\text{Conditional Adversarial Loss}}$$

4   Dataset

A new dataset is created using the CUB dataset [25] with text descriptions written in Hindi. Our dataset contains 2,933 test pictures and 8,855 training photos, each with ten text descriptions in Hindi. Google Translate was used to convert English captions into Hindi but revealed several issues: 1) The translations often lose their intended meaning due to the lack of context-aware translation methods, and literal translations could be grammatically incorrect in different situations. Moreover, 2) the accuracy was inconsistent, as it depended on specific source and target language combinations. Consequently, human annotators had to step in to correct the inaccuracies in Google's translations.

5   Implementation

The Hindi text description is given as input to the Word2Vec tokenizer to convert into word tokens and encoded tokens. Pre-trained LSTM [24] on the Hindi dataset is used for converting text description into sentence vector 256-dimensional feature vector where each word is of the 18-dimensional feature vector. The proposed model contains three stages. At stages 2 and 3, channel-wise and spatial attention are applied, and images are generated at the three scales: 64×64, 128×128, and 256×256 at each stage, respectively. The content loss is computed at Relu2_2 of VGG-16 [25], which is a pre-trained model on the ImageNet dataset, to compute perceptual loss. The learning rate is 0.0002, and the Adam optimiser [26] is used for training the GAN network. For the Hindi dataset, we found the best values for hyper-parameter of λ_1, λ_2, λ_3 and λ_4 are 0.75, 1.2, 1.2, and 6, respectively (Table 1).

Table 1. Perceptible Comparison on Inception Score and L_2 reconstruction error.

Method	Inception Score	L_2 Error
StackGAN++	4.04 ± 3.72	0.29
AttnGAN	4.36 ± 4.43	0.26
ControlGAN	4.58 ± 0.92	0.18
HGAN (Our)	5.38 ± 0.22	**0.15**

6   Result

The resulting image's quality and variety are evaluated using Inception Score [28]. A statistic for assessing the image's quality is the inception score. The proposed model shows a considerable increase in the inception score on the Hindi dataset, which is 20.3%. This indicates that the suggested model can generate higher-quality, more diversified pictures that are accurate as per the text description.

Reconstruction error [29, 30], L2 is estimated between the picture produced by the original text and the image derived from the updated text. The proposed model has produced lower reconstruction errors on the Hindi dataset than the English dataset, with a decrease of 16.6% (Table 1).

Figure 2 shows that every result seems semantically correct and follows the description appropriately.

7   Component Analysis

The proposed method combines spatial and channel-wise attention in the generator to create realistic pictures. The interlude results and related attention

इस पक्षी की एक छोटी सी चाँच और एक छोटी काली आँख होती है।

चमकीले पीले सिर, काले और पीले पंख, और छाती पर भूरी धारियों वाला एक रंगीन छोटा पक्षी।

चमकदार लाल सिर और गले, भूरे और लाल पंख, काली पूंछ-टिप और सफेद पेट वाला एक छोटा सुंदर लाल पक्षी।

पक्षी के पास एक छोटा सफेद शरीर होता है जिसमें नीले और हरे रंग का मुकुट और आवरण होता है।

Fig. 2. Results: The text in the first row is given as input to the HGAN model, and below is the output produced by the model.

maps at different stages are observed to understand the effectiveness of attention processes. It is found that spatial attention is primarily focused on colour descriptions, and channel-wise attention is directly tied to the semantic elements of objects. The corresponding intermediate results are shown in Fig. 3.

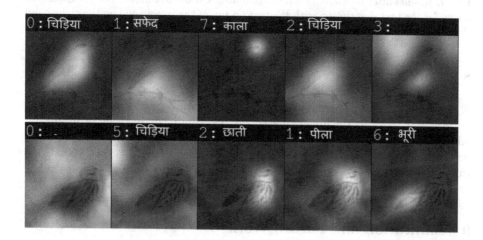

Fig. 3. Intermediate results

The channel-wise attention assigns significant correlation values to channels that are semantically meaningful to the term characterising areas of a bird. Without channel-wise attention, the method could fail to yield reasonable findings, even after the change in the terminology used to describe bird parts. On the other hand, the adopted approach with channel-wise attention can result in more controllable outcomes. In Fig. 4, some sample results are presented, representing the generated image before and after modifying a text by changing the colour.

Fig. 4. Results after modifying certain attributes: On the left-hand side, the generated image for the text description with the text on top of the image, and on the right-hand side, the regenerated image after text modification (changing the colour attribute).

8　Conclusion

A new Hindi dataset is created, and a generative adversarial network, HGAN, is proposed to generate images based on the Hindi language description and allow the manipulation of certain attributes of an image by changing the text description. The proposed model successfully disentangles distinct visual features, allowing precise manipulation of sections of the synthetic image while preserving the generation of other components.

9　Limitation and Future Work

This work creates a new dataset for visual content generation from Hindi descriptions. Due to the scarcity of linguistic resources, the dataset is limited to a very small domain, which could be extended to more objects and complex images in future.

References

1. Singh, V., Agrawal, P., Tiwary, U.S.: Scene description with context information using dense-LSTM. J. Intell. Fuzzy Syst. **44**(5), 7553–7565 (2023)
2. Garg, K., Singh, V., Tiwary, U.S.: Textual description generation for visual content using neural networks. In: Kim, J.H., Singh, M., Khan, J., Tiwary, U.S., Sur, M., Singh, D. (eds.) IHCI 2021. LNCS, vol. 13184, pp. 16–26. Springer, Cham (2021). https://doi.org/10.1007/978-3-030-98404-5_2
3. Singh, V., Khushaboo, K., Singh, V.K., Tiwary, U.S.: Describing images using CNN and object features with attention. In: 2023 International Conference on Information Technologies (InfoTech), pp. 1–6. IEEE (2023)
4. Singh, V., et al.: Performance analysis of GANs for de-noising images. In: 2023 International Conference on Information Technologies (InfoTech), pp. 1–7. IEEE (2023)
5. Denton, E.L., Chintala, S., Szlam, A., Fergus, R.: Deep generative image models using a Laplacian pyramid of adversarial networks. In: Advances in Neural Information Processing Systems, pp. 1486–1494 (2015)
6. Goodfellow, I., et al.: Generative adversarial nets. In: Advances in Neural Information Processing Systems, pp. 2672–2680 (2014)
7. Bahdanau, D., Cho, K., Bengio, Y.: Neural machine translation by jointly learning to align and translate. arXiv preprint arXiv:1409.0473 (2014)
8. Dong, H., Yu, S., Wu, C., Guo, Y.: Semantic image synthesis via adversarial learning. In: Proceedings of the IEEE International Conference on Computer Vision, pp. 5706–5714 (2017)
9. Reed, S., Akata, Z., Yan, X., Logeswaran, L., Schiele, B., Lee, H.: Generative adversarial text to image synthesis. arXiv preprint arXiv:1605.05396 (2016)
10. Reed, S.E., Akata, Z., Mohan, S., Tenka, S., Schiele, B., Lee, H.: Learning what and where to draw. In: Advances in Neural Information Processing Systems, pp. 217–225 (2016)
11. Singh, V., Tiwary, U.S.: Visual content generation from textual description using the improved adversarial network. Multimedia Tools Appl. **82**(7), 10943–10960 (2023)
12. Xu, T., et al.: AttnGAN: fine-grained text to image generation with attentional generative adversarial networks. In: Proceedings of the IEEE Conference on computer vision and pattern recognition, pp. 1316–1324 (2018)
13. Brock, A., Lim, T., Ritchie, J.M., Weston, N.: Neural photo editing with introspective adversarial networks. arXiv preprint arXiv:1609.07093 (2016)
14. Johnson, J., Alahi, A., Fei-Fei, L.: Perceptual losses for real-time style transfer and super-resolution. In: Leibe, B., Matas, J., Sebe, N., Welling, M. (eds.) ECCV 2016. LNCS, vol. 9906, pp. 694–711. Springer, Cham (2016). https://doi.org/10.1007/978-3-319-46475-6_43
15. Mansimov, E., Parisotto, E., Ba, J.L., Salakhutdinov, R.: Generating images from captions with attention. arXiv preprint arXiv:1511.02793 (2015)
16. Nguyen, A., Clune, J., Bengio, Y., Dosovitskiy, A., Yosinski, J.: Plug & play generative networks: Conditional iterative generation of images in latent space. In: Proceedings of the IEEE Conference on Computer Vision and Pattern Recognition, pp. 4467–4477 (2017)
17. Oliva, A., Torralba, A., Castelhano, M.S., Henderson, J.M.: Top-down control of visual attention in object detection. In :Proceedings of International Conference on Image Processing (Cat. No. 03CH37429), vol. 1, pp. 253–256. IEEE (2003)

18. Lin, T.-Y., et al.: Microsoft COCO: common objects in context. In: Fleet, D., Pajdla, T., Schiele, B., Tuytelaars, T. (eds.) ECCV 2014. LNCS, vol. 8693, pp. 740–755. Springer, Cham (2014). https://doi.org/10.1007/978-3-319-10602-1_48

19. Xu, K., et al.: Show, attend and tell: neural image caption generation with visual attention. In: International Conference on Machine Learning, pp. 2048–2057 (2015)

20. Simonyan, K., Zisserman, A.: Very deep convolutional networks for large-scale image recognition. arXiv preprint arXiv:1409.1556 (2014)

21. Li, X., Wu, B., Song, J., Gao, L., Zeng, P., Gan, C.: Text-instance graph: exploring the relational semantics for text-based visual question answering. Pattern Recogn. **124**, 108455 (2022)

22. Salimans, T., Goodfellow, I., Zaremba, W., Cheung, V., Radford, A., Chen, X.: Improved techniques for training GANs. In: Advances in Neural Information Processing Systems, pp. 2234–2242 (2016)

23. Szegedy, C., et al.: Going deeper with convolutions. In: Proceedings of the IEEE Conference on Computer Vision and Pattern Recognition, pp. 1–9 (2015)

24. Huang, X., Lin, N., Li, K., Wang, L., Gan, S.: HinPLMs: pre-trained Language Models for Hindi. In: International Conference on Asian Language Processing (IALP), pp. 241–246. IEEE (2021)

25. Wah, C., Branson, S., Welinder, P., Perona, P., Belongie, S.: The Caltech-Ucsd Birds-200-2011 dataset (2011)

26. Kingma, D.P., Ba, J.: Adam: a method for stochastic optimization. arXiv preprint arXiv:1412.6980 (2014)

27. Russakovsky, O., et al.: Imagenet large scale visual recognition challenge. Int. J. Comput. Vision **115**(3), 211–252 (2015)

28. Kingma, D.P., Mohamed, S., Rezende, D.J., Welling, M.: Semi-supervised learning with deep generative models. In: Advances in Neural Information Processing Systems, pp. 3581–3589 (2014)

29. Radford, A., Metz, L., Chintala, S.: Unsupervised representation learning with deep convolutional generative adversarial networks. arXiv preprint arXiv: 1511.06434 (2015)

30. Schuster, M., Paliwal, K.K.: Bidirectional recurrent neural networks. IEEE Trans. Signal Process. **45**(11), 2673–2681 (1997)

Adopting Pre-trained Large Language Models for Regional Language Tasks: A Case Study

Harsha Gaikwad[(✉)], Arvind Kiwelekar, Manjushree Laddha,
and Shashank Shahare

Department of Computer Engineering, Dr. Babasaheb Ambedkar Technological
University, Lonere 402 103, Maharashtra, India
{harsha.gaikwad,awk,mdladdha,srshahare}@dbatu.ac.in

Abstract. Large language models have revolutionized the field of Natural Language Processing. While researchers have assessed their effectiveness for various English language applications, a research gap exists for their application in low-resource regional languages like Marathi. The research presented in this paper intends to fill that void by investigating the feasibility and usefulness of employing large language models for sentiment analysis in Marathi as a case study. The study gathers a diversified and labeled dataset from Twitter that includes Marathi text with opinions classified as positive, negative, or neutral. We test the appropriateness of pre-existing language models such as Multilingual BERT (M-BERT), indicBERT, and GPT-3 ADA on the obtained dataset and evaluate how they performed on the sentiment analysis task. Typical assessment metrics such as accuracy, F1 score, and loss are used to assess the effectiveness of sentiment analysis models. This research paper presents additions to the growing area of sentiment analysis in languages that have not received attention. They open up possibilities for creating sentiment analysis tools and applications specifically tailored for Marathi-speaking communities.

Keywords: Natural Language Processing · Large Language Models · Sentiment Analysis

1 Introduction

Large language models are a remarkable advancement in the fields of artificial intelligence and natural language processing [23]. These models are distinguished by their enormous dimensions and intricate designs. Large language models (LLMs) have a wide range of applications, including language generation, language translation, sentiment analysis, question answering, chatbot development, text summarization, and text paraphrasing [3]. These models have been extensively evaluated and applied to various tasks in the English language.

However, there is a notable research gap when it comes to the application of Large Language Models to regional and low-resource languages like Marathi [15].

B. J. Choi et al. (Eds.): IHCI 2023, LNCS 14531, pp. 15–25, 2024.
https://doi.org/10.1007/978-3-031-53827-8_2

Although these models have demonstrated outstanding performance in English, there is still much to learn about how to adapt and use them in regional languages. Our knowledge of how LLMs can be used to support linguistic diversity and assist non-English speaking communities is constrained by the lack of research in this area. Further investigation and development of LLMs for regional and low-resource languages like Marathi could open up new avenues for improving natural language processing capabilities and accessibility for a broader range of users. This paper explores the case study of Marathi sentiment Analysis using Large Language Models.

Sentiment analysis, also known as opinion mining or emotion AI, is a prominent branch of natural language processing (NLP) that focuses on extracting and understanding sentiments or emotions expressed in textual data [22]. The ability to automatically analyze sentiments from text has many applications, including understanding customer feedback, monitoring public opinion, social media analysis, and market research [12].

While sentiment analysis in significant languages such as English has seen substantial progress, there remains a scarcity of research in sentiment analysis for low-resource languages, including Marathi [1]. Marathi, an Indo-Aryan language primarily spoken in the Indian state of Maharashtra, has a rich literary and cultural heritage [8]. It boasts a sizable native-speaker population and is widely used in various domains, including media, literature, business, and social communication.

The unique linguistic characteristics and limited digital resources for Marathi present challenges for natural language processing tasks like sentiment analysis [15]. Developing effective sentiment analysis models for Marathi requires overcoming the scarcity of labeled data, handling linguistic nuances, and adapting large language models to this specific language.

In recent years, the emergence of large language models, such as GPT and BERT, has revolutionized the field of NLP [16]. These models leverage deep learning techniques and are pre-trained on vast amounts of textual data, enabling them to capture complex language patterns and representations [5,10,17]. Large language models have showcased impressive results in various NLP tasks, making them promising candidates for sentiment analysis [16].

This study investigates the use of large language models for sentiment analysis in Marathi. The study looks into the viability of using pre-trained models already in use, such as mBERT, IndicBert, and GPT-3 ADA. It explores the possible gains that might be made by these models using a tagged dataset of Marathi text. The following are the study's primary contributions:

1. **Creation of Dataset:** A diversified and labeled collection of Marathi texts with sentiments classified as positive, negative, or neutral is developed for the study's purposes. Future Marathi sentiment analysis research will greatly benefit from this dataset as a useful resource.
2. **Fine-Tuning Analysis:** Investigated if domain-specific modifications could improve model performance by examining the effect of large language models on Marathi sentiment analysis.

Table 1. Marathi Words showing sentiments

Sentiment	Marathi Words
Positive	प्रफुल्लीत, जलसा, आनंदोत्सव, जोश, तत्पर, उत्तेजित, उमेदी, उत्सुक, नामांकित, विजयी, अतिसुखी, रसिक, सुखी, आशावादी, उत्साही, स्वाभिमान, उल्हास, समाधानी, हुरूप, हौशी.
Negative	दुर्बळ, झिडकारणे, एकटे पडणे, तिरस्कृत, दुर्लक्षित, दुःखी, धडकी, लज्जास्पद, निराश, नकार, भयग्रस्त, चिंताग्रस्त, सुरक्षित, भंगलेला.
Neutral	स्तब्ध, साशंक होणे, पेचात पडणे, संभ्रमित, चकित, समजेनासे होणे.

3. **Evaluation of Large Language Models:** Performance assessment of state-of-the-art large language models, like GPT-3 ADA, IndicBERT, and mBERT, on Marathi sentiment analysis This evaluation will shed light on the inherent capabilities of these models for low-resource languages.

The rest of the paper is organized as follows. The Sect. 2 discusses the Pre-Trained Language models such as IndicBERT, mBERT, GPT-3 ADA. Data collection and preprocessing techniques are described in Sect. 3. The Sect. 4 discusses the fine-tuning of the large language models and the results obtained through our analysis. This study endeavors to pave the way for a deeper understanding of sentiment analysis in Marathi and contribute to the advancement of NLP in low-resource language contexts. The related work in the Marathi sentiment analysis is described in Sect. 5 and, finally, concludes the paper's findings in Sect. 6.

2   Pre-trained Language Models

The Marathi sentiment analysis problem has been tackled using various approaches such as rule-based techniques, machine learning algorithms, and deep learning-based techniques. However, this paper proposes a novel approach that leverages the power of large language models.

Large language models represent a recent breakthrough in deep learning techniques for handling human language. They can comprehend and generate text in a manner that resembles human-like understanding. Our research explores applying large language models such as IndicBert, mBert, and GPT-3 ADA to address the Marathi sentiment analysis problem. These transformer-based models understand the language context [9]. This paper employed the following advanced large language models to achieve accurate and robust Marathi sentiment analysis.

Table 2. Examples from the dataset

Sr. No.	Sentence	Translation	Sentiment
1.	त्यांचा वैज्ञानिक दृष्टिकोन पुढील पिढ्यांसाठी प्रेरणादायी आहे.	His scientific approach is an inspiration to the next generations.	Positive
2.	समृद्धी महामार्गाच्या तिसऱ्या टप्प्याचे काम सुरु असतांना शहापूर इथे क्रेन कोसळून कामगारांचा मृत्यू झाल्याची बातमी अत्यंत दुःखद आहे.	The news is very sad that while the work of third phase of Samriddhi highway is going on, the workers died when the crane collapsed in Shahapur.	Negative
3.	हे जपलं पाहिजे.	This should be preserved.	Neutral
4.	लाज वाटायला हवी या राजकारण्यांना उगाचच आग करून घेत आहेत	Shame on these politicians	Negative
5.	खूप छान चित्रपट	Very nice movie	Positive
6.	कलाकार यांना योग्य उत्पन्न होणे अपेक्षित आहे या साठी चित्रपत आणि वाहिनी क्षेत्रात उद्योजक यांची भांडवली गुंतवणूक गरजेचे आहे.	Capital investment by entrepreneurs in the film and channel sector is necessary for artists to expect a decent income.	Neutral

a) **IndicBert:** BERT, an abbreviation for Bidirectional Encoder Representations from Transformers, utilizes Transformers, a deep learning model with interconnected output and input elements, and their dynamic weightings are calculated based on these connections [7]. It is an open-source machine learning framework tailored for natural language processing (NLP).

BERT's primary aim is to enhance computers' comprehension of ambiguous language in textual data by considering the context provided by surrounding text. The framework was initially pre-trained using Wikipedia text and can be further customized for specific tasks using question-and-answer datasets. IndicBERT is a unique multilingual ALBERT model specifically designed for the Indian context, as it is trained exclusively on 12 major Indian languages [11].

This model undergoes pre-training using our novel monolingual corpus, which comprises approximately 9 billion tokens. Subsequently, it undergoes rigorous evaluation on a diverse range of tasks. Remarkably, IndicBERT achieves superior performance. The 12 languages skillfully covered by IndicBERT include Assamese, Bengali, English, Gujarati, Hindi, Kannada, Malayalam, Marathi, Oriya, Punjabi, Tamil, and Telugu.

b) **mBert:** mBERT stands for "Multilingual BERT", a type of pre-trained language model developed by Google Research. mBERT is an extension of the original BERT model designed to handle multiple languages [18]. Instead of being trained on a single language, mBERT is trained on a large corpus of text from multiple languages simultaneously. This enables the model to learn language-agnostic representations that can be fine-tuned for specific NLP tasks in any supported language.

c) **GPT-3 ADA:** GPT stands for Generative Pre-trained Transformer. It is an auto-regressive language model. Its full version has the capacity of 175 billion machine learning parameters [25].

The GPT-3 model is not a single model but a family of models. Each model in the family has a different number of trainable parameters. GPT-3 family of models is based on the same transformer (decoder) based architecture of GPT-2. ADA is Apt for basic assignments, typically the swiftest model among the GPT-3 series, and available at the most affordable price [13].

3   Preparation of Sentiment Analysis Dataset

This section elaborates on data collection, preprocessing, and preparation of datasets to fine-tune the language models.

a) **Data Collection:** The dataset was obtained from Twitter and consisted of 1,500 tweets. These tweets were manually annotated into three categories: positive, negative, and neutral, based on the emotions present in each tweet. A tweet is classified as "Positive" if it conveys a favorable emotional connotation, expresses positivity, and shows agreement or approval.

Positive sentences in the tweets often include words such as "congrats", "applauding", "warm wishes", "praises", and "appreciation". On the other hand, a tweet is labeled as "Negative" if it carries an overwhelmingly negative meaning or contains a higher proportion of negative comments than other emotions. Negative sentences in the tweets typically involve terms like "boycott", "punishing", and "assessing". These statements may comprise a negative word paired with a positive adjective.

Finally, a tweet is categorized as "Neutral" if the statement's meaning is uncertain, indicated by words like "maybe" or "may not be". Neutral statements in the tweets neither express disagreement nor complete agreement. The manual annotation process involved analyzing the content of each tweet to determine its emotional connotation, leading to a balanced dataset of 500 positive, 500 negative, and 500 neutral tweets. This dataset can be used for various natural language processing tasks, sentiment analysis, or emotion recognition applications.

In Table 1, we have enlisted some words that show emotions in the Marathi language. प्रफुल्लीत, जलसा, आनंदोत्सव (cheerfully, happy) shows positive emotion. दुर्बळ, झिडकारणे, एकटे पडणे (Weak, stubborn, lonely) showing negative emotion. संभ्रमित, चकित, समजेनासे होणे (confused, surprised) shows neutral emotion.

In Table 2, we discussed some examples from the dataset below:

(i) **Sentence 1:** "यांचा वैज्ञानिक दृष्टिकोन पुढील पिढ्यांसाठी प्रेरणादायी आहे." (His scientific approach is an inspiration to the next generations.) - This sentence carries a positive sentiment due to the word "प्रेरणादायी" (inspiration).

Table 3. Hyper-Paremeter Tuning for LLM

Sr. No.	Parameters	indicBERT	mBERT	GPT-ADA
1.	Epochs	10	10	4
2.	Learning Rate	e^{-5}	e^{-5}	0.05
3.	Batch Size	30	30	20% of Training Data

(ii) **Sentence 2:** "समृद्धी महामार्गाच्या तिसऱ्या टप्प्याचे काम सुरु असतांना शहापूर इथे क्रेन कोसळून कामगारांचा मृत्यू झाल्याची बातमी अत्यंत दुःखद आहे." (The news is very sad that while the work of the third phase of Samriddhi highway is going on, the workers died when the crane collapsed in Shahapur.) - This sentence conveys a negative emotion due to the phrase "अत्यंत दुःखद" (very sad).

(iii) **Sentence 3:** "हे जपलं पाहिजे." (This should be preserved.) - This sentence is neutral, offering instruction without any specific positive or negative emotional connotation.

(iv) **Sentence 4:** This sentence carries a negative emotion due to the word "लाज" (shame).

(v) **Sentence 5:** This sentence has a positive sentiment because of the word "छान" (nice).

b) **Data Preprocessing and Data preparation:** This section discusses the data preparation and processing phase for several language models.

For IndicBERT and mBERT:
- **Text Cleaning:** Removing stopwords, hashtags, numbers, special characters, punctuation marks, URLs, and emojis.
- **Label Encoding:** Assigning integer labels (1 for positive, 0 for negative, and 2 for neutral) to each sentence based on their sentiment.

For GPT-3 ADA:
- **Text Cleaning:** Like BERT, perform text cleaning by removing unwanted elements like stopwords, hashtags, numbers, special characters, punctuation marks, URLs, and emojis.
- **JSONL Format:** Convert the dataset into a JSONL (JSON Lines) format. Each line in the JSON file represents a data instance as a JSON object.
  - "prompt" field: Represents the sentence.
  - "completion" field: Indicates the sentiment, either "positive", "negative", or "neutral".

Here's an example of how a data instance might look in the JSONL format for ADA:

e.g. "prompt": "सरकारला ❑गोवर' ची परिस्थिती नीट सांभाळता येईना यांनी 'कोरोना'ची परिस्थिती काय सांभाळली असती.\n\n###\n\n" , "completion":"negative"

Table 4. Results of fine-tuned Model

Sr. No.	Model	Accuracy	Loss	F1 Score
1.	**IndicBERT**	0.7387	0.2658	0.7335
2.	**mBERT**	0.8291	0.0522	0.8283
3.	**ADA**	0.65	0.015992	0.65

4   Fine Tuning of Language Models

We fine-tuned the language models on our custom dataset. On the parameters, we performed hyper-parameter tuning as shown in Table 3. We determined that IndicBERT and mBERT achieve their highest accuracy with an epoch value of 10. Conversely, the ADA model employs a default epoch value of 4 [20]. Subsequent experimentation led us to establish a learning rate of $1e^{-5}$ for both IndicBERT and mBERT, while the GPT-3 ADA model adheres to the default learning rate of 0.05, 0.2, 0.1 depending on the batch size.

Regarding batch size, the BERT models utilize a value of 30, whereas ADA's batch size defaults to 0.20% of the training dataset size. We have used accuracy loss and F1 score matrices to evaluate the model's performance. Experimental results show that mBERT gives the best performance accuracy of 0.8291 and a loss of 0.0522 for the given dataset of Marathi sentiment analysis. In contrast, ADA gives the least accuracy of 0.65 and a loss of 0.015992, while indicBERT gives an accuracy of 0.7387 and a loss of 0.26. Figure 1 Shows the Accuracy graphs of GPT-3 ADA, mBERT, and IndicBERT Models.

As a powerful language model, ADA gives the least accuracy for the given dataset. The probable reasons might be

i) The small size of the dataset. Observations indicate that doubling the size of the dataset results in a proportional enhancement of model quality [21].
ii) Sentiment analysis models often rely on sentiment lexicons or dictionaries that contain words and phrases associated with positive or negative sentiment. Maybe the model needs a comprehensive Marathi sentiment lexicon; it could struggle to predict accurately.
iii) Sentiment analysis in Marathi is influenced by cultural context as well. Specific expressions of sentiment can be viewed as positive or negative in one cultural setting but not necessarily in another.

Table 4 shows the experimental results for several fine-tuned model parameters.

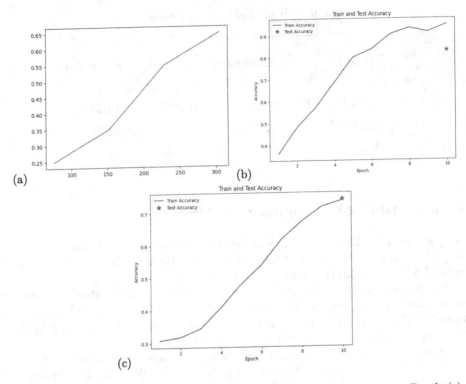

Fig. 1. (a) GPT-3 ADA Model Accuracy Graph (b) mBERT Accuracy Graph (c) IndicBERT Accuracy Graph

5   Related Work

Sentiment analysis is primarily a classification problem, which can be addressed using two main approaches: lexicon-based and machine learning. The lexicon-based technique can be further divided into two methods: the corpus-based method and the dictionary-based method [4]. The corpus for the Sinhala Language is generated from English Senti-Wordnet, where keywords are mapped to analyze the sentiment. The lexicon-based approach is used to find out the sentiment of the sentence.

A dataset of 200 sentences of the Marathi language is created: 100 sentences are positive, and 100 sentences are negative [6]. The system operates successfully due to the implementation of the lexicon approach. A dataset contains a collection of positive and negative words [24]. A machine learning approach is famous for solving the sentiment classification problem. A comprehensive dataset was assembled with 4248 political tweets. The dataset is manually labeled with positive and negative sentiments.

Data is cleaned during the data preprocessing stage by eliminating hash-tags, special symbols, emojis, etc. Various machine learning techniques, such as Support Vector Machines (SVM) with both linear and RBF kernel, Multinomial

Naïve Bayes, Logistic Regression, and Random Forest, are utilized to train classifiers. These classifiers use the Term Frequency vs. Inverse Document Frequency (TF-IDF) as features to distinguish the tweets as either positive or negative. Notably, the Multinomial Naïve Bayes algorithm demonstrates exceptional performance in this task [19].

A dataset comprising approximately 300 Marathi documents sourced from various social media platforms, including Chats, Tweets, and YouTube comments, was utilized for analysis. KNN, SVM, and Naive Bayes are used for sentiment classification for the Marathi language [2]. A dataset of 16,000 tweets written in the regional languages (i.e. Marathi) has been meticulously created and manually annotated with labels for positive, negative, and neutral sentiments. Different variations of models, such as CNN, LSTM, IndicBERT, and mBERT, have been employed for sentiment classification. A dedicated team conducted the annotation process. Notably, IndicBERT outperformed the other models, achieving the highest accuracy. A sample of this dataset is available for reference [14].

6 Conclusion

Sentiment Analysis, called opinion mining, constitutes a pivotal research domain within natural language processing. This paper introduces Marathi-senti, a seminal publicly accessible dataset tailored for Marathi Sentiment Analysis, encompassing approximately 1500 unique tweets. The document elucidates the annotation policy employed to label the entire dataset meticulously. Leveraging this dataset, we conducted a comprehensive 3-class sentiment classification. In the study, we harnessed prominent language models, including mBERT, IndicBERT, and GPT-3 ADA models, to conduct our analysis. Among these, the mBERT model demonstrated the highest accuracy. Notably, the ADA model's performance yielded sub-optimal results, prompting exploration of the plausible factors contributing to its unsatisfactory performance.

Data and Model Availability. The dataset and working models for the proposed article are available on the GitHub repository. The link to the GitHub repository is https://github.com/CompDbatu/MarathiSentimentAnalysis.

References

1. Agüero-Torales, M.M., Salas, J.I.A., López-Herrera, A.G.: Deep learning and multilingual sentiment analysis on social media data: an overview. Appl. Soft Comput. **107**, 107373 (2021)
2. Ansari, M.A., Govilkar, S.: Sentiment analysis of mixed code for the transliterated Hindi and Marathi texts. Int. J. Nat. Lang. Comput. (IJNLC) **7** (2018)
3. Bender, E.M., Gebru, T., McMillan-Major, A., Shmitchell, S.: On the dangers of stochastic parrots: can language models be too big?. In: Proceedings of the 2021 ACM Conference on Fairness, Accountability, and Transparency, pp. 610–623 (2021)

4. Chathuranga, P., Lorensuhewa, S., Kalyani, M.: Sinhala sentiment analysis using corpus based sentiment lexicon. In: 2019 19th International Conference on Advances in ICT for Emerging Regions (ICTer), vol. 250, pp. 1–7. IEEE (2019)
5. Deshmukh, R., Kiwelekar, A.W.: Deep convolutional neural network approach for classification of poems. In: Kim, J.-H., Singh, M., Khan, J., Tiwary, U.S., Sur, M., Singh, D. (eds.) IHCI 2021. LNCS, vol. 13184, pp. 74–88. Springer, Cham (2022). https://doi.org/10.1007/978-3-030-98404-5_7
6. Deshmukh, S., Patil, N., Rotiwar, S., Nunes, J.: Sentiment analysis of Marathi language. Int. J. Res. Publ. Eng. Technol. [IJRPET] **3**, 93–97 (2017)
7. Devlin, J., Chang, M.W., Lee, K., Toutanova, K.: BERT: pre-training of deep bidirectional transformers for language understanding. arXiv preprint arXiv:1810.04805 (2018)
8. Dhumal Deshmukh, R., Kiwelekar, A.: Deep learning techniques for part of speech tagging by natural language processing. In: 2020 2nd International Conference on Innovative Mechanisms for Industry Applications (ICIMIA), pp. 76–81 (2020)
9. Gillioz, A., Casas, J., Mugellini, E., Abou Khaled, O.: Overview of the transformer-based models for NLP tasks. In: 2020 15th Conference on Computer Science and Information Systems (FedCSIS), pp. 179–183. IEEE (2020)
10. Han, X., et al.: Pre-trained models: past, present and future. AI Open **2**, 225–250 (2021)
11. Jain, K., Deshpande, A., Shridhar, K., Laumann, F., Dash, A.: Indic-transformers: an analysis of transformer language models for Indian languages. arXiv preprint arXiv:2011.02323 (2020)
12. Khan, R., Shrivastava, P., Kapoor, A., Tiwari, A., Mittal, A.: Social media analysis with AI: sentiment analysis techniques for the analysis of twitter COVID-19 data. J. Crit. Rev **7**(9), 2761–2774 (2020)
13. Kublik, S., Saboo, S.: GPT-3. O'Reilly Media, Inc. (2022)
14. Kulkarni, A., Mandhane, M., Likhitkar, M., Kshirsagar, G., Joshi, R.: L3CubeMahaSent: a Marathi tweet-based sentiment analysis dataset. arXiv preprint arXiv:2103.11408 (2021)
15. Lahoti, P., Mittal, N., Singh, G.: A survey on NLP resources, tools, and techniques for Marathi language processing. ACM Trans. Asian Low-Resour. Lang. Inf. Process. **22**(2), 1–34 (2022)
16. Min, B., et al.: Recent advances in natural language processing via large pre-trained language models: a survey. ACM Comput. Surv. (2021)
17. Naseem, U., Razzak, I., Khan, S.K., Prasad, M.: A comprehensive survey on word representation models: from classical to state-of-the-art word representation language models. Trans. Asian Low-Resour. Lang. Inf. Process. **20**(5), 1–35 (2021)
18. Nozza, D., Bianchi, F., Hovy, D.: What the [mask]? Making sense of language-specific BERT models. arXiv preprint arXiv:2003.02912 (2020)
19. Patil, R.S., Kolhe, S.R.: Supervised classifiers with TF-IDF features for sentiment analysis of Marathi tweets. Soc. Netw. Anal. Min. **12**(1), 51 (2022)
20. Sawicki, P., et al.: On the power of special-purpose GPT models to create and evaluate new poetry in old styles (2023)
21. Smith, S., et al.: Using deepspeed and megatron to train megatron-turing NLG 530B, a large-scale generative language model. arXiv preprint arXiv:2201.11990 (2022)
22. Soong, H.C., Jalil, N.B.A., Ayyasamy, R.K., Akbar, R.: The essential of sentiment analysis and opinion mining in social media: introduction and survey of the recent approaches and techniques. In: 2019 IEEE 9th Symposium on Computer Applications & Industrial Electronics (ISCAIE), pp. 272–277. IEEE (2019)

23. Torfi, A., Shirvani, R.A., Keneshloo, Y., Tavaf, N., Fox, E.A.: Natural language processing advancements by deep learning: a survey. arXiv preprint arXiv:2003.01200 (2020)
24. Vidyavihar, M.: Sentiment analysis in Marathi language. Int. J. Recent Innov. Trends Comput. Commun. 5(8), 21–25 (2017)
25. Zhou, C., et al.: A comprehensive survey on pretrained foundation models: a history from BERT to ChatGPT. arXiv preprint arXiv:2302.09419 (2023)

Text Mining Based GPT Method for Analyzing Research Trends

Jeong-Hoon Ha[1,2(✉)], Dong-Hee Lee[1,2], and Bong-Jun Choi[1,2]

[1] Dongseo University, 47 Jurye-ro, Sasang-gu, Busan 47011, Korea
logo8044@gmail.com, bongjun.choi@dongseo.ac.kr
[2] Department of Software, Dongseo University, 47 Jurye-ro, Sasang-gu, Busan 47011, Korea

Abstract. In recent years, keywords are often extracted using text mining techniques when analyzing research trends to start a new study. In order to check whether the extracted keywords are relevant to the sentences, it is necessary to analyze the accuracy by using patterns, trends, etc. when extracting. In this study, we extract keywords using text mining techniques for a specific domain and compare and analyze the sentences generated by using the extracted keywords as input to GPT fine-tuned with data from the domain. By comparing the keywords extracted using text mining techniques with the sentences generated using GPT, it is verified that the results are relevant to the domain data. Through this process, it is expected that more accurate sentences can be generated by using the keywords extracted using text mining techniques as the result of GPT, unlike when only existing text mining techniques are used.

Keywords: Text mining · GPT (Generative Pre-trained Transformer) · Accuracy

1 Introduction

When analyzing past studies to start a new one, you have to sift through a large amount of research data. Manually categorizing them is time-consuming and inefficient. To compensate for this, many researchers have turned to text mining techniques to extract keywords and make it easier to understand research trends.

Text mining is a technique that aims to process unstructured and semi-structured text data by extracting useful information based on natural language processing techniques. By utilizing this technique, it is possible to extract meaningful information from a large amount of domain text data, check its relevance to the domain, and obtain various results. As a result, the domain category of the text can be found and analyzed to reproduce the information.

However, if the text mining process simply extracts preprocessed text documents, it is difficult to analyze or evaluate whether the extracted keywords are relevant to the domain data. Therefore, it is necessary to analyze patterns and trends in the extracted results to check for accuracy. After these steps, the extracted keywords can be fed as input to a GPT trained with domain data to compare and analyze the generated sentences.

© The Author(s), under exclusive license to Springer Nature Switzerland AG 2024
B. J. Choi et al. (Eds.): IHCI 2023, LNCS 14531, pp. 26–31, 2024.
https://doi.org/10.1007/978-3-031-53827-8_3

2 Related Research

Existing research using text mining is used to analyze trends by extracting keywords from various domain data when analyzing text data among unstructured data, which is increasing exponentially. [1] In this related work, we introduce text mining techniques, similarity and pattern recognition methods, and explain the techniques used in this system. In addition, we explain how GPT is utilized, how it differs from the existing GPT, and where it is used similarly.

2.1 Text Mining

Text mining is a technique that extracts meaningful information from text and is used in applications such as classification, summarization, and clustering. Classification is the process of analyzing the content of a given text and assigning predefined categories to it appropriately. It can automatically categorize documents without the need for manual understanding of the text's content. Classification techniques include similarity of text data and pattern recognition. Figure 1 shows the process of text mining techniques.

Fig. 1. Text mining process

   Summarization is the process of extracting a representative part from a large amount of text data. It helps reduce the time required to access specific documents by providing an understanding of their content. Summarization techniques include surface-level approaches and object-level approaches.

   Clustering is the process of dividing text data into small groups based on content similarity. It is often used to improve information retrieval performance by exploring smaller groups within large amounts of data. Clustering can be further divided into static clustering and dynamic clustering.

   In general, structured data can be directly processed by computers, while text data, also known as unstructured data, consists of natural language and cannot be directly processed by computers. Therefore, structuring textual data is essential for analyzing it effectively. In this study, we use frequency analysis to extract keywords"[2].

3   Research Trend Analysis System Utilizing GPT Based on Text Mining

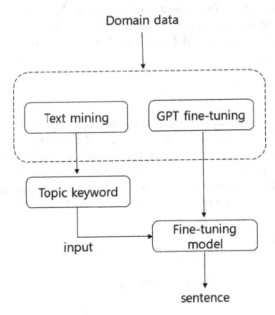

Fig. 2. System concept

In the text mining-based GPT-enabled research trend analysis system proposed in this paper, the user inserts the domain data to be analyzed into the system. The inserted data is extracted as many top keywords as the user wants using text mining techniques, and the keywords are used as prompt values for the GPT model trained to match the extracted domain data. The result is a sentence generated by inserting the keywords as input to the GPT model. Figure 2 shows the overall concept of the system proposed in this paper.

The system that preprocesses specific domain data, extracts keywords using text mining techniques, and generates sentences as inputs to GPT for comparison utilizes the Scopus API to collect specific domain data. Scopus API is an API provided by Elsevier and can be used to find a large number of articles published annually or to identify and organize overall trends. After preprocessing the collected domain data, the system can be constructed using text mining techniques.

3.1 Keyword Extraction

"Assuming that a user is analyzing research trends in a specific domain, this system uses text mining techniques to extract keywords from the preprocessed data. These keywords are extracted from the topic keywords using scoring through the aforementioned frequency-based analysis [3].

Existing text mining techniques can identify trends in research but have limitations when it comes to analyzing differences in trends between years or observing changes over time. To compensate for this, our system employs frequency-based analysis. This is a weighted statistical method that quantifies the importance of a particular keyword across different years" [4].

3.2 Training GPT Models

The system pre-trains GPT on specific domain data and compares the sentences generated by using the resulting keywords extracted by text mining techniques as input to GPT. A fine-tuning process is required to train GPT on the collected domain data and generate sentences.

The GPT model used in this system is the GPT2 model supported by Hugging Face. This model has 1.5 billion parameters and has the advantage that you do not need to declare layer, model, etc. or implement a learning script when using it. Using these models, we fine-tuned them to generate sentences related to domain data. The domain data used for fine-tuning consists of yearly articles collected using the aforementioned Scopus API. The papers collected data from 2001 to 2021 on automated guided vehicles (AGV), which are machines that automatically transport cargo unattended. GPT was pre-trained on the test data and generated sentences similar to the content of the trained paper data. For the same domain data, the results of the text mining technique and the results generated using GPT were compared and analyzed. Two keywords were extracted for each domain by year using text mining techniques, and sentences were generated using GPT by entering test prompts for the same domain data.

Table 1. Text mining results

domain/years	keywords
AGV/2001	control, dispatching, multiple, conflict, CMM
AGV/2019	AGV, dynamic, optimization, intermittent, binary
AGV/2020	AGVS, matrix, DABC, positioning, heuristic

Table 2. GPT results

domain/years	keywords	sentences
AGV/2001	dispatching	Mathematical model evaluates AGVS vehicle requirement base on dispatching rule and traffic management impact
AGV/2019	optimize	This paper optimizes sectional power supply path for logistics AGV using adaptive weight value multi-objective genetic algorithm and validates it through simulation
AGV/2020	DABC	This paper proposes a DABC for solving the multiple automatic guided vehicle dispatching problem in a matrix manufacturing workshop

Tables 1 and 2 above show the respective results for sentence generation using text mining techniques and GPT. Table 2 shows the sentences generated by selecting one of the keywords extracted from Table 1 for a specific domain by year and applying it as a prompt. Before building the system we finally propose in this study, the above experiments allow us to compare the problems and advantages of using text mining techniques and GPT with the same domain data.

As shown in Table 1, when keywords are extracted by text mining techniques from preprocessed domain data, it is possible to know the most used words in the text and to exclude unnecessarily repeated words because of the processing of stop words and the use of scoring through frequency-based analysis. However, it is not easy to understand the overall trend of the data based on the extracted words alone, and it is difficult to find the relevance of the extracted keywords to the domain. In this study, we experimented with classification techniques using text similarity and pattern recognition to compensate for these shortcomings.

As shown in Table 2, when sentences are generated using pre-trained GPT by entering prompts related to domain data, the results are good, but it is necessary to check whether the generated sentences are related to the domain and whether the keywords extracted using text mining techniques are in the sentences.

3.3 Generate Sentences

The keywords and domain data extracted using text mining techniques are inserted into the fine-tuned GPT model as input to generate sentences. To generate consistent sentences from the GPT model, the max_length value, which can adjust the maximum length of sentence generation, was set to 90, and the temperature, which sets the degree of freedom of the sentence, was set to 0.8.

4   Conclusion

The data of 2001, 2019, and 2020 for AGVs were applied to the system proposed in this paper. The top keywords were extracted from the preprocessed data of each domain through text mining techniques, and these keywords were used as inputs to GPT, and the results are shown in Table 2.

Table 3. Text mining results

domain/years	domain data/generated sentences	
AGV/2001	Mathematical model evaluates AGVS vehicle requirement base on dispatching rule and traffic management impact	Based on the effect of dispatching rule and traffic management method on system performance, a mathematical model was put forward, with which the number of vehicles that an AGVS needs can be easily evaluated
AGV/2019	This paper optimizes sectional power supply path for logistics AGV using adaptive weight value multi-objective genetic algorithm and validates it through simulation	Modern flow cytometry technology has enabled the simultaneous analysis of multiple cell markers at the single-cell level, and it is widely used in a broad field of research
AGV/2020	This paper proposes a DABC for solving the multiple automatic guided vehicle dispatching problem in a matrix manufacturing workshop	a discrete artificial bee colony algorithm (DABC) is presented together with some novel and advanced techniques for solving the problem

The results of manually classifying and analyzing the data to see if the sentences shown in Table 2 can generate sentences similar to the yearly data of each domain are shown in Table 3 above. By directly comparing the main content of each data and the sentences generated by GPT through keywords, we found that the generated sentences were consistent with the content of the domain data.

Acknowledgement. Following are results of a study on the "University innovation" project, supported by the Ministry of Education and National Research Foundation of Korea.

References

1. Jo, T.-H.: Concepts and applications of text mining. Knowl. Inf. Infrastruct. **5**, 76–85 (2001)
2. Kim, J-S.: A review of big data utilization and related technologies. J. Korean Content Soc. **10**(1), 34–40
3. Lee, K., Choi, H.: Comparative analysis of keywords in winter olympics related studies using text mining techniques. Korean Soc. Sports Meas. Eval. **9**, 67–76 (2018)
4. Kim, S., Cho, H., Kang, J.: The status of using text mining in academic research and analysis methods. J. Inf. Technol. Architect. **13**(2), 317–329 (2016)

Effect of Speech Entrainment
in Human-Computer Conversation:
A Review

Mridumoni Phukon(✉) and Abhishek Shrivastava

Indian Institute of Technology Guwahati, Guwahati, India
{mridumonip,shri}@iitg.ac.in

Abstract. The phenomenon of entrainment in conversation is the process where participants become more similar to each other in terms of different verbal and non-verbal aspects such as acoustic-prosodic, lexical, syntactic, pitch, and speech rate. This process of becoming similar to each other is the key to effective human-human conversation. To replicate the effectiveness observed in human-human conversation, it is equally critical to explore the occurrence of entrainment within human-machine conversation. This review article examines the various non-verbal and verbal aspects that machines are able to adapt for improved entrainment in human-machine conversation. Initially, we categorize the specific verbal and non-verbal behaviors of human users that machines are capable of adapting. Subsequently, we analyze the likely challenges that have prevented the speech technology sector from enabling smooth, natural interactions between humans and machines. These obstacles have hindered the industry's ability to leverage the phenomenon of entrainment for more fluid and intuitive human-machine conversation. Finally, we advocate for a mechanomorphic design strategy in human-machine conversation, outlining the rationale for its potential efficacy.

Keywords: Human-Computer Interaction · Conversational
Interfaces · Entrainment · Voice User Interfaces · Turn-Taking

1 Introduction

Entrainment, also known as adaptability, is the process where conversational partners engaged in conversation gradually exhibit similar behavior [30]. Although the concept originated within cognitive psychology, recently it has attracted the attention of professionals from the Human-Computer Interaction (HCI) fields. Utilizing different approaches such as data-driven methodologies and objective metrics, these researchers have explored the implications of entrainment in both human-human and human-computer interactions. Their findings have revealed different aspects of entrainment that lead to efficiency of conversation such as user satisfaction and task efficiency. However, despite their

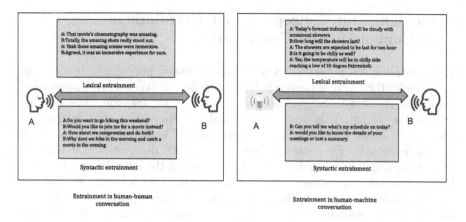

Fig. 1. Different types of entrainment in human-human and human-machine conversation

research to explore entrainment in human-computer conversation and technology advancement, the speech industry however unable to deliver intuitive, voice user interfaces capable of handling smooth, fluent conversation. Existing Voice user interfaces that use wake word (Hey Siri/Alexa)- one shot user launched command, are not up to the user's expectations [16,17,20] (Fig. 1).

In this paper, we mainly analyze the following two research questions.

1) **What are the different entrainment strategies used for fluid and smooth conversation?**
2) **what are the potential benefits of incorporating entrainment in human-machine conversation?**

In the subsequent sections, this paper presents the fundamental concepts about entrainment in human-human conversation (Sect. 2). Section 3 describes different entrainment strategies employed in voice user interfaces and their impact on user experience and satisfaction. Section 4 explains the potential benefits of incorporating entrainment in human-machine conversation. Section 5 concludes by stating the summary and states a couple of directions where we think more research is needed (Table 1).

Table 1. Overview of Research Questions

#	Research Question
RQ1	What are the different entrainment strategies employed in human-machine conversation for smooth and fluid communication? (Sect. 3)
RQ2	What are the potential benefits of incorporating entrainment into human-machine conversation? (Sect. 4)

2 Fundamental Concept About Entrainment in Human-Human Conversation

Speech entrainment is the tendency of conversational partners to become similar to each other during conversation [37]. Different studies have consistently shown that conversational partners often adjust their speaking style during a conversation [21,38]. This occurs across multiple dimensions such as tone, linguistic style, speed of speech, pitch, voice quality, loudness, and even specific phonetic attributes. For instance, Natale et al. [22] discovered that an interviewer's vocal intensity can directly influence that of the interviewee. Heldner et al. [23] observed that speakers often align their pitch with their conversational partners when providing a backchannel response. Similarly, linguistic styles like word choice and structure also tend to align among people interacting via computer chats, as noted by previous research.

Pardo's [24] work highlights that conversational partners even start to mimic each other's pronunciation over time. Further nuanced findings by Levitan et al. indicate that such synchronization happens not only during specific conversational turns but also throughout the course of the entire conversation. Levitan and colleagues [7] also found that mixed-gender pairings show a higher level of entrainment on various acoustic-prosodic factors during computer game interactions. Lubold et al. [25] corroborated this by revealing that in collaborative learning scenarios, people often adapt to each other's acoustic intensity. One recent study by Wynn et al. [27] investigated speech entrainment in neurotypical early adolescents and found that the ability to align speech patterns with interlocutors improves throughout adolescence but does not fully reach adult levels. The study revealed that speech entrainment is predictive of conversational quality and efficiency, suggesting the developmental importance of this skill and its potential impact on successful communication. Again, another study by Wynn et al. [28] found that in order to have a successful speech interaction, the prosodic entrainment that involves rhythmic adaption should occur.

3 Entrainment/Alignment Strategies Employed in Voice-Based Human-Machine Conversation and Their Impact on User Experience and Satisfaction

In this section, we have analyzed research papers that experimented the phenomenon of entrainment during voice-based human-machine conversation. Tables 2 and 3 list the summary of the papers considered for analysis.

Table 2. Summary of studies on different types of entrainment

Paper title	Types of entrainment	Language	Impact on User Experience and Satisfaction
[1]	Acoustic-prosodic	English	positive association between entrainment and an avatar's perceived reliability and likability
[2]	Acoustic-prosodic entrainment and disentrainment in SDSs, focusing on speech rate, pitch, and intensity	Argentine Spanish	Acoustic-prosodic adaptation in Speech Dialogue Systems (SDSs) affects user trust, with intensity entrainment increasing it, and pitch entrainment decreasing it
[3]	Acoutsic-prosodic	slovac	Making avatars more engaging can increase people's trust in them. A person's gender can influence their trust in avatars. Women generally prefer advice from less engaging avatars
[31]	Acoustic-Prosodic Entrainment, specifically pitch and intensity	English	Entrainment fosters emotional positivity and learning, and can improve cooperation in team settings like board games
[32]	Prosodic entrainment in Mandarin conversations, specifically focusing on proximity, convergence, and synchrony over tones, turns, and conversational levels	Mandarin	The paper implies that studying entrainment in Mandarin may enhance communication and inform the design of speech systems for human-computer interaction
[38]	Global and local, synchrony, constant, convergent, backchannel-inviting cues	Examines speakers of Standard American English and Mandarin Chinese	Improved user perception of conversational agents, increased naturalness in human-computer interactions
[43]	speech-rate entrainment	English	Users slowed their speech rate during interaction with the social bot, especially in the in-lab context
[39]	Lexical entrainment	English	Lexical entrainment leads to more successful and engaging conversations, facilitates shared understanding, and reduces ambiguity. Lack of lexical entrainment in conversational systems may lead to sub-optimal responses and user dissatisfaction
[5]	Lexical alignment	Chinese (Mandarin)	Stronger alignment in HCI, influenced by social skill in HHI
[6]	Lexical Alignment	Not specified	Alignment increases positive evaluation of agent competence

Table 3. Summary of studies on different types of entrainment

Paper title	Types of Entrainment	Language	Impact on User Experience and Satisfaction
[35]	Lexical Entrainment	Not specified	Improved recognition of referred-to objects by robots
[9]	syntactic alignment	Not specified	Enhanced syntactic alignment with the presence of an interlocutor, influencing communicative efficiency
[36]	Syntactic Alignment	Not specified	No effect on syntactic alignment, significant effect on user satisfaction with different voice types

3.1 Acoustic-Prosodic Entrainment

When people converse, they often start to match each other's way of speaking in terms of loudness, speed, and pitch, as well as the words and sentence structures they use. This is because they want to either get along better or show that they are different from each other. Studies have shown that this matching can lead to fluent and smooth interactions during human-human conversation [22,42].

Acoustic-prosodic entrainment, the phenomenon where conversational partners adapt to each other's speech patterns, has been shown to significantly affect user interaction with conversational avatars and AI interfaces. Levitan et al. [1] demonstrated how this entrainment can be implemented in avatars for more natural interaction. Gálvez et al. [2]) furthered this by empirically establishing a link between entrainment and the perceived trustworthiness of avatars, a sentiment echoed by Beňuš et al. [3] in their examination of trust in human-computer interaction. Paletz et al. [31] expanded the scope by exploring how team personality composition affects entrainment, suggesting broader implications for collaborative AI tools (Small Group Research). Cohn et al. [43] their study showed that people speak slower to an Amazon Alexa socialbot than usual, in order to be clear. They also match their speed to the bot's last speech rate, showing adaptation. They found that changes were greater in a lab setting and didn't change much with speech recognition errors or how users felt about the talk. Collectively, these studies underscore the importance of prosodic patterns in AI system design for improved user trust and interaction, pointing to a need for further research on individualized user experiences and effectiveness in team settings.

3.2 Lexical Entrainment

Lexical entrainment is when people in a conversation start using the same words or phrases [33]. This usually happens without participants even realizing it and helps make the conversation flow more smoothly. Hyuang et al. [4] in their study on conversational systems highlighted the importance of lexical entrainment in making digital assistants seem more in tune with users, demonstrating that current response generation models need to better incorporate this phenomenon for more natural interactions. They also examined the impact of social skills on lexical alignment and found that in Human-human interaction, individuals' social skills, particularly those related to considering others' social standing, can predict the degree of lexical alignment. Interestingly they found, that this alignment was stronger in human-computer interaction, suggesting that users adapt their language more when interacting with computers than with humans.

In the context of Human-Robot interaction Lio et al. in their study [35], showed that robots could induce lexical entrainment, leading to two identified types: term-specific and type-specific entrainment. This capability has significant implications for improving robot comprehension and interaction with humans, especially in recognizing and referring to objects.

The study by Huiyang et al. [5] delved into the psychological effects of lexical (non-) alignment by virtual agents on human users. The study found that users rated the agent more competent when it shifted from non-alignment to alignment, whereas the shift from alignment to non-alignment negatively affected the agent's evaluation. This highlights the dynamic nature of alignment in affecting user perceptions.

Across these studies, a common theme is that lexical entrainment is not static but rather a dynamic process influenced by multiple factors, including the type of interlocutor (human or artificial), the context of the interaction, and the social skills of the human participants. While the research predominantly shows positive effects of lexical entrainment on user experience and satisfaction, the complex nature of communication and individual differences means that the outcomes of lexical entrainment can vary widely. Moreover, the studies suggest that incorporating lexical entrainment into the design of conversational agents, robots, and virtual agents can improve the naturalness and effectiveness of interactions, potentially leading to more positive user experiences. However, the exact mechanisms, including the psycho-physiological effects, are still areas requiring further exploration to fully understand how alignment processes impact human-machine interactions.

3.3 Syntactic Alignment

Syntactic alignment indicates the mirroring of grammatical structures between conversational partners [41]. In a study by Schhot et al. [9] reported how much people tend to copy the sentence structures they've just heard-a phenomenon known as "syntactic priming" or "structural persistence". They found that this copying behavior is stronger when people are actually talking to someone else, compared to when they're just listening to a recorded voice. However, the study didn't find evidence that the extent to which one person copies sentence structures affects how much their conversation partner does the same. The study by Cowan et al. [36] in their study found that the voice type of a computer interlocutor does not significantly affect syntactic alignment. However, user satisfaction varied significantly with voice type; a basic computer voice was rated significantly lower in satisfaction compared to human-like and advanced computer voices. This suggests that while voice type influences user satisfaction, it does not impact linguistic alignment in human-computer dialogue. The study also challenges the use of syntactic alignment as a metric for interaction satisfaction, as no significant correlation was found between the two.

The three entrainment strategies—acoustic-prosodic, lexical, and syntactic alignment—offer distinct pathways to enhance human-computer conversation. While acoustic-prosodic focuses on mimicking sound attributes, lexical and syntactic alignments work on word and sentence structure synchronization. Entrainment/alignments are not only reciprocal but also adaptive, catering to specific interaction conditions. The strategic utilization of these alignments can lead to a more natural and effective conversation flow, although complexities and nuances in their application should be carefully considered.

4 Potential Benefits of Incorporating Entrainment in Voice-Based Human-Machine Conversation

Incorporating entrainment in voice-based human-machine conversation offers a multitude of benefits for creating more natural and effective interactions between humans and computer systems. The following elucidates the potential advantages:

4.1 Naturalness

Naturalness is important as it bridges the gap between human-human and human-machine interaction. A conversation that provides naturalness reduces the cognitive load on users. Nankova et al. [10] checked out two different sets of dialogues to see if entrainment actually leads to smoother, more successful conversations. Their result reported that when people do start to sound like each other by using common words, the conversation tends to be natural and tasks get completed more successfully. Benus et al. [3] in their study show that voice entrainment enhances learning outcomes, and perceived naturalness suggests a positive correlation with trust and effective communication.

4.2 Enhanced Trustworthiness

Conversational entrainment, particularly when it concerns voice alignment, plays a significant role in shaping perceptions of rapport and trustworthiness during interactions. Galvez et al. [2] in their study see how mimicking human voice qualities (like tone and rhythm) can make spoken dialog systems) better. Their results show that when these systems match certain voice aspects with the user, people tend to trust them more and think they're more capable. Interestingly, matching the loudness level helped gain trust, but mimicking the pitch of the voice didn't work as well; in fact, it had a negative significant effect. Benus et al. [3] in their study to see the relationship between speech entrainment and trust develop a Harry Potter-inspired game where players interact with two owl-like avatars that offer advice, with their voices either mimicking the player's speech patterns or doing the opposite. Players then decide whose advice to follow, which gauges their trust in each avatar. Their findings show that intensifying entrainment may improve trust, and that a player's gender influences how they react to entrainment. Their study found that women actually prefer advice from avatars that don't mimic their voices. Another study by Linneman et al. [10] examined how users perceive adaptive and non-adaptive spoken dialogue systems regarding trustworthiness and usability. Their results revealed that when the spoken dialogue system (SDS) does not align with users, they perceive higher cognitive demand and on the other hand they feel that aligned SDS have more integrity and likable.

5   Summary and Future Prospect

As the review has shown, entrainment is an innate aspect of dialogue, operating subconsciously across several verbal dimensions like acoustic-prosodic features, lexical choices, and syntactic structures. Scholars from a range of disciplines have exploited this phenomenon to enhance the quality of human-machine inter-actions. By capitalizing on entrainment in human-machine dialogues, several enhancements have been observed for the human participants in human-machine conversation, including 1) natural conversational flow, and 2) increased levels of trust. 3) low cognitive load.

However, during the review, we were unable to find studies on entrainment that enhance the experience of turn-taking in human-machine interfaces. The underlying reason for this may be many- the majority of the studies exploit entrainment in an anthropomorphic design where the machine is designed to adapt human user's verbal behaviors (acoustic-prosodic, lexical, syntactic), etc. Entraining the aspects of turn-taking in human-machine dialogues presents a complex challenge. Humans use complex cues like syntax, temporal, and prosody, which are difficult for machines to adapt. Adding to this are individual variabil-ity and the need for real-time processing. Contextual factors further complicate the machine's ability to adapt. Ethical concerns, such as the potential for manip-ulation of temporal aspects (since the existing VUIs in the market do not allow manipulation of temporal aspects of turn-taking), add another layer of complex-ity. These challenges highlight the need for multi-disciplinary research to advance conversational systems.

In the rest of the section, we will discuss some potential future research on mechanomorphic entrainment.

5.1   Implementing Entrainment for Mechanomorphic Design

While conducting the literature review, we observed that a significant portion of the existing studies focus on the anthropomorphic design of VUIs- where the interfaces are designed to mimic human-like characteristics or behaviors instead of human users mimicking machine's behaviour. The prevailing rationale behind this approach is the aspiration to emulate human-like qualities in VUI interactions. This perspective is fueled by the belief that anthropomorphism fosters more engaging and natural communication between humans and machines/interfaces. However, the anthropomorphic design approach also has its drawbacks. Specifically, it tends to foster a mismatch between the capabilities and expectations of human users and the functionalities offered by current technology. This dissonance leads to what has been termed a "habitability gap" [7]. Essentially, as these systems strive to be more human-like, they create an illusion of capabilities that they cannot fully deliver, ultimately leading to user dissatisfaction and disengagement. Moreover, the anthropomorphic design has the risk of evoking the "uncanny valley" [11]. Uncanny Valley- a situation where designing something that feels almost like a human, can make people feel weird. Contrary to the prevalent anthropomorphic trend, a nascent body of literature suggests a counter-intuitive, mechanomorphic approach to VUI design. Studies indicate that mechanomorphic designs show promise for delivering improved task success [12,13]. Mechanomorphism, being devoid of the complications of simulating human qualities, enables a more straightforward user-machine interaction, thus sidestepping the uncanny valley effect [11]. In addition, a specific investigation into prosodic entrainment between humans and machines found that mechanomorphic designs have enabled users to more naturally adapt to the speaking rate of simulated dialogue systems [2].

While these insights are illuminating, there remains a notable scarcity of research that systematically compares the effectiveness, user satisfaction, and entrainment capabilities of anthropomorphic versus mechanomorphic VUIs. This lack of comparative studies represents a significant research gap. Addressing this gap is crucial for providing empirical evidence to guide future VUI design strategies.

5.2 Manipulation of Temporal Aspects of the Machine to Provide Enhanced Turn-Taking

In conversation between humans, the time it takes to respond is influenced by various factors, such as cognitive workload, individual personality traits, and the specific context of the conversation [14]. Replicating such naturalistic turn-taking timing in anthropomorphic designs presents a significant challenge due to its context-sensitive nature. A novel method by Phukon et al. [15] employed a mechanomorphic Voice User Interface (VUI) prototype, specifically altering its temporal characteristics to investigate whether users would synchronize their turn-taking behavior with the machine. Their findings indicate that human participants did adjust their timing, resulting in a more fluid and intuitive interaction experience.

5.3   Manipulation of Machine's Cues for Better Entrainment

Research has shown that entrainment happens in all different verbal aspects of conversation (acoustic, prosodic, syntactic) [1–3,5,6,9] which provides various benefits such as increased trust, naturalness etc. However, despite all these developments speech industry is unable to provide us with a human-machine interface that provides smooth intuitive interaction between humans and machines. Existing interfaces are quite unnatural with one-shot user launch command/wake word hiding its capabilities. This nature of existing interfaces provide user no cues about their inabilities and capabilities leading to frustration [16]. Thus we proposed to design an interface that itself reveals its capabilities/inabilities through its cue so that human users adapt to its behaviour accordingly leading to a smooth intuitive conversation.

References

1. Levitan, R., et al.: Implementing Acoustic-Prosodic Entrainment in a Conversational Avatar. Interspeech. vol. 16 (2016)
2. Gálvez, R.H., et al.: An empirical study of the effect of acoustic-prosodic entrainment on the perceived trustworthiness of conversational avatars. Speech Commun. **124**, 46–67 (2020)
3. Beňuš, Š., et al.: Prosodic entrainment and trust in human-computer interaction. In: Proceedings of the 9th International Conference on Speech Prosody. Baixas, France: International Speech Communication Association (2018)
4. Iio, T., et al.: Lexical entrainment in human robot interaction: do humans use their vocabulary to robots? Int. J. Soc. Robot. **7**, 253–263 (2015)
5. Huiyang, S., Min, W.: Improving interaction experience through lexical convergence: the prosocial effect of lexical alignment in human-human and human-computer interactions. Int. J. Hum. Comput. Interact. **38**(1), 28–41 (2022)
6. Nuñez, T.R., et al.: Virtual agents aligning to their users. Lexical alignment in human-agent-interaction and its psychological effects. Int. J. Hum Comput Stud. **178**, 103093 (2023)
7. Levitan, R., et al.: Entrainment and turn-taking in human-human dialogue. In: 2015 AAAI Spring Symposium Series (2015)
8. Linnemann, G.A., Jucks, R.: can i trust the spoken dialogue system because it uses the same words as i do?-influence of lexically aligned spoken dialogue systems on trustworthiness and user satisfaction. Interact. Comput. **30**(3), 173–186 (2018)
9. Schoot, L., Hagoort, P., Segaert, K.: Stronger syntactic alignment in the presence of an interlocutor. Front. Psychol. **10**, 685 (2019)
10. Hirschberg, J.B, Nenkova, A., Gravano, A.: High frequency word entrainment in spoken dialogue (2008)
11. Mori, M., MacDorman, K.F., Kageki, N.: The uncanny valley [from the field]. IEEE Robot. Autom. Mag. **19**(2), 98–100 (2012)
12. Lubold, N., Pon-Barry, H., Walker, E.: Naturalness and rapport in a pitch adaptive learning companion. In: 2015 IEEE Workshop on Automatic Speech Recognition and Understanding (ASRU). IEEE (2015)
13. Balentine, B.: It's Better to Be a Good Machine Than a Bad Person: Speech Recognition and Other Exotic User Interfaces in the Twilight of the Jetsonian Age. ICMI Press (2007)

14. Strömbergsson, S., et al.: Timing responses to questions in dialogue. Interspeech. vol. 2013 (2013)
15. Phukon, M., Shrivastava, A., Balentine, B.: Can VUI turn-taking entrain user behaviours? voice user interfaces that disallow overlapping speech present turn-taking challenges. In: Proceedings of the 13th Indian Conference on Human-Computer Interaction (2022)
16. Goetsu, S., Sakai, T.: Different types of voice user interface failures may cause different degrees of frustration. arXiv preprint arXiv:2002.03582 (2020)
17. Goetsu, S., Sakai, T.: Voice input interface failures and frustration: developer and user perspectives. In: Adjunct Proceedings of the 32nd Annual ACM Symposium on User Interface Software and Technology (2019)
18. Kim, J., Jeong, M., Lee, S.C.: Why did this voice agent not understand me?" error recovery strategy for in-vehicle voice user interface. In: Proceedings of the 11th International Conference on Automotive User Interfaces and Interactive Vehicular Applications: Adjunct Proceedings (2019)
19. Jiang, J., Jeng, W., He, D.: How do users respond to voice input errors? Lexical and phonetic query reformulation in voice search. In: Proceedings of the 36th International ACM SIGIR Conference on Research and Development in Information Retrieval (2013)
20. Luger, E., Sellen, A.: like having a really bad PA the gulfbetween user expectation and experience of conversational agents. In: Proceedings of the 2016 CHI Conference on Human Factors in Computing Systems, pp. 5286–5297 (2016)
21. Tannen, D.: Talking Voices: Repetition, Dialogue, and Imagery in Conversational Discourse, vol. 26. Cambridge University Press, Cambridge (2007)
22. Natale, M.: Convergence of mean vocal intensity in dyadic communication as a function of social desirability. J. Pers. Soc. Psychol. **32**(5), 790 (1975)
23. Heldner, M., Edlund, J., Hirschberg, J.B.: Pitch similarity in the vicinity of backchannels (2010)
24. Pardo, J.S.: On phonetic convergence during conversational interaction. J. Acoust. Soc. Am. **119**(4), 2382–2393 (2006)
25. Lubold, N., Pon-Barry, H.: Acoustic-prosodic entrainmentand rapport in collaborative learning dialogues. In: Proceedings of the 2014 ACMworkshop on Multimodal Learning Analytics Workshop and Grand Challenge, pp. 5–12 (2014)
26. Manson, J.H., et al.: Convergence of speech rate in conversation predicts cooperation. Evol. Hum. Behav. **34**(6), 419–426 (2013)
27. Wynn, C.J., et al.: Speech entrainment in adolescent conversations: a developmental perspective. J. Speech, Lang. Hearing Res. 1–19 (2023)
28. Wynn, C.J., Barrett, T.S., Borrie, S.A.: Rhythm perception, speaking rate entrainment, and conversational quality: a mediated model. J. Speech Lang. Hear. Res. **65**(6), 2187–2203 (2022)
29. Borrie, S.A., et al.: Syncing up for a good conversation: a clinically meaningful methodology for capturing conversational entrainment in the speech domain. J. Speech, Lang. Hearing Res. **62**(2), 283–296 (2019)
30. Coupland, J., Coupland, N., Giles, H.: Accommodation theory. communication, context and consequences. Contexts Accommodation, 1–68 (1991)
31. Paletz, S.B.F., et al.: Speaking similarly: team personality composition and acoustic-prosodic entrainment. Small Group Res. 10464964231178748 (2023)
32. Xia, Z., Hirschberg, J., Levitan, R.: Investigating prosodic entrainment from global conversations to local turns and tones in mandarin conversations. Speech Commun. **153**, 102961 (2023)

33. Brennan, E.S.: Lexical choice and conceptual pacts in conversation. J. Exp. Psychol. Learn. Mem. Cogn. **22**, 1482–1493 (1996)
34. Shen, H., Wang, M.: Effects of social skills on lexical alignment in human-human interaction and human-computer interaction. Comput. Hum. Behav. **143**, 107718 (2023)
35. Iio, T., et al.: Lexical entrainment in human-robot interaction: can robots entrain human vocabulary?. In: 2009 IEEE/RSJ International Conference on Intelligent Robots and Systems. IEEE (2009)
36. Cowan, B.R., Branigan, H.P., Beale, R.: Investigating the impact of interlocutor voice on syntactic alignment in human-computer dialogue. In: The 26th BCS Conference on Human Computer Interaction, vol. 26 (2012)
37. Beňuš, Š.: Social aspects of entrainment in spoken interaction. Cogn. Comput. **6**, 802–813 (2014)
38. Levitan, R.: Acoustic-Prosodic Entrainment In Human-human and Human-computer Dialogue. Columbia University, New York (2014)
39. Levitan, R., et al.: Acoustic-prosodic entrainment and social behavior. In: Proceedings of the 2012 Conference of the North American Chapter of the Association for Computational Linguistics: Human language Technologies (2012)
40. Shi, Z., Sen, P., Lipani, A.: Lexical Entrainment for Conversational Systems. arXiv preprint arXiv:2310.09651 (2023)
41. Giles, H., Powesland, P.F.: Speech Style and Social Evaluation. Academic Press, Cambridge (1975)
42. Levitan, R., Hirschberg, J.B.: Measuring acoustic-prosodic entrainment with respect to multiple levels and dimensions (2011)
43. Cohn, M., et al.: Speech rate adjustments in conversations with an Amazon Alexa socialbot. Front. Commun. **6**, 671429 (2021)

HUCMD: Hindi Utterance Corpus for Mental Disorders

Shaurya Prakash[1], Manoj Kumar Singh[1]([✉]), Uma Shanker Tiwary[2], and Mona Srivastava[3]

[1] Banaras Hindu University, Varanasi 221005, India
{shprakash,manoj.dstcims}@bhu.ac.in
[2] Indian Institute of Information Technology, Prayagraj 211015, India
ust@iiita.ac.in
[3] Institute of Medical Sciences, Banaras Hindu University, Varanasi 221005, India
mona.srivastava1@bhu.ac.in

Abstract. As our knowledge, there is no dialog system for mental health-care domain in Hindi. This may be due to unavailability of user utterances corpora in Hindi for this domain. In this paper, we propose a novel algorithmic approach for user utterance generation in Hindi by considering dialects, linguistic attributes, symptoms, frequency of symptoms, and intensity of symptoms and history of symptoms. We use nine symptoms (anger, emptiness, fear, irritation, restlessness, suicide, sadness, tension, worry) as given in DSM5, ICD-11, and WHO guideline. These symptoms were used for generation of utterances and validation of the generated utterances for different type of mental diseases. We collected utterances by interviewing patients in clinic and found that it closely match to the utterance generated by proposed algorithm. The generated utterance corpus is also validated using machine learning methods in the framework of CNN, Bi-LSTM and Dense.

Keywords: Frequency Words · Intensity Words · History Words · Utterance · Slot · Intent · Formal · Semi-Formal · kernel · max pooling

1 Introduction

India has a labyrinth of geographical, cultural and linguistic diversity along with plethora of beliefs and dialects that predominates in different pocket/domains of this diversity. Providing quality mental health care to its huge population is very challenging. For example, in the State of Jharkhand, despite having world class facility like Central Institute of Psychiatry, there is a huge treatment gaps for different types of mental disorders, such as the treatment gaps of 76.1% for common mental disorders, 80.1% for depressive disorders and 33.3% for severe mental disorders. On the population of 100,000 only 103 psychiatrists, 19 clinical psychologists and 8 psychiatric social workers are available. It was also recommended to develop solutions like near-to-home evidence based electronic decision support system to improve mental health framework [21]. In order to provide awareness, privacy and mental healthcare, dialog system [1] seems to be a good choice for those who refrain themselves from seeing a doctor either due to social stigma or disease borne hesitations or sometimes in cases of non-availability of mental health facilities/institutions. Hindi language based mental health assistant can make its reach to approx 600 millions Hindi speaking population via mobile phones and personal computers. To develop such systems most basic requirement is user utterance data. To the best of our knowledge, no such mental disorder-based utterance repository is available in Hindi.

2 Challenges

Hindi is a low-resource language which means limited availability and access to natural language processing (NLP) resources like datasets, corpora, parts of speech taggers, vocabularies etc. Without knowing the history of medical or mental conditions, interviewing a mental patient may lead to undesirable outcomes. Supervision of medical expert ensures safe and sound interview process but their availability to supervise such interviews is limited. Availability of patient who are ready to interact with interviewer is another issue. Due to free word order [2], cultural influences [21], popular figures of speech and slangs which varies region to region, a statement in one dialect appears to be totally different in the others. A person prefers privacy and refrains herself/himself from indulging in any such activities (interviews/interactions) which she/he thinks encroaches her/his personal space. This keeps researchers from getting data from patients. Interviewing a patient requires ethical and medical clearances at different levels which requires time and efforts, and the governing body (committee/ board/ authority) may or may not approve such activities. If data collection forms are floated on internet to gather data, result is non-natural language data format i.e., records.

3 Related Work

ATIS (Air Travel Information System) [11] by Microsoft CNTK, has BIO (Beginning, Inside, Out) scheme to labeled user utterances. SNIP [10] uses similar approach; the only difference between ATIS and SNIP is the domain of interest. The SNIP dataset is mostly used for benchmarking purposes for tasks like slot filling and intent detection. The TamilATIS [12] is a translated version of ATIS; the original dataset has been translated to Tamil to provide a natural language resource in Dravidian languages. HDRS [4] is a restaurant search dataset in Hindi that is based on the WOZ (Wizard of Oz) paradigm and has a one-to-one multi-turn conversation to facilitate the belief state tracking models [9]. The ATIS and SNIP datasets appear to be more inspiring for utterance generation in a limited resource environment. Transformers capture sequence features as well as utterance features simultaneously that enable joint intent detection and slot filling tasks [3]. The JointBert [5] has used transformer networks for joint intent detection and slot filling tasks using the ATIS and SNIP datasets. The DSM-5 [15], ICD-11 [16], and WHO report [17] provide detailed information about symptoms and disease classifications. Difficulties and challenges are mentioned in the NIMHANS report [21].

4 Proposed Approach for HUCMD

Despite having different dialects, figures of speech, slangs, and word orders, communication between the doctor and the patient takes place in formal (the nationwide accepted standard form of Hindi in India) or semi-formal (the few most prominent dialects of Hindi, either used by a very large population or spoken over significant geographies in India). Capturing the dialect-based variations and incorporating linguistic features in first-person, one-to-one conversation form, sentences are subdivided into substructures like symptom words, intensity words, frequency words, history words, trailers, and actor words. By analysis, experience, and observation, the following are the suggested substructures:

4.1 Symptom Words

A set of 13 commonly spoken symptom words was chosen that express symptoms of mental disorders, along with their inflectional derivatives and synonyms. Example: घबराहट (Restlessness), घबरा, घबराया,etc.

4.2 Intensity Words

The extent of the feeling of depressive symptoms provides partial information about the depression stages. So far, the total number of such words or figures of speech is 59. These are further subdivided into three categories: high, medium, and low. Examples: B_MediumIntensity, I_Medium Intensity tags were assigned to tag medium-intensity words. थोड़ा-अधिक ,थोड़ा-ज्यादा ,थोड़ी-अधिक, औसत, मामुली, सामान्यetc.

4.3 Frequency Words

The total number of most-used 45 words that express the frequency of the occurrence of a symptom has been identified and subdivided into three sub-categories: Always, Often and Sometimes. Example: High: बराबर (B_Always), हमेशा (B_Always), सदैव, हरवक्त(B_AlwaysI_Always), ज्यादातर (B_Often), आमतौरसे(B_OftenI_OftenI_Often), अक्सर(B_Often), कभी-कभार(B_Sometimes), यदाकदा(B_Sometimes) etc.

4.4 History Words

To establish the symptom occurrence pattern, whether they are occurring or have occurred in the past, history words are used. Example: है, हैं, हूं, थी, था, थे.

4.5 Actor Words

A word denoting a subject in an utterance. Example: मैं, मुझे, मुझको, हम.

4.6 Trailer Words

These words serve as sentence endings. More precisely, these constructs are not only end sentences but also provide partial information about the history of symptoms. Trailers: 'से घिरा रहता हूँ', 'हो जाया करती हैं', 'हो जाया करती हूं', 'का अनुभव करते हैं','का अनुभव करते थे'.

4.7 Utterance Generation Algorithm

There are four types of sentences that were frequently used by the patients during interactions. We chose four main kinds of sentence structures to develop this repository.

Basic Sentences: Sentences with only symptoms and no frequency or intensity information. Example: हम उदास रहते हैं.

Prepend Sentences: Sentences where frequency information comes first. Example: कभी कभी हम उदास रहते हैं.

Append Sentences: Frequency information follows the actor-words... Example: कभी कभी उदास रहते हैं.

Headless Sentences: Sentences without actor-words. Example: कभी कभी उदास रहते हैं

Algorithm 1: Generation Natural Sentences
Inputs   : actor-words , symptoms , trailers
Output   : Natural Hindi Sentences

Step1: generateList = [(actor words$_1$, symtopms$_1$, trailers$_1$)........]
Step2: Generator(generationList) :
Step3: for actor-words$_i$, symtopms$_i$, trailers$_i$ in generationList :
          Basic Sentences = [actor-words$_1$+ symtopms$_1$ + trailers$_1$]
Step4: for each frequency_list in [freq_high, freq_med, freq_low] , do
          for frequency word in frequency_list :
              prepend = [frequency word + actor-words$_1$+ symtopms$_1$ + trailers$_1$]
              append = [actor-words$_1$+frequency words+ symtopms$_1$ + trailers$_1$]
              headless = [frequency words + symtopms$_1$ + trailers$_1$]
              frequency_list_merge = [basic,prepend,append,headless]
          end for
Step5: for each intensity_list in [inten _high, inten_medium, inten__low] , do
          for intensity word in intensity _list :
              prepend = [intensity word + actor-words$_1$+ symtopms$_1$ + trailers$_1$]
              append = [actor-words$_1$ + intensity words+ symtopms$_1$ + trailers$_1$]
              headless = [intensity words + symtopms$_1$ + trailers$_1$]
          low_intensity   = merge [basic, prepend, append, headless],
          medium_intensity = merge [basic, prepend, append, headless],
          high_intensity   = merge [basic, prepend, append, headless]
          end for

4.8 Generated Utterances

The algorithm discussed above is used to generate utterances. Utterances have been generated for a total of nine symptoms: three dialects, two sexes, and 16 sub-symptoms in JSON format. In the first version, a total of 5019739 tagged sentences have been generated for nine symptoms so far. These utterances are generated with the help of dictionaries containing frequency, intensity, and historical information. Once utterances are generated, they are further tagged with corresponding tags (Table 1).

Table 1. Utterances generated by Algorithm 1.

Sex	Dialect	Symptom	Sentence Type	Utterances
Female	Formal	Anger	Append	'मैं अक्सर गुस्से में रहा करती हूँ'
Female	Formal	Emptiness	Prepend	'गाहे ब गाहे जबरदस्त खोखलापन को महसूस करती हूं'
Male/Female	Semi-Formal	Worry	Append	'हमको यदा कदा बेहिसाब चिंता लगी रहती थी'
Female	Formal	Sad	Headless	'दुःखी रहती थी'
Male	Formal	Fear	Prepend	'बहुत ज्यादा डर जाया करता हूँ'

5 Dataset Validation

5.1 Bidirectional Long Short Term Memory (Bi-LSTM)

A bidirectional long-short memory network [19] remembers the information for a long period of time. It does not suffer from any disadvantages of RNN, like exploding gradients and vanishing gradients. Bi-LSTM is often used for sequence labelling tasks because it collects two types of features: one in the forward direction and the other in the backward direction. Finally, these two features are merged into a single feature vector. Bi-LSTM also performs well for classification tasks.

5.2 Convolution Neural Network

In a convolutional neural network (CNN) [18], a sentence is represented as a sequence of words:

$$S_i = x_1, x_2, \cdots, x_n, \tag{1}$$

where x_i is a word. Given that embedding vector w_i, x_i is transformed into sequence of p-dimensional word embeddings where, $w_i \in R^p$, hence embedded sentence:

$$S_i = w_1, w_2 \cdots w_n, \tag{2}$$

has dimension of $R^{n \times p}$. A convolution filter of dimension moves over embedded word vectors of sentences in sliding window fashion to produce a feature map by capturing words. When the sliding window finishes one pass, it generates the convolution feature map for words, which is represented by

$$c = [c_1, c_2 .... c_{n-h+1}].\qquad(3)$$

For character-level embeddings, vocabulary is defined by a dictionary of 127 Hindi Unicode characters and some other characters (e.g., {, -}). Every character in the vocabulary was encoded into q-dimensional character embeddings. A word is represented as a sequence of characters,

$$w = c_1, c_2, \ldots c_n,\qquad(4)$$

and has dimension of $c_i \in R^{n \times q}$, which was treated exactly the same way as mentioned above. A convolution filter F' of dimension h' moves over embedded characters-vectors of a word in sliding window fashion to produce feature map c_i' by capturing $c_{i:i-h'+1}$ characters. When sliding window finishes one pass, it generates the convolution feature map for characters, which is represented by

$$c_1', c_2' \cdots c_{n-h'+1}'\qquad(5)$$

5.3 Network Architecture

In this work, the network architecture consists of three types of neural networks. At layer 1, CNN takes input in the form of word embeddings (intent detection) or character embeddings (slot filling). After convolving, max pooling, and concatenation operations, it gives the output feature vector. This output feature vector serves as input to the BILSTM network, which isolates the forward and backward features and concatenates them into a single feature vector. Finally, layer 3 is a fully connected network that does the classification by looking at the output features of BILSTM (Table 2).

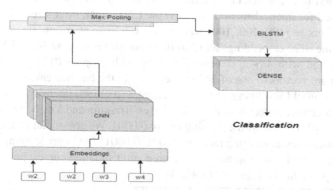

Fig. 1. Representation of network architecture used in Intent Detection framework

Table 2. Training Parameters

Learning Rates	Optimizer	Loss Function	Weight Decay	Task	Library
0.01	SGD	Cross Entropy	0.0001	Slot Filling	Pytorch
0.001	Adam	Cross Entropy	Default	Intent detection	Pytorch

5.4 Sampling Algorithm

To sample data for a particular symptom, we use a parameter called sample_size. Each symptom has been divided into 16 sub-symptoms. The sample_size parameter samples randomly shuffled data from each subsymptom and finally merges them to give the final symptom utterances.

Algorithm 2 : **SampleSentences**(sample_Size, symptom_file_path) :
Inputs   : symptoms_file_path
Output   :Natural Hindi Sentences

Step 1: JsonData = FileRead(symptom_file_path)
Step 2: for field in JsonData.Fields do :
Step 3:   utterances = getDataFrom(fields)
Step 4:   if Sample_size > utterances :
Step 5:   shuffledData = shuffle(utterances)
Step 6:   Data   = shuffledData[:sample_size]
Step 7:   else:
Step 8:   Data   = utterances
Step 9: return Data

5.5 Intent Detection

Intent detection is a sentence classification task. In this work, word-level embeddings have been used for intent detection tasks. At the CNN layer, filter size = [3, 4, 5] and number of filters = [100, 100, 100] were applied, which, after concatenation, finally transformed the word embeddings into 300-dimensional vectors. At the Bi-LSTM layer, we experimented with 4/6-layer networks along with dropout (0.2). The final layer maps the feature vectors to one of 144 classes. Due to the batched environment, Adam Optimizer was used for training. By concatenating symptoms, intensity, and frequency information together, we get a total of 144 classes. Training and loss curves show convergence after 400 epochs with a learning rate of 0.0001. We got similar results on {200, 300, 400} epochs and at learning rates of {0.001, 0.0001}. At lower learning rates, like 0.1, there was a significant gap between training and validation curves; hence, training was stopped prematurely. Example (Table 3):

Input: 'ज्यादातर हम काफी अधिक तनाव का अनुभव करते हैं'
Output: TENSION_often_high

Table 3. Performance Evaluations on CNN + BiLSTM + DENSE (Intent Detection)

Samples (Train, Valid, Test)	Performance Metric (Precision, Recall, F1)	Accuracies (Train, Valid, Test)
154544, 51548, 51567	**0.98, 0.98, 0.98**	**98.20, 99.38, 99.0**
63828, 21291, 21303	0.86, 0.85, 0.84	90.78, 89.64, 85.0
32240, 10771, 10759	0.99, 0.99, 0.99	97.46, 97.63, 99.0

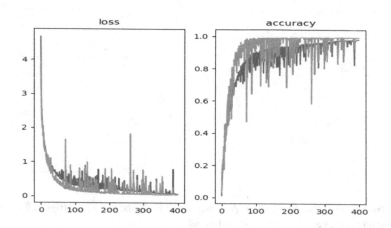

Fig. 2. Training Progress -Intent Detection: Loss and Accuracy curves for CNN + BiLSTM + DENSE network , fluctuations in the learning curves are due to use of dropout.

5.6 Slot Prediction

Slot prediction is a sequence-labelling task. We have chosen a work flow that is similar to the named entity recognition task. Here, character-level embeddings were used, and every sentence is considered a batch of words. Since each sentence serves as a batch of words, learning takes place at every utterance in the training set. Therefore, the stochastic gradient descent optimizer was used in this setup. At the CNN layer, filter size = [3, 4, 5] remains the same, but there was a change in the number of filters = [150, 150, 150] and filters = [150, 150, 150] and dropout = 0.5 was introduced to avoid over-fitting. With the increasing number of layers in BiLSTM validation, accuracy was not changing. As the number of filters increased and the model depth increased by adding more linear layers, the gap between training and validation curves reduced significantly. Example (Table 4):

Input: 'मैं गाहे ब गाहे कमतर कोप महसूस करती हूं'
Output: O B_Sometimes I_Sometimes I_Sometimes B_Lowintensity B_Anger O O B_Existingsymptom

Table 4. Performance Evaluations on CNN + BiLSTM + DENSE (Slot Filling)

Samples (Train, Valid, Test)	Performance Metric (Precision, Recall, F1)	Accuracies (Train, Valid, Test) %
6712, 2242, 2246	**0.89, 0.86, 0.85**	**90.02, 90.02, 92.78**
32240, 10759, 10771	0.90, 0.88, 0.87	89.63, 89.81, 88.32
79428, 26491, 26509	0.91, 0.88, 0.87	93.73, 88.63, 87.60

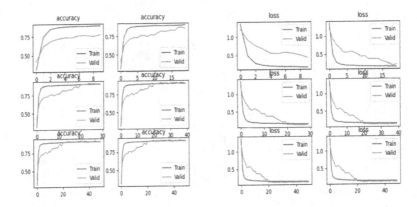

Fig. 3. Training Progress: Slot Filling: Accuracy curve (left) and loss curve (right); fluctuations in the learning curves are due to the use of dropout.

6 Conclusion

In this work, linguistic attributes like symptoms of mental disorders, frequency of occurrence of symptoms, intensity of symptoms, and history of illness were used to generate sequence-labelled utterances for prominent dialects of Hindi. A cross-validation step was included to validate the data against the clinical interviews of patients under the supervision of highly trained doctors and psychiatrists at the Department of Psychiatry, Institute of Medical Sciences, Banaras Hindu University, Varanasi. The CNN + BiL-STM + Dense neural network was deployed to estimate the performance of the system in terms of weighted precision, recall, and F1 scores for intent detection and slot filling tasks for nine symptoms collected from the DSM-5, ICD-11, and WHO guidelines. It was found that the generated utterances closely match the patient utterances - symptom words, frequency words, intensity words, and history words- used in generated utterances closely match patient vocabulary. Using different dialects gives another advantage, as the HUCMD corpus maps patients with different educational backgrounds, cultures, and geographical regions well. This type of cross-border mapping of the corpus (educational, religious, cultural, and geographical) makes it more general in terms of application.

References

1. Mesnil, G., et al.: Using recurrent neural networks for slot filling in spoken language understanding. In: IEEE/ACM Transactions on Audio, Speech, and Language Processing, vol. 23, pp. 530–539(2015)
2. Hanjung, L.: The Emergence of the Unmarked Order in Hindi. Northeast Linguistics Society: vol.30, article 6 (2000). https://scholarworks.umass.edu/nels/vol30/iss2/6. Accessed 14 Sept 2023
3. Chen, Q., Zhuo, Z., Wang, W.: BERT for Joint Intent Classification and Slot Filling. ArXiv (2019)
4. Malviya, S., Mishra, R., Barnwal, S.K., Tiwary, U.S.: HDRS: Hindi dialogue restaurant search corpus for dialogue state tracking in task-oriented environment. IEEE/ACM Trans. Audio, Speech, Lang. Process. **29**, 2517–2528 (2021)
5. Lee, H., Lee, J., Kim, T.Y.: SUMBT: slot-utterance matching for universal and scalable belief tracking. In: Proceedings of the 57th Annual Meeting of the Association for Computational Linguistics, pp. 5478–5483. Association for Computational Linguistics Florence, Italy (2019)
6. Ma, Z., Sun, B., Li, S.: A two-stage selective fusion framework for joint intent detection and slot filling. IEEE Trans. Neural Netw. Learn. Syst. (2022). https://doi.org/10.1109/TNNLS.2022.3202562
7. Wu, J., Harris, I. G., Zhao, H., Ling, G.: A graph-to-sequence model for joint intent detection and slot filling. In: 2023 IEEE 17th International Conference on Semantic Computing (ICSC), pp. 131–138. Laguna Hills, CA, USA (2023)
8. Ma, X., Hovy, E.: End-to-end sequence labeling via Bi-directional LSTM-CNNs-CRF. In: Proceedings of the 54th Annual Meeting of the Association for Computational Linguistics (Volume 1: Long Papers), pp. 1064–1074. Berlin, Germany. Association for Computational Linguistics (2016)
9. Mrkšić, N., Séaghdha, D.O., Wen, T.H., Thomson, B., Young, S.: Neural belief tracker: data-driven dialogue state tracking. In: Proceedings of the 55th Annual Meeting of the Association for Computational Linguistics vol. 1: Long Papers, pp. 1777–1788, Vancouver, Canada. Association for Computational Linguistics (2017)
10. Coucke, A., et al.: Snips voice platform: an embedded spoken language understanding system for private-by-design voice interfaces. arXiv:1805.10190 (2018)
11. Hemphill, C.T., Godfrey, J.J., Doddington, G.R.: The ATIS spoken language systems pilot corpus. Speech and natural language. In: Proceedings of a Workshop Held at Hidden Valley, Pennsylvania, June 24–27 (1990)
12. Ramaneswaran, S., Vijay, S., Srinivasan, K.: TamilATIS: dataset for task-oriented dialog in Tamil. In: Proceedings of the Second Workshop on Speech and Language Technologies for Dravidian Languages (2022)
13. Kane, B., Rossi, F., Guinaudeau, O., Chiesa, V., Quénel, I., Chau, S.: Joint intent detection and slot filling via CNN-LSTM-CRF. In: 2020 6th IEEE Congress on Information Science and Technology (CiSt), Agadir - Essaouira, Morocco, pp. 342–347. (2020)
14. Lafferty, J., McCallum, A., Pereira, F.C.: Conditional random fields: probabilistic models for segmenting and labeling sequence data. In: ICML 2001: Proceedings of the Eighteenth International Conference on Machine Learning, pp. 282–289 (2001)
15. American Psychiatric Association.Diagnostic and statistical manual of mental disorders (5th ed., text rev.) (2022)
16. International Classification of Diseases, Eleventh Revision (ICD-11), World Health Organization (WHO) https://icd.who.int/browse1 International Classification of Diseases, Eleventh Revision (ICD-11), World Health Organization (WHO) (2019/2021)

17. World Health Organization: Doing What Matters in Time of Stress -An Illustrated Guide. World Health Organization (WHO) (2020)
18. Zhang, X., Zhao, J., LeCun, Y.: Character-level convolutional networks for text classification Courant Institute of Mathematical Sciences, New York University 719 Broadway, 12th Floor, New York, NY 10003 (2015)
19. Hochreiter, S., Schmidhuber, J.: Long short term memory. Neural Comput. 9(8), 1735–1780 (1997)
20. Kim, Y., Jernite, Y., Sontag, D., Rush, A.M.: arXiv:1508.06615 (2015)
21. National Institute of Mental Health and Neurosciences Bengaluru, National Mental Health Survey of India, 2015–16: Mental Health Systems. (2015–2016)

Classification of Cleft Lip and Palate Speech Using Fine-Tuned Transformer Pretrained Models

Susmita Bhattacharjee[1]([✉]) [iD], H. S. Shekhawat[1] [iD], and S. R. M. Prasanna[2] [iD]

[1] Indian Institute of Technology Guwahati, Guwahati, India
{sbhattacharjee,h.s.shekhawat}@iitg.ac.in
[2] Indian Institute of Technology Dharwad, Dharwad, India
prasanna@iitdh.ac.in

Abstract. Cleft lip and palate speech (CLP) is a cranio-facial disorder which leads to spectro-temporal distortions in the speech of an individual. This makes accessibility of CLP speakers to speech enabled applications which require Human-computer interaction (HCI) such as voice assistants very challenging. Recently the availability of pretrained models have made the constraint of low resource language very convenient. Recent findings have proven that pretrained transformer models perform way ahead of traditional classifiers. In this paper, with an aim to achieve high end classification results, pretrained Transformer models fine-tuned on CLP data are used. The results obtained from the transformer models such as Wav2Vec2, SEW, SEW-D, UniSpeechSat, HuBERT, DistilHu-BERT showed a comparative performance of the models and specially DistilHuBERT showed a significant improvement in the accuracy being close to 100%.

Keywords: Cleft lip and palate speech · Wav2Vec2 · SEW · SEW-D · UniSpeechSat · HuBERT · DistilHuBERT

1 Introduction

Speech disorder or speech intelligibility is the result of deformation of structure and function of the articulatory system. Disordered speech is found to degrade both in quality and intelligibility [1–3]. As a result this type of speech becomes difficult to perceive by other normal listeners [4]. One such speech disorder is the Cleft Lip and Palate speech. It is one of the most common congenital disorder which affects the cranio-facial region [5–7]. This impairment in the articulatory system leads to other speech disorders due to oro-nasal fistula, velopharyngeal dysfunction and mislearning [8]. The disorders include hyponasality, hypernasality, voice disorder and misarticulation errors [5,6]. Although clinical interventions which include prosthetics, surgery and speech therapy were able to decrease the distortion to a significant extent but the CLP speakers still produce speech that is far from being intelligible to human.

© The Author(s), under exclusive license to Springer Nature Switzerland AG 2024
B. J. Choi et al. (Eds.): IHCI 2023, LNCS 14531, pp. 55–61, 2024.
https://doi.org/10.1007/978-3-031-53827-8_6

Classification task is necessary due to various reasons. The first and foremost being the diagnosis at the comfort of the patients and secondly to use these classification techniques after distortion enhancement tasks which usually use Mel Cepstral Distortion (MCD) and Mean opinion score (MOS) for estimating the enhancement [9]. A recent work included classification of CLP and normal data by extracting low level embeddings from Wav2Vec 2.0 model and then classifying using ML based method [10]. In this work we hypothesized that using pretrained models which are usually trained on large data and then fine-tuning it on desired classification task will help the low resource pathological speech data achieve better classification performance for same amount of inference time.

The rest of the paper is organized as follows. The database used for the experiments is discussed in Sect. 2. Section 3 discusses the methods of fine-tuning the pretrained models. Section 4 elaborates the experiments and observations. Eventually, the paper is concluded in Sect. 5.

2   Dataset

The dataset NMCPC-CLP was obtained from New Mexico Cleft Palate centre by Anil et al. [11]. This private dataset comprises 41 CLP speakers out of which 22 male and 19 females speakers are present. Also a set of 32 normal speakers speaking the similar utterances are present. The total number of utterances were 76 spoken in english. The age group of the CLP speakers is 9.2 ± 3.3 years. The CLP speakers were classified into mild, moderate and severe. The Table 1 shows the number of subjects along with the severity of pathology and their corresponding utterances. In Fig. 1a, the distribution of length of utterances are plotted. The minimum being 1 s and maximum is 4.5 s. Figure 1b shows the imbalance of CLP and normal data. Label 0 shows the CLP data and label 1 the normal data.

Table 1. Description of NMCPC dataset [11]

Type of audio Data	Severity of Speaker	# Subjects	# Utterances
Normal	–	32	406
CLP	*Mild*	11	357
	Moderate	14	373
	Severe	16	406

3   Methods

Pathological speech classification is important from an engineering perspective as a wide number of HCI applications now-a days require audio input for various tasks such as security or speech recognition. Deformed pathological speech makes

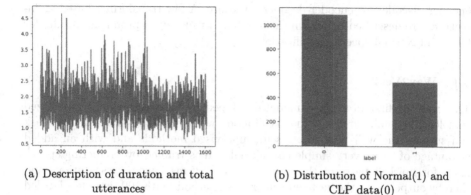

(a) Description of duration and total
utterances

(b) Distribution of Normal(1) and
CLP data(0)

Fig. 1. Description and Distribution of Normal and CLP data in NMCPC-CLP dataset

it difficult for the applications to perceive the speech utterances. Classification task in speech pathology has gained momentum as it allows ease of detection of pathology at clinical level but also for further enhancement of the deviant speech for accessing HCI applications and adapting the devices for working with pathological speech in the future. In the past decade, deep learning has taken over the research due to high-end performance in maximum cases. However it becomes difficult while dealing with less data or low resource languages as there is a risk of overfitting.

Transfer Learning. While training a neural network, the most important requirement is to find out the correct weights for the underlying network by forward backward iterations. One way is to train Deep learning models on very large datasets which require a lot of compute resource and is very time consuming the other way out is to use a pre-trained model which have been trained on large datasets and are task specific. Transfer Learning allows to use the architecture and the weights obtained from a pretrained model and apply it to the desired task at hand. The reason why transfer learning works is because of the fact that the layers that learn to identify classes which are task specific are the final layers which need the training or in other words called as fine-tuning. Transfer learning using transformer models are gaining momentum nowadays due to their advantages in performances over machine learning and deep learning techniques. Transformers are a pretrained state-of-the-art models for audio and speech processing tasks [12–16]. Most transformer models are a variant of an encoder, decoder, or encoder-decoder structure. A few transformer based architectures are described below which were found to produce promising results for CLP and Normal speech classification.

3.1   Wav2Vec2

This was the first pretrained model developed by facebook which emphasized that learning embeddings from speech along with fine-tuning on a desired task can surpass the well performing semi-supervised methods and with an added advantage of being very simple conceptually. It is trained by predicting speech units for masked parts of the audio by using 25 ms long representations and thereby outperforms the best-semisupervised methods with 100 times less labeled training data [12].

3.2   SEW

SEW or squeezed and Efficient Wav2vec is a pre-trained architecture which has high improvement as compared to wav2vec2 in terms of both performance, efficiency and speed. The architecture of SEW has three major modifications as compared to Wav2Vec2. The first is the introduction of a compact feature extractor which makes the model faster without hampering performance by allocating computation more evenly across the layers. Secondly, a squeeze context network to downsample the audio thereby reducing computation and memory usage and allowing usage of larger model without having to sacrifice inference speed. Thirdly, during pre-training the introduction of an MLP predictor heads improve the performance [13].

3.3   SEW-D

Squeezed and Efficient Wav2Vec with Disentangled attention replaces the normal self-attention with disentangled self-attention and hence with half the number of parameters obtains a better performance and quite a significant decrement in both inference time and memory footprint [13].

3.4   UniSpeechSat

Universal Speech Representation Learning With Speaker Aware Pre-Training. Here multi-task learning is applied and for better speaker discrimination, an utterance mixing strategy was applied for data augmentation. In the strategy mentioned above, additional overlapped utterances were created unsupervisedly and incorporated during training thereby achieving a significant performance [14].

3.5   HuBERT

The Hidden-Unit BERT (HuBERT) is a self-supervised speech representation learning approach which provides aligned target labels similar to a BERT-like prediction loss by using an offline clustering step. The prediction loss is applied over the masked regions only which forces the model to learn a combined acoustic and language model over the continuous inputs. Cross-entropy loss is used instead of contrastive loss and diversity loss as used by wav2vec 2.0 which makes training easier and more stable [15].

3.6   DistilHuBERT

This framework distills knowledge from HuBERT in a layer-wise manner. It retains most of HuBERT's performance and that too at 25% of its size. Moreover it reduces HuBERT's size by 75% and making it 73% faster [16].

4   Experimentation and Results

Experiments are implemented with NMCPC-CLP dataset. Before starting the model training, the dataset is divided into train and test parts. While fine-tuning of the models using the NMCPC dataset, the train test split is done by stratification in the same ratio as the dataset distribution. Before feeding the audio samples to the model, preprocessing is required which is done by Hugging Face Transformers Feature Extractor which also resamples the inputs. Next for feature extraction, Auto Feature Extractor is used which ensures that the Feature Extractor loaded by the respective prompt corresponds to the model architecture used. Also the models are trained for 10 epochs each with optimal hyperparametres. After training the model, the classes are predicted for audio samples in the test set. From the Table 2, it can be seen that DistilHuBERT gives the best performance along with the minimal time taken. The F1 score is 0.9986 with an accuracy of 99.38%.

Table 2. Table showing performances of various models and stipulated time to run on NMCPC-CLP dataset.

TRANSFORMER MODELS	ACCURACY (%)	TIME TAKEN (min)
Wav2vec2	67.28	24
SEW	85.18	20
SEW-D	82.71	27
UniSpeechSat	74.07	23
HuBERT	99.38	29
DistilHuBERT	**99.38**	**22**

5   Conclusion

This work focusses on how the pretrained models have made the limitation of classification tasks related to pathological speech very convenient. Classification task in speech pathology is essential at clinical level for classifying enhanced speech output and analyze the enhancement in HCI applications. The experiments show that pretrained transformer models fine-tuned on CLP classification task performs very well in classifying CLP speech from Normal speech. Distil-HuBERT performs the best among the models with an accuracy of almost 100% and in the least time thereby saving computation time.

References

1. Kummer, A.W.: Cleft Palate and Craniofacial Anomalies: Effects on Speech and Resonance (2007)
2. Peterson-Falzone, M. A. H.-J. S. J., Karnell, M.P.: Cleft palate speech (2001)
3. Grunwell, P., Sell, D.: Speech and cleft palate/velopharyngeal anomalies. Management of Cleft Lip and Palate. Whurr, London (2001)
4. Whitehill, T.: Assessing intelligibility in speakers with cleft palate: a critical review of the literature. Cleft Palate-Craniofacial Journal : Official Publication of the American Cleft Palate-craniofacial Association, vol. 39, pp. 50–8 (2002)
5. Zajac, D.J., Vallino, L.: Evaluation and management of cleft lip and palate: a developmental perspective (2017)
6. Lohmander, A., Olsson, M.: Methodology for perceptual assessment of speech in patients with cleft palate: a critical review of the literature. Cleft Palate Craniofac. J. **41**, 64–70 (2004)
7. Stengelhofen, J.: Cleft palate: The nature and remediation of communication problems Churchill Livingstone (1993)
8. Hsu, C.-C., Hwang, H.-T., Wu, Y.-C., Tsao, Y., Wang, H.: Voice conversion from unaligned corpora using variational autoencoding Wasserstein generative adversarial networks. In: INTERSPEECH (2017)
9. Bhattacharjee, S., Sinha, R.: Sensitivity analysis of maskcyclegan based voice conversion for enhancing cleft lip and palate speech recognition, pp. 1–5 (2022)
10. Baumann, I., et al.: Influence of utterance and speaker characteristics on the classification of children with cleft lip and palate. In: INTERSPEECH 2023 (2022)
11. Javid, M.H., Gurugubelli, K., Vuppala, A.K.: Single frequency filter bank based long-term average spectra for hypernasality detection and assessment in cleft lip and palate speech. In: ICASSP 2020–2020 IEEE International Conference on Acoustics, Speech and Signal Processing (ICASSP), pp. 6754–6758 (2020)
12. Baevski, A., Zhou, H., Mohamed, A., Auli, M.: Wav2vec 2.0: a framework for self-supervised learning of speech representations. In: Proceedings of the 34th International Conference on Neural Information Processing Systems, ser. NIPS 2020. Red Hook, NY, USA: Curran Associates Inc. (2020)
13. Wu, F., Kim, K., Pan, J., Han, K.J., Weinberger, K.Q., Artzi, Y.: Performance-efficiency trade-offs in unsupervised pre-training for speech recognition. In: ICASSP 2022–2022 IEEE International Conference on Acoustics, Speech and Signal Processing (ICASSP), pp. 7667–7671 (2021)

14. Chen, S., et al.: Unispeech-sat: universal speech representation learning with speaker aware pre-training. In: ICASSP 2022–2022 IEEE International Conference on Acoustics, Speech and Signal Processing (ICASSP), pp. 6152–6156 (2021)
15. Hsu, W.-N., Bolte, B., Tsai, Y.-H.H., Lakhotia, K., Salakhutdinov, R., Rahman Mohamed, A.: Hubert: self-supervised speech representation learning by masked prediction of hidden units. IEEE/ACM Trans. Audio, Speech, Lang. Process. **29**, 3451–3460 (2021)
16. Chang, H.-J., wen Yang, S., Yi Lee, H.: Distilhubert: speech representation learning by layer-wise distillation of hidden-unit Bert. In: ICASSP 2022–2022 IEEE International Conference on Acoustics, Speech and Signal Processing (ICASSP), pp. 7087–7091 (2021)

Affective Computing and Human Factors

Visual-Sensory Information Processing Using Multichannel EEG Signals

Kamola Abdurashidova[1]([✉]) [iD], Farkhat Rajabov[1] [iD], Nozima Karimova[2] [iD],
and Shohida Akbarova[2] [iD]

[1] Department of Computer Systems, Tashkent University of Information Technologies Named After Muhammad Al-Khwarizmi, Tashkent, Republic of Uzbekistan
kamolabdurashidova@gmail.com
[2] Department of of Information Technologies, Tashkent State Technical University Named After Islam Karimov, Tashkent, Republic of Uzbekistan

Abstract. Brain-computer interfaces (BCIs) are an application of EEG that is still relatively young but is expanding. BCI is a technology that enables brain-computer intercommunication. A computer system called a brain-computer interface (BCI) takes in brain signals, processes them, and then transforms them into orders that are delivered to an output device to carry out the intended operation. Currently, there is growing interest in the analysis of the processes occurring in the cerebral cortex during the reception and processing of sensory information using multichannel signals of its electrical activity. In this case, special attention is paid to identifying typical scenarios for the involvement of various parts of the brain in the process of emotional perception. Purpose: to study the features of the spatio-temporal and frequency-temporal structure of signals of electrical activity of the brain during the processing of visual sensory information. As a result, based on the methods of time-frequency analysis, the processes of changes in the energy of electrical activity of neurons in the frequency ranges of 8–12 Hz (alpha activity) and 15–30 Hz (beta activity) in various environments of the brain were studied. Perception and processing of a visual stimulus has been shown to cause a decrease in alpha energy and an increase in beta energy. Moreover, these processes are observed in different parts of the brain. It was also noted that characteristic patterns appear on the back of the head and forehead, and then include neurons in the central and frontal regions of the brain. The resulting templates are separated into classes and run using the necessary instructions. It recognizes and categorizes these patterns into various jobs, such as controlling a wheelchair or a computer program (such as cursor movement). This article focuses on the creation of a software application that uses the aforementioned techniques and tools to analyze and categorize electrical multichannel EEG signals produced by the brain as it processes visual sensory data.

Keywords: BCI · Multichannel EEG Signals · Visual Perception · Time-Frequency Analysis · Sensory Processing

B. J. Choi et al. (Eds.): IHCI 2023, LNCS 14531, pp. 65–75, 2024.
https://doi.org/10.1007/978-3-031-53827-8_7

1 Introduction

At present, interest is growing in the analysis of the processes occurring in the cerebral cortex during the reception and processing of sensory information using multichannel signals of its electrical activity obtained through surface skin sensors. At the same time, special attention is paid to identifying typical scenarios for the involvement of various parts of the brain in the process of emotional perception.

2 Main Part

Analysis of the processes occurring in the cerebral cortex during the perception and processing of sensory data is not only an urgent task of the physical and mathematical sciences, neurophysiology, but also the task of information technology and artificial intelligence [5, 6]. Interest in solving this problem is associated with the possibility of identifying scenarios of neural activity that are characteristic of the perception of a large amount of sensory information under conditions of high cognitive load [6].

The cerebral cortex's neural network is a distributed computing system with the ability to flexibly rearrange its layout for effective processing of sensory data and decision-making, as is well known. The brain activates tiny groups of neurons in the cortex when doing straightforward activities that don't call for the processing and interpretation of a significant quantity of sensory data, according to the findings of neurophysiological investigations. When addressing a problem involves a lot of effort, the brain simultaneously stimulates several peripheral nerve structures as well as the connections between them [1, 2]. The global thought field theory is how this hypothesis is referred to in scientific literature.

It is specifically known that the activation of the principal center of the visual analyzer, situated in the posterior part of the cerebral cortex, and the center of visual attention, located in the parietal area, are related to the perception of visual information [2]. According to the findings of neurophysiological investigations, the frontal area is activated at the same time as the occipital-spinal region when a visual task is performed for a longer period of time or when the complexity of the visual stimuli presented increases.

Despite many works devoted to the analysis of the properties of neural activity associated with the processing of sensory information, the mechanisms responsible for the formation of such a spatially distributed structure in the neural network of the brain remain unknown [7, 8]. Thus, the mechanisms that allow neurons located in different areas of the brain to interact with each other are not well understood. One of the effective non-invasive methods for analyzing the dynamics of the brain neuron network is to determine the characteristic time-frequency and spatio-temporal characteristics of electrical activity by recording and analyzing multichannel electroencephalograms (EEG). In this regard, the use of methods of statistical and spectral analysis to study the characteristics of EEG signals makes it possible to model various scenarios of the activity of ensembles (groups) of brain neurons at the microscopic level [5].

Considering the above, in this work we classify and process multichannel EEG signals in the process of perception of visual information. Based on the results obtained, the scenario of the activity of brain neurons associated with the perception and processing of visual sensory information is described.

2.1 Significance of Analysis and Classification of EEG Signals

When measuring EEG, we often have a large amount of data of different categories, especially when recordings are made over a long period of time. To obtain information from such a large amount of data, automated methods for analyzing and classifying data using an appropriate algorithm are needed. Although EEG recordings provide valuable information about brain activity, procedures for classifying and evaluating these signals are not yet well developed. The evaluation of the EEG recording is usually carried out by experienced neurologists who visually scan the EEG recordings. Such a visual study of EEG signals is not a satisfactory procedure, since there are no standard evaluation criteria and it is a laborious process, often subject to errors by the interpreting neurologist (translator) [9]. Therefore, there is a need to develop automatic systems for classifying recorded EEG signals. Because BCIs attempt to translate brain activity into commands to control an external device that acts as a communication device, BCI systems need to correctly and efficiently recognize brain desires using appropriate classification algorithms.

The creation of a useful classification system is challenging due to the considerable variability of EEG data in the presence of different external perturbing influences. Distinct computer analysis techniques have been developed recently to extract pertinent data from EEG recordings and identify distinct EEG data types. The majority of the procedures given have a low likelihood of success, according to the research. To do the necessary computing work, some approaches take more time, while others are exceedingly challenging to apply. A huge number of EEG recording data points were represented by certain approaches using small sample set (SSS) data points. They were often insufficiently representative to categorize EEG signals [10]. However, despite the fact that the parameters had a major influence on classification performance, the described techniques frequently did not choose their parameters using the appropriate approaches.

EEG signals play a significant role in biological science, hence rigorous analysis of EEG recordings is required to reveal important data and further understanding. The most precise categorization of time-varying electroencephalographic (EEG) data is one of the objectives of contemporary biomedical research. Using common EEG signal patterns, it has been claimed that a number of categorization systems have been created to recognize different neurological ailments as well as different mental problems in the impaired. Currently, there isn't enough research on how to classify EEG data during epileptic activity and BCI using BI tasks (brain interface) [11]. It will need a significant amount of neurological study to do this. Therefore, our goal in this research project is to create systems for categorizing brain activity.

2.2 EEG Signal Processing Based on Brain Computer Interface

A relatively new but growing field of EEG application is brain-computer interfaces (BCIs). BCI is a technology that provides communication between the brain and the computer [12]. BCI is a computer system that receives brain signals, analyzes them and translates them into commands that are sent to an output device to perform the desired action. The main goal of MCT research is to create a new communication path that allows you to directly transmit messages from the brain by analyzing the mental activity of the brain in patients with severe neuromuscular diseases, as well as control various household electrical appliances using remote control (Fig. 1). To measure the EEG signal, a headgear with electrodes (EEG helmet) is put on the user's head [13]. To control a machine (computer), the user imagines a specific task, such as moving limbs or forming words. These mental tasks act on patterns of EEG signals and determine characteristic states using special recognition algorithms based on fast Fourier transform (FFT) and neural networks [14, 15]. The resulting templates are broken down into classes and executed as required commands. For example, to control a computer application (such as cursor movement) or a machine (such as a wheelchair), it detects and classifies these patterns into different tasks.

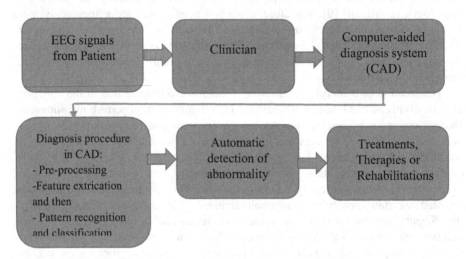

Fig. 1. Scheme of computer diagnostics of EEG.

Since BCIs don't involve real physical movement, they could be the sole way for patients with severe mobility limitations to communicate. Additionally, BCIs aid in easing the signs and symptoms of stroke, autism, mental distress, and attention deficit disorder. BCI comes in two flavors: invasive, which relies on signals gathered by electrodes implanted in the cerebral cortex (and requires surgery), and non-invasive, which relies on signals gathered by electrodes applied to the scalp (and requires no surgery). Non-invasive EEG has emerged as the method of choice in recent investigations [16].

In general, BCI systems (such as human-machine interfaces controlled or driven by myoelectric impulses) enable humans to interact with the outside world by deliberately managing their thoughts rather than muscular contractions. According to Mason and Birch (2003), a typical feedback process for a BCI system entails the following steps: measuring brain activity, preprocessing, feature extraction, classification, access to instructions, and replay (feedback). The categorization outcome enables signal control by external devices. Another feature of BCI systems is that while brain signals are being recorded and analyzed, the user is exposed to stimuli (visual, aural, or tactile) and/or is engaged in mental activities. EEG signals may be used to identify a variety of events or actions, depending on the stimulus or task the user is performing.

Measurement of brain activity. Effective measurement of brain activity is an essential step for BCI communication. Human desires modulate electrical signals, which are measured using various types of electrodes, and then these signals are digitized.

This scientific work is based on the analysis of the formation of characteristic stable EEG patterns in the obtained points based on the projection of certain parts of the brain to the organs by returning the EEG signal to the BCI TGAM device connected via the Bluetooth wireless interface.

This "Thought" program is designed to record an EEG signal on an BCI TGAM device connected via a Bluetooth wireless interface in order to determine and display the state of the brain (Fig. 2).

The purpose of this work is to develop a subsystem for computer acquisition, storage and analysis of EEG. The program must perform preliminary processing of the digitized EEG, determine the parameters of the encephalogram, and based on them make assumptions about the patient's health status: whether it is normal or whether there are pathologies. Since the EEG is a time-varying quantity, that is, a signal, the task of EEG analysis is reduced to digital signal processing (Fig. 2).

Based on the functional diagram of the computer EEG system (Fig. 2) and the algorithms given in the previous sections, application software for the computer EEG "Thought" was developed. Application software "Thought" is created on the basis of an object-oriented integrated system, visually programmed and designed to work in Microsoft Windows operating systems. User interfaces have been developed for the 3-channel and 12-channel versions of the EEG "Thought".

The "Thought" program is designed to determine the general state of the brain by EEG signals taken from a certain point, and display all the necessary information, as well as to notify about this by sending information to a wireless printer when a specified limit is reached. Therefore, this program is intended for scientific experimental measurements and research.

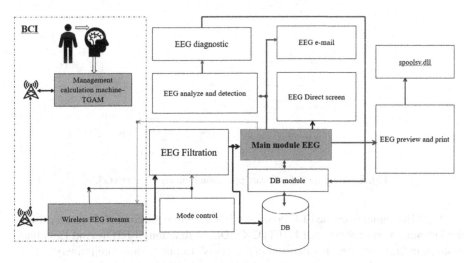

Fig. 2. General architecture of the "Thought" system.

1. Algorithm of operation of the software package running Windows OS.

In accordance with the general rules for constructing interface (window) programs under the Windows operating system, windows are built using the constructor (Windows OS function subroutine) Create. As can be seen from the block diagram (Fig. 3), in our case there are three such windows (forms). These are MainForm (main window), DM (database management window) and Ritm Form (rhythmogram analysis form). The main window is responsible for the existence of this program in the Windows OS environment and therefore closing this window using the Close method will terminate the operation of this program.

As is known in the Windows environment, windows can be in several states: active or passive, visible or invisible, full screen or initial, or minimized. Changing the state of the window will cause events that are indicated by special codes, the so-called messages, generated by the operating system kernel. These messages are transmitted to all currently active programs, and a special processing loop containing active programs determines whether these messages apply to them or not. This cycle is called the message processing cycle and in our case is called after the main window is built using the Application. Run method (Fig. 3). The created software "Thought" provides for receiving, displaying, storing in a database, loading, analyzing and printing EEG signal results.

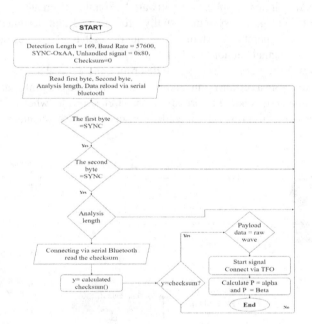

Fig. 3. Algorithm of software operation running Windows OS.

The EEG signal from the EEG controller is loaded into the computer memory using the Dynamic Load Library (DLL (FTD2XX.DLL)) function of FTDI. And EEG data is saved in files with the off extension in the "as is" format without preliminary digital processing. The advantage of storing the EEG signal in this form is that you can quickly change the signal scale in time or amplitude, and quickly apply various digital filters. In addition, as a result of incorrect digital processing, the necessary information is not lost, and it becomes possible to quickly return to the original signal. It will also be possible to test new mathematical algorithms on the original signal.

The program can be in 4 modes:

1. Initial mode (Initial) – the EEG display area and most control and processing functions are not available. This mode is transitional and when loading the program will be in this mode;   .
2. Monitoring mode - the real-time encephalogram display area shows the c encephalogram curve against the background of graph paper. This mode is used mainly when the encephalograph is used as a bedside monitor or to control the quality of the recorded encephalogram;
3. View mode – in this mode, the captured encephalogram is displayed in the encephalogram display area. It is possible to view various sections of encephalograms, delete unnecessary sections, make copies to the clipboard, clear encephalograms, measure time and frequency characteristics, view the spectrum of the encephalogram and print;
4. Recording mode - in this mode, the recorded encephalogram is displayed in the encephalogram display area while simultaneously recording it into the database for subsequent processing. Recording can be made for any duration, and the current recording time is shown in the status and status indicator. In the monitoring mode, without leaving it, you can arbitrarily switch between different groups of leads, turn on/off the Filter and switch to recording mode, if this mode is available. At the same time, if the signal - the cardiac curve - is of insufficient magnitude, you can use the sensitivity regulator to increase this signal and vice versa. You can also change the speed of EEG measurement using the speed controller.

When the cardiac curves go beyond the measured range due to transient processes, you can use the quick recovery key by pressing the keyboard spacebar. In this case, a high-pass filter with a cutoff frequency of ~2 Hz is turned on for a period of 1 s, which allows accelerating transient processes.

Now, having launched the "Thought" program (Fig. 4.), the same memorized COM port number (in the example COM12) should be set in the list above the "Communication" button in the upper right part of the screen, and this completes the process of establishing contact with. If necessary, you can also connect a bluetooth printer by repeating the seven steps described above, but instead of its pin code, you need to enter "1234", and return the created COM port to the list on the left side of the "Check" window. Printer" (Fig. 4.).

2. Classification of the main elements of the interface of the "Thought" program.
In order for the program to start receiving information from MCI TGAM, you must press the "Communication" button on the screen (Fig. 4.). After that, if the device is turned on and configured (see the first section), the red inscription "Synchronization" will appear, and the number of received packets will begin to be displayed in the "Receive:" line.

At the same time, the received information begins to be expressed in real time in a three-dimensional graphical form.

The first graphic image called "Status Symbols" is located at the very top of the screen (Fig. 5). It is on this graph that the quality of the signal in the form of colored

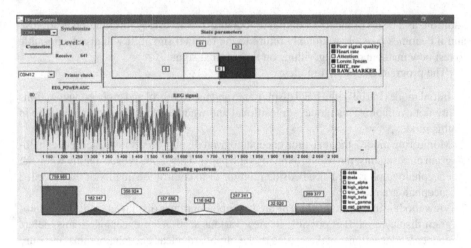

Fig. 4. Main program window.

bars and its quantity are recognized, as well as two general states of the brain based on a special analysis, i.e. concentration of thought (attention) or imagination.

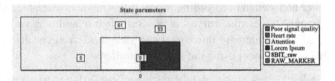

Fig. 5. Graph depicting the state of the brain.

The middle graph shows the temporal distribution (oscillogram) of the repeated EEG signal (Fig. 6.). Based on this graph, the quality of the received EEG signal is evaluated and, if necessary, the electrical contact of the connected electrodes with the scalp is corrected. In addition, these received EEG signals serve as inputs that can be used in other new algorithms. If the graphical representation of the signal deviates from the specified limits, you can move the diagram up or down using the "+" and "−" buttons on the right side of the graph so that the EEG signal is drawn in the middle of the graph.

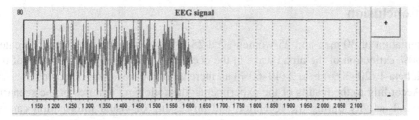

Fig. 6. A graph of the initial recorded EEG signal.

The last plot shows the spectrum of the EEG signal determined from the Fast Fourier Transform (FFT) (Fig. 7.). Here, the frequency axis represents the energy distribution of the EEG signal from the delta range to the gamma range, adopted in the analysis of the EEG.

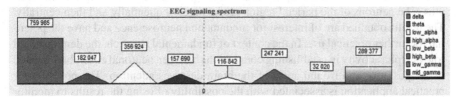

Fig. 7. A graph of the EEG signal spectrum.

Also, if a wireless Bluetooth printer is connected to the system (see the first section), it can be activated by clicking the "Check Printer" button. If everything is connected correctly, then every time you press the button, the thermal paper in the printer will say "Checking... Good!!!" but if the printer is not connected, the error message "the printer is not connected" appears on the screen.

If the printer is started normally, then when the state of concentration of the brain reaches 80 units, the paper will say "You are focused!!!" will appear. If a similarly defined delusion score reaches 80, "You're distracted!!!" printed from the printer.

BCI can identify and classify certain patterns of activity in continuous brain signals associated only with specific tasks or events. What the BCI user must do to create these templates is determined by the mental strategy that the BCI system uses. Mental strategy is at the heart of any interaction between the brain and the computer. The mental strategy defines what the user must do in order to create brain patterns that the BCI can interpret. The most common mental strategies are motion imagery (MI) and selective attention. MI is the imagination of an action without actually performing it. On the other hand, BCIs based on selective attention require external stimuli provided by the BCI system. Stimuli can be auditory and somatosensory. In this study, we are working on MI for BCI systems.

3   Conclusion

Examination of 30 men and 25 women aged 25 to 60 years using the device revealed 70–90% extinction of the alpha wave in the occipitalis and frontalis projections of the head, beta-1 (20–30%) beta 2 (40–60%) an increase in the wave was detected.

According to the results of the experiments, the following features of the scenario of neural activity during the reception and processing of visual information can be distinguished:

- During the presentation of a visual stimulus, the visual center is excited directly on the back of the head. Currently, electrical activity in this area is characterized by a decrease in alpha activity and an increase in beta activity.
- 0.3 s after the presentation of the stimulus (stimulus), it is processed by activating the distributed fronto-parietal neural network. At this point, the electrical activity in the region is characterized by an increase in beta activity.
- Activation of neurons in the alpha range occurs gradually within 0.3 s after the stimulus. The neurons of the parietal region are involved sequentially and then centrally. The results obtained are of interest for fundamental neuroscience and have the potential for further practical use. In the context of fundamental research, the demonstrated role of alpha activity in establishing connections between neuronal ensembles located in the posterior and frontal areas of the brain deserves attention. The potential for practical application is associated with the possibility of using the results to monitor human cognitive activity in the processing of emotional information, a passive brain-computer interface that monitors and controls neural activity during human activities associated with a high cognitive load.

Studies show that the perception and processing of a visual stimulus causes a decrease in the energy of alpha activity and an increase in the energy of beta activity. Moreover, these processes are observed in different parts of the brain. It has also been noted that characteristic patterns appear on the back of the head and forehead, and then turn on neurons in the central and frontal regions of the brain.

References

1. Sanei, S., Chambers, J.A.: EEG Signal Processing. John Wiley & Sons Ltd., Hoboken (2007)
2. Leong, W.Y.: EEG signal processing: feature extraction, selection and classification methods. MAHSA University, Petaling Jaya, Malaysia, p. 297. Travis, Jeffrey (2019)
3. Djumanov, J.X., Rajabov, F.F., Abdurashidova, K.T., Tadjibaeva, D.A., Atadjanova, N.S.: Development of the method, algorithm and software of a modern non-invasive biopotential meter system. In: Singh, M., Kang, D.-K., Lee, J.-H., Tiwary, U.S., Singh, D., Chung, W.-Y. (eds.) IHCI 2020. LNCS, vol. 12615, pp. 95–103. Springer, Cham (2021). https://doi.org/10.1007/978-3-030-68449-5_10
4. Djumanov, J., Abdurashidova, K., Rajabov, F., Akbarova, S.: Determination of characteristic points based on wavelet change of electrocardiogram signal. In: 2021 International Conference on Information Science and Communications Technologies (ICISCT), pp.1–5 (2021)
5. Milton, A., Pleydell-Pearce, C.W.: The phase of pre-stimulus alpha oscillations influences the visual perception of stimulus timing. Neuroimage **133**, 53–61 (2016). https://doi.org/10.1016/j.neuroimage.2016.02.065

6. Shourie, N.: Cepstral analysis of EEG during visual perception and mental imagery reveals the influence of artistic expertise. J. Med. Signals Sens. **6**(4), 203 (2016)
7. Nasimova, N., Muminov, B., Nasimov, R., Abdurashidova, K., Abdullaev, M.: Comparative analysis of the results of algorithms for dilated cardiomyopathy and hypertrophic cardiomyopathy using deep learning. In: 2021 International Conference on Information Science and Communications Technologies (ICISCT), pp. 1–5 (2021)
8. Djumanov, J.X., Rajabov, F.F., Abdurashidova, K.T.: Development of a multifunctional medical diagnostic system based on modern element base. Tashkent TUIT -BULLETIT **2**(58), 160–166 (2021)
9. Singh, D., Singh, M., Hakimjon, Z.: Signal Processing Applications Using Multidimensional Polynomial Splines. Springer, Singapore (2019)
10. Rajabov, F.F., Abdurashidova, K.T.: Typical solutions for the construction of modern electrocardiographs (ECG). Tashkent, TUIT -BULLETIT **2**(46), 42–55 (2018)
11. Rajabov, F.F., Abdurashidova, K.T., Salimova, H.R.: The issues of creating a computer biomeasuring and noise suppression methods. Muhammad al-Khwarizmi Descendants. Sci.-Pract. Inf. Anal. J. Tashkent **1**(3), 23–27 (2019)
12. Djumanov, J., Rajabov, F., Abdurashidova, K., Xodjaev, N.: Autonomous wireless sound gauge device for measuring liquid level in well. In: E3S Web of Conferences, vol. 401, p. 01063. EDP Sciences (2023)
13. Kuchkorov, T., Ochilov, T., Gaybulloev, E., Sobitova, N., Ruzibaev, O.: "Agro-field boundary detection using mask R-CNN from satellite and aerial images. In: 2021 International Conference on Information Science and Communications Technologies (ICISCT), Tashkent, Uzbekistan, pp. 1–3 (2021). https://doi.org/10.1109/ICISCT52966.2021.9670114
14. Kuchkorov, T., Khamzaev, J., Allamuratova, Z., Ochilov, T.: Traffic and road sign recognition using deep convolutional neural network. In: 2021 International Conference on Information Science and Communications Technologies (ICISCT), Tashkent, Uzbekistan, pp. 1–5 (2021). https://doi.org/10.1109/ICISCT52966.2021.9670228
15. Kuchkorov, T., Urmanov, S., Kuvvatova, M., Anvarov, I.: Satellite image formation and preprocessing methods. In: 2020 International Conference on Information Science and Communications Technologies, ICISCT 2020 (2020)

Vision Transformer-Based Emotion Detection in HCI for Enhanced Interaction

Jayesh Soni[1]($\boxtimes$), Nagarajan Prabakar[2], and Himanshu Upadhyay[3]

[1] Applied Research Center, Florida International University, Miami, FL, USA
jsoni@fiu.edu
[2] Knight Foundation School of Computing and Information Sciences, Florida International University, Miami, FL, USA
prabakar@cis.fiu.edu
[3] Electrical and Computer Engineer, Florida International University, Miami, FL, USA
upadhyay@fiu.edu

Abstract. Emotion recognition from facial expressions is pivotal in enhancing human-computer interaction (HCI). Spanning across diverse applications such as virtual assistants, mental health support, and personalized content recommendations, it promises to revolutionize how we interact with technology. This study explores the effectiveness of the Vision Transformer (ViT) architecture within the context of emotion classification, leveraging a rich dataset. Our research methodology is characterized by meticulous preprocessing, extensive data augmentation, and fine-tuning of the ViT model. For experiment purposes, we use the Emotion Detection FER-2013 dataset. Rigorous evaluation metrics are meticulously employed to gauge the model's performance. The research underscores the potential for enhancing user experiences, facilitating mental health monitoring, and navigating the ethical considerations inherent in emotion-aware technologies. Our model achieved a testing accuracy of 70%. As we chart new horizons in HCI, future endeavors should focus on fine-tuning model accuracy across various emotion categories and navigating the complexities of real-world deployment challenges.

Keywords: Emotion Recognition · Vision Transformer (ViT) · Emotion Detection FER-2013 Dataset · Human-Computer Interaction (HCI)

1 Introduction

Emotions represent a fundamental aspect of human communication and interaction, underpinning our ability to connect, empathize, and respond effectively in social contexts. Our innate recognition and interpretation of emotions conveyed through facial expressions are central to this process. However, in recent years, rapid advancements in computer vision and artificial intelligence have unveiled a transformative opportunity: automating the intricate process of emotion recognition from facial cues. This research explores the formidable potential at the intersection of HCI and emotion recognition. Specifically, we delve into applying the ViT [1] architecture to tackle the complex emotion classification task. By doing so, we seek to harness the power of ViT in the realm of HCI and illuminate its implications across various domains.

B. J. Choi et al. (Eds.): IHCI 2023, LNCS 14531, pp. 76–86, 2024.
https://doi.org/10.1007/978-3-031-53827-8_8

The significance of emotion classification in HCI cannot be overstated. As HCI systems evolve, they aim to become responsive and intuitive. Emotion-aware systems, capable of detecting and responding to users' emotional states, are poised to redefine how we interact with technology. Imagine virtual assistants that adapt their responses based on your emotional state, video conferencing platforms that gauge participant engagement in real-time, or mental health applications that monitor and offer timely interventions. Emotion classification holds the key to realizing these transformative possibilities [2].

Our research unfolds with a clear and ambitious objective: to assess the prowess of the ViT architecture in the nuanced task of classifying emotions from facial expressions within the confines of the provided dataset. We endeavor to ascertain ViT's proficiency in distinguishing between the seven emotion categories, uncover its areas of excellence, and dissect the challenges it faces in achieving consistent and accurate emotion recognition. In so doing, we strive to contribute to the ever-expanding landscape of emotion recognition, particularly within the purview of HCI.

Through this comprehensive study, we aim to offer insightful revelations into the capabilities and limitations of state-of-the-art computer vision models when applied to emotion classification. These insights are poised to serve as the foundation for designing more emotionally intelligent HCI systems, thereby elevating user experiences and propelling advancements in fields as diverse as mental health and personalized technology.

The forthcoming sections are structured as follows: Sect. 2 delves into related work, providing an overview of existing research. Section 3 offers a comprehensive description of the dataset employed in this study. Section 4 describes the high-level overview of the research framework. Metrics assessed in the research are illustrated in Sect. 5. Experimental results are thoroughly analyzed and discussed in Sect. 6. Lastly, Sect. 7 serves as the conclusion discussing potential avenues for future research.

2 Related Work

Facial Emotion Recognition (FER) is a multifaceted challenge, and researchers have explored diverse classic machine-learning techniques to address it effectively. Methods like support vector machines and K-nearest neighbors have been pivotal in advancing this field. Lee et al. [4] pursued a different path by implementing a boosting technique for classification. They also harnessed the potential of the contourlet transform and a specialized wavelet transform tailored for 2D feature extraction. This method enabled the extraction of intricate spatial and structural details from facial images, augmenting the overall discriminative capacity of the system. A localized approach was adopted in a significant study by Feng et al. [5]. They generated local binary pattern histograms from multiple small segments of facial images, effectively capturing essential texture information. To enhance the accuracy of emotion classification, a sophisticated amalgamation of local features was expertly crafted into a unified feature histogram. The classification task was approached by applying linear programming techniques, highlighting the potential of intelligent feature selection and integration for optimizing emotion recognition. Drawing inspiration from face recognition, Chang and Huang [5] introduced face recognition methodologies into the realm of FER, aiming to enhance the identification of individual emotional expressions. Their classification approach [6, 7] involved implementing a radial basis function neural network, which displayed the potential to address the complex and nuanced aspects of emotional expressions effectively. Building on this foundation, Xiao-Xu and Wei [8] introduced an innovative facet to FER by seamlessly incorporating wavelet energy features into facial images, thereby elucidating spatial frequency characteristics.

Deep learning has firmly established itself as a prevailing paradigm in machine learning dedicated to recognizing emotions. This methodology has garnered substantial interest, with numerous investigations delving into utilizing Convolutional Neural Networks (CNNs). As an illustration, Pranav et al. [9] ventured into the arena of FER utilizing self-captured emotional facial images. Their methodology revolved around a conventional CNN framework, encompassing two convolutional pooling layers tailored for extracting pertinent facial characteristics. Furthermore, Pons et al. [10] pushed the frontiers of emotion recognition by pioneering an ensemble comprising 72 distinctive CNNs. Their ingenious approach entailed training individual CNNs with varying filter dimensions within convolutional layers and diverse quantities of neurons, thereby embracing a wide range of architectural configurations. This ensemble-based strategy vividly exemplified the escalating intricacy and refinement of CNN-based models within emotion recognition research. In a distinctive effort, Ronak et al. [11] introduced the concept of "Facial Affect Analysis Based on Graphs," seamlessly integrating two distinct approaches for representing emotions. These approaches encompassed 26 distinct emotional categories and the continuous valence, arousal, and dominance dimensions. They employed various CNN models to ensure robust emotion recognition within this framework.

Additionally, Mengting et al. [12] pioneered the Attention-Based Magnification-Adaptive Network (AMAN), a network designed to learn adaptive magnification levels for microexpression (ME) representation. This development introduces an innovative perspective on the nuanced interpretation of emotions, highlighting the evolving landscape of emotion recognition research. Remarkably, Li et al. [13] bolstered the effectiveness of this framework by incorporating transfer learning techniques, thus enhancing its overall capabilities. Meanwhile, Soumya et al. [14] engineered a conversational emotion recognition system that integrated multiple modalities encompassing speech and text data. This project involved the development of a bidirectional GRU network augmented with self-attention mechanisms. These enhancements enabled the distinct processing of textual and auditory representations within the system.

This research aims to contribute to this evolving landscape by evaluating the efficacy of the Vision Transformer (ViT) architecture in emotion classification. Our findings not only enrich the field of emotion recognition but also offer practical insights for HCI, ultimately fostering more empathetic and responsive technology interfaces.

3   Dataset

In this research, we leverage the "Emotion Detection FER-2013" dataset [15], a comprehensive resource for facial emotion recognition. This dataset comprises a diverse collection of 35,685 grayscale images, each measuring 48x48 pixels. The emotion categories encompass Angry, Disgusted, Fearful, Happy, Neutral, Sad, and Surprised, covering the spectrum of human emotional expressions.

Due to its size, diversity, and relevance to human-computer interaction, this dataset represents a critical asset in advancing the understanding and application of emotion recognition in real-world scenarios. Table 1 describes the dataset. Figure 1 and 2 shows the sample images from the training and testing set.

Table 1. Dataset Categories

Emotion	Train	Test
Angry	3995	958
Disgusted	436	111
Fearful	4097	1024
Happy	7215	1774
Neutral	4965	1233
Sad	4830	1247
Surprised	3171	831

Fig. 1. Training Set

Fig. 2. Testing Set

4   Methodology

In this section, we outline the comprehensive methodology underpinning our research endeavors. We begin by explaining the significance and workings of the ViT architecture within the context of computer vision for emotion recognition. Furthermore, we detail the preprocessing steps executed for our dataset, encompassing essential techniques such as data augmentation and normalization. To provide a holistic understanding, we explore the training and evaluation procedures, addressing critical elements such as loss functions, optimization methods, and metrics employed to assess model performance thoroughly. Figure 3 shows the research framework.

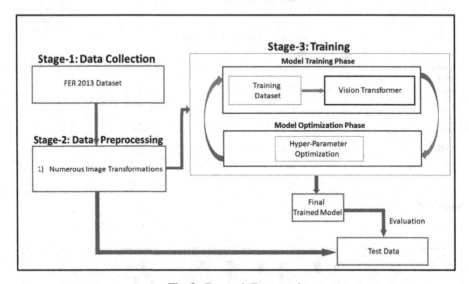

Fig. 3. Research Framework

Stage 1: Data Collection:
The FER-2013 dataset is available from Kaggle and described in the previous section.

Stage 2: Data Preprocessing:
Robust model performance hinges on the effective preprocessing of our dataset, a crucial preparatory step in our research endeavor. This section details the comprehensive preprocessing pipeline, encompassing data augmentation techniques to introduce diversity and normalization procedures to standardize pixel values. These preprocessing steps are instrumental in fortifying our model's capacity to generalize seamlessly across a spectrum of emotional expressions.

Data Augmentation: To enrich the dataset and ensure the model's resilience to variations in facial expressions, we employ data augmentation techniques. Augmentation introduces controlled variations into the training data, expanding its diversity without requiring additional labeled samples. We create augmented versions that simulate a broader range of facial expressions by applying rotation, scaling, and horizontal flipping to the original images. This augmentation strategy equips the model with the adaptability needed to effectively recognize emotions under various conditions, enhancing its overall performance.

Normalization: Normalization of pixel values is a fundamental preprocessing step that plays a pivotal role in ensuring consistency and aiding the training process. We standardize the pixel values across all images to bring them within a standard scale. This standardization typically involves mean subtraction and division by the standard deviation, yielding a distribution centered around zero with unit variance. Normalization enhances convergence during training and ensures that the model's learning process is stable and efficient.

Stage 3: Algorithm.

Vision Transformer (ViT) Architecture: In our methodology, the ViT architecture stands as a cornerstone, playing a pivotal role in shaping the framework for our research. ViT offers transformative capabilities for processing visual data effectively. By harnessing the principles of Transformers [16], initially developed for natural language processing tasks, ViT demonstrates remarkable adaptability in efficiently handling visual information.

ViT: Adapting Transformers for Computer Vision: ViT reimagines images as sequences of patches, breaking down visual data into smaller, manageable units. This approach, often called "patch processing," enables ViT to harness Transformers' potent attention mechanisms [17] to analyze and understand complex visual patterns. Its self-attention mechanism allows it to weigh the importance of different patches when encoding information:

$$Attention(Q, K, V) = softmax\left(\frac{QK^T}{\sqrt{d_k}}\right)V$$

$Q, K,$ *and* V represent the query, key, and value matrices. The softmax function normalizes the scores, and $\sqrt{d_k}$ is a scaling factor. Another essential equation within the ViT architecture is the calculation of the multi-head attention output:

$$MultiHead(Q, K, V) = Concat(head_1, head_2, \ldots, head_n s)W^o$$

where $head_i$ represents the i-th attention head, and W^o is the output weight matrix.

Furthermore, ViT employs positional embeddings to impart spatial information to the model. The position embeddings can be represented as:

$$PE_{(pos, 2i)} = \sin(\frac{pos}{1000^{\frac{2i}{d_{model}}}})$$

$$PE_{(pos, 2i+1)} = \cos(\frac{pos}{1000^{\frac{2i}{d_{model}}}})$$

These equations allow ViT to account for the relative positions of patches in the image.

5 Evaluation Metrics

The metrics provide a comprehensive assessment of our model's capabilities, facilitating a more insightful interpretation of the results.

Accuracy: A fundamental metric that quantifies the proportion of correctly classified instances from the total number of instances in the dataset. It provides a high-level overview of our model's overall performance in recognizing emotions from facial expressions.

$$Accuracy = \frac{TP + TN}{TP + TN + FP + FN}$$

where:

- **True Positives (TP):** These are instances correctly classified as belonging to a specific emotion category.
- **True Negatives (TN):** These instances were correctly classified as not belonging to a specific emotion category.
- **False Positives (FP):** These instances were incorrectly classified as belonging to a specific emotion category when they do not.
- **False Negatives (FN):** These instances were incorrectly classified as not belonging to a specific emotion category when they did.

Precision quantifies the accuracy of positive predictions made by the model. It is particularly valuable when we want to minimize false positives, ensuring that when the model predicts an emotion, it is highly likely to be correct.

$$Precision = \frac{TP}{TP + FP}$$

Recall measures the model's ability to identify positive instances correctly. It is crucial to ensure that the model does not miss any instances of a specific emotion.

$$Recall = \frac{TP}{TP + FN}$$

The F1-score is a harmonic mean of precision and recall and is particularly useful in cases where there is an imbalance between different emotion categories.

$$F1 - Score = \frac{2 * Precision * Recall}{Precision + Recall}$$

6   Experimental Results

We use the ViT Base model with 12 layers, 786 hidden dimensions, and 12 heads. The total size of the model is 86M. The experiments were conducted using the Hugging Face Transformers library [18], a powerful tool for natural language processing and computer vision tasks, with the underlying deep learning framework PyTorch [19] serving as the foundation. Figure 4 represents the loss at each epoch. Adam optimizer is employed for gradient optimization with a batch size of 32. The learning rate is set to 0.001. The x-axis represents the number of training epochs, while the y-axis represents the loss values. The training loss consistently decreases with each epoch, indicating that the model effectively learns from the training data. The final loss achieved is 0.0844.

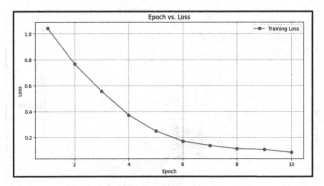

Fig. 4. Epoch Vs. Loss

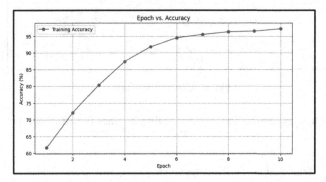

Fig. 5. Epoch Vs. Accuracy

Figure 5 represents the accuracy at each epoch. The x-axis represents the number of training epochs, while the y-axis represents the model's accuracy on the training dataset. The graph demonstrates that the model's accuracy steadily improves as training progresses, highlighting its ability to correctly classify emotions in facial expressions. The confusion matrix for the test dataset is shown in Fig. 6. Table 2 shows the evaluated metrics on the training and testing dataset. Table 3 shows the precision, recall and F1-score for individual categories for the test dataset.

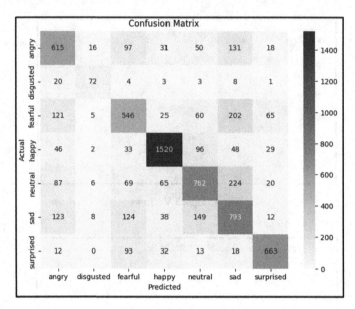

Fig. 6. Confusion Matrix

Confusion Matrix depicts the model's ability to classify different emotional states accurately.

Table 2. Evaluated metrics on the training and testing dataset

Metric	Training Dataset	Testing Dataset
Accuracy	0.982	0.701
Precision	0.982	0.697
Recall	0.982	0.692
F1 Score	0.982	0.694

Table 3. Evaluated metrics on testing dataset on each emotion

Emotions	Precision	Recall	F1-Score
Angry	0.66	0.54	0.59
Disgusted	0.57	0.80	0.66
Fearful	0.53	0.56	0.54
Happy	0.88	0.86	0.87
Neutral	0.63	0.69	0.6
Sad	0.57	0.62	0.59
Surprised	0.85	0.73	0.78

The model excelled in recognizing happiness, achieving a Precision of 88%. The model was able to recall 86% of genuine happy expressions, resulting in a notable F1-Score of 87%. Table 4 compares the results with existing work.

Table 4. Comparative Analysis

Work	Accuracy
Akash Saravanan et al. [20]	60%
Yijun Gan et al. [21]	64.2%
Wang et al. [22]	67.7%
Our Proposed Work	**70.1%**

The model also demonstrated strong performance in recognizing surprised expressions with a precision of 85%, with the F1-Score of 78%.

7 Conclusion

In this research, we trained the innovative ViT architecture for emotion classification using facial expressions. The ViT model exhibits promising performance in capturing the complexity of emotions. As we navigate the landscape of HCI, the potential applications of our research are far-reaching. The ability to accurately detect emotions from facial expressions holds promise in various domains, including virtual therapy, customer support, and educational technologies, where understanding and responding to user emotions are pivotal for enhancing user experiences.

While our results are promising, there remains room for further exploration. Future research may refine the model's performance for specific emotions and extend its applicability to real-time interactive systems. Additionally, incorporating multimodal data sources like voice and text could enhance the model's robustness and broaden its applications.

References

1. Han, K., et al.: A survey on vision transformer. IEEE Trans. Pattern Anal. Mach. Intell. **45**(1), 87–110 (2022)
2. Barling, J., Slater, F., Kelloway, E.K.: Transformational leadership and emotional intelligence: an exploratory study. Leadersh. Org. Dev. J. **21**(3), 157–161 (2000)
3. Lee, C.-C., Shih, C.-Y., Lai, W.-P., Lin, P.-C.: An improved boosting algorithm and its application to facial emotion recognition. J. Ambient. Intell. Humaniz. Comput. **3**, 11–17 (2012)
4. Feng, X., Pietikainen, M., Hadid, A.: Facial expression recognition based on local binary patterns. Pattern Recognit Image Anal. **17**, 592–598 (2007)
5. Chang, C.-Y., Huang, Y.-C.: Personalized facial expression recognition in indoor environments. In: Proceedings of the 2010 International Joint Conference on Neural Networks (IJCNN), Barcelona, Spain, pp. 1–8 (2010)

6. Peddoju, S.K., Upadhyay, H., Soni, J., Prabakar, N.: Natural language processing-based anomalous system call sequences detection with virtual memory introspection. Int. J. Adv. Comput. Sci. Appl. **11**(5) (2020)
7. Soni, J., Prabakar, N., Upadhyay, H.: Deep learning approach to detect malicious attacks at system level: poster. In: Proceedings of the 12th Conference on Security and Privacy in Wireless and Mobile Networks, pp. 314–315 (2019)
8. Xiao, X.Q., Wei, J.: Application of wavelet energy feature in facial expression recognition. In: Proceedings of the 2007 International Workshop on Anti-Counterfeiting, Security, and Identification (ASID), Xiamen, China, pp. 169–174 (2007)
9. Pranav, E., Kamal, S., Chandran, C.S., Supriya, M.: Facial emotion recognition using deep convolutional neural network. In: Proceedings of the 2020 6th International Conference on Advanced Computing and Communication Systems (ICACCS), Coimbatore, India, pp. 317–320 (2020)
10. Pons, G., Masip, D.: Supervised committee of convolutional neural networks in automated facial expression analysis. IEEE Trans. Affect. Comput. **9**, 343–350 (2018)
11. Kosti, R., Alvarez, J.M., Recasens, A., Lapedriza, A.: Context-based emotion recognition using EMOTIC dataset. IEEE Trans. Pattern Anal. Mach. Intell. **42**, 2755–2766 (2020)
12. Wei, M., Zheng, W., Zong, Y., Jiang, X., Lu, C., Liu, J.: A novel micro-expression recognition approach using attention-based magnification-adaptive networks. In: Proceedings of the ICASSP 2022—2022 IEEE International Conference on Acoustics, Speech, and Signal Processing (ICASSP), Singapore, pp. 2420–2424 (2022)
13. Li, J., et al.: Facial expression recognition by transfer learning for small datasets. In: Yang, C.-N., Peng, S.-L., Jain, L.C. (eds.) SICBS 2018. AISC, vol. 895, pp. 756–770. Springer, Cham (2020). https://doi.org/10.1007/978-3-030-16946-6_62
14. Dutta, S., Ganapathy, S.: Multimodal transformer with learnable frontend and self attention for emotion recognition. In: Proceedings of the ICASSP 2022—2022 IEEE International Conference on Acoustics, Speech, and Signal Processing (ICASSP), Singapore, pp. 6917–6921 (2022)
15. https://www.kaggle.com/datasets/msambare/fer2013
16. Han, K., Xiao, A., Wu, E., Guo, J., Xu, C., Wang, Y.: Transformer in transformer. Adv. Neural. Inf. Process. Syst. **34**, 15908–15919 (2021)
17. Shaw, P., Uszkoreit, J., Vaswani, A.: Self-attention with relative position representations (2018). arXiv preprint arXiv:1803.02155
18. Wolf, T., et al.: Huggingface's transformers: State-of-the-art natural language processing (2019). arXiv preprint arXiv:1910.03771
19. Paszke, A., et al.: Automatic differentiation in pytorch (2017)
20. Saravanan, A, Perichetla, G, Gayathri, D.K.: Facial emotion recognition using convolutional neural networks (2019). arXiv preprint arXiv:1910.05602
21. Gan, Y.: Facial expression recognition using convolutional neural network. In: Proceedings of the 2nd International Conference on Vision, Image and Signal Processing, pp. 1–5 (2018)
22. Wang, Y., Li, Y., Song, Y., Rong, X.: Facial expression recognition based on auxiliary models. Algorithms **12**(11), 227 (2019)

GenEmo-Net: Generalizable Emotion Recognition Using Brain Functional Connections Based Neural Network

Varad Srivastava[1], Ruchilekha[2(✉)], and Manoj Kumar Singh[3]

[1] Quantitative Analytics, Barclays, London, UK
[2] DST- Centre for Interdisciplinary Mathematical Sciences, Banaras Hindu University, Varanasi, India
ruchilekha.cims@bhu.ac.in
[3] Department of Computer Science, Banaras Hindu University, Varanasi, India
manoj.dstcims@bhu.ac.in

Abstract. The aim of this research is to construct a generalizable and biologically-interpretable emotion recognition model utilizing complex electroencephalogram (EEG) signals for realizing emotional state of human brain. In this paper, the spatial-temporal information of EEG signals is used to extract brain connectivity-based feature, i.e., phase-locking value (PLV), that incorporates phase information between a pair of signals. These functional features are then fed as input to our proposed model (GenEmo-Net), which encompasses of Graph Convolutional Neural Network (GCNN) and Long Short-Term Memory Network (LSTM). It is able to dynamically learn the adjacency matrix that resembles functional connections in the brain, and are combined with the temporal features learnt by LSTM. To validate the generalization ability of our model, the experimental setup combines three emotion databases, namely DEAP, DREAMER, and AMIGOS, which increases variability and reduces biasness among subjects and trials. We evaluated the performance of our proposed model on the combined dataset, which achieved a classification accuracy of 70.98 ± 0.73, 65.47 ± 0.56, and 70.09 ± 0.37 for discrimination of valence, arousal, and dominance, respectively. Notably, our generalized model gives more robust results for emotion recognition tasks when compared to other methods. In addition, the biological interpretation of GenEmo-Net is tested via the final adjacency matrix, learnt at the end of training, for VAD processing units. Above results demonstrate the efficacy of the GenEmo-Net for recognizing human emotions and also highlight substantial variations in the spatial and temporal brain characteristics across distinct emotional states.

Keywords: Emotion recognition · Phase-Locking Value · Graph neural network · LSTM · GCNN · Brain connectome

B. J. Choi et al. (Eds.): IHCI 2023, LNCS 14531, pp. 87–98, 2024.
https://doi.org/10.1007/978-3-031-53827-8_9

1   Introduction

Emotion recognition is a multidisciplinary endeavor that merges aspects of psychology, neuroscience, and engineering, aiming to discern emotional states through various modalities such as facial expression, voice tone, and physiological signals. Among these modalities, electroencephalogram (EEG) signals present a unique and compelling avenue for emotion recognition. Unlike other methods, which often require observable expressions or reactions that may be intentionally masked or altered, EEG offers an objective measure of neural activities directly related to emotional experiences. EEG has advantages such as high temporal resolution which allows us to detect rapid changes in emotions, non-invasiveness, and being relatively low cost, making it a viable option for real-time applications. Additionally, EEG has the capability to record brain activity related to both overt and covert emotional processes, providing a holistic view of emotional states. In contrast to emotions deduced from physiological indicators of the Autonomic Nervous System (such as heart rhythm or skin conductivity), which can be affected by noise, those obtained directly from the Central Nervous System (like EEG) portray emotional experiences from their source. This has ignited a plethora of research in the domain of emotion recognition via EEG, with the goal to leverage EEG's potential to deepen our comprehension of emotions and set the stage for real-world uses.

The conceptual framework for emotion recognition generally falls into two main categories: discrete and dimensional models. The discrete model identifies fundamental emotions such as happiness, sadness, anger, and fear, assuming these to be universally experienced and biologically based. Conversely, the dimensional model represents emotions along continuous axes, typically arousal (indifference to excitement), valence (negative to positive), and dominance (submissive to dominance), which allows for a more nuanced capture of emotional states. In this work, we explore the latter, as it allows for comparatively better quantitative investigations.

While substantial progress has been made in the field of emotion recognition using EEG data, existing methodologies still exhibit some limitations. Our research primarily deals with a couple of limitations:

1. Generalizability: There is a gap in the existing literature about generalizable models for emotion recognition from EEG. Previous works have devised models that are finely tuned to specific datasets, and hence, specific types of EEG hardware, or emotional stimuli, thereby lacking generalizability required to be applicable across diverse datasets.
2. Biologically-Interpretable: Most of the existing models use either univariate features like PSD (Power Spectral Density) which lack spatial information, or multivariate features like PLV (Phase-Locking Value) which represent functional connectivity in the brain, but this is constructed rather than learnt by the model. This leads to lacking of a more interpretable representation of emotional states, which could be closer to brain.

Therefore, the objective of this research is to develop a model architecture for emotion recognition using EEG data that is not only generalizable (able to adapt to different subjects, EEG acquisition systems, and emotional stimuli), but also closer to the brain's natural processing mechanisms. This would make it suitable for real-time deployment in various application scenarios, from clinical settings to everyday interactions with technology.

To achieve these objectives, we propose a new model - GenEmo-Net, which utilises the combined capabilities of Graph Convolutional Neural Network and LSTM to learn spatial brain functional connectivity features over time for each emotional state. This model is able to learn the underlying functional connectivity of emotion states in the brain across major public datasets - DEAP, DREAMER and AMIGOS [6,7,10], resulting in state-of-the-art performance across them, making it both biologically-interpretable as well as generalizable.

This research contributes to the literature by presenting a model that strikes a balance between the neural realism and generalizability, necessary for real-world applications.

2 Related Work

The application of EEG data for emotion recognition has garnered substantial attention in recent years. Various machine learning and signal processing techniques have been employed, such as Support Vector Machines (SVM) and Convolutional Neural Networks (CNN) [2,9,13]. However, most of these approaches are dataset-specific, focusing on individual datasets like DEAP [7], DREAMER [6], or AMIGOS [10], limiting their generalizability across multiple datasets. The challenge of building a generalized model applicable across diverse datasets is under-explored in the current literature.

Traditional approaches often use univariate features extracted from EEG data, such as power spectral density, entropy and statistical measures, which provide valuable insights but lack spatial information, that could provide a more comprehensive understanding of brain activity related to emotional states. Additionally, models are rarely designed to closely mimic neurobiological mechanisms underlying emotional processing. To fill these gaps, some works have explored functional connectivity metrics like Coherence, Mutual information, and Granger causality [5,13,14], but have not reached a consensus on a standardized feature set. Functional connectivity in the brain refers to the temporal correlation between activity in different regions. Phase Locking Values (PLVs) offer an effective measure of functional connectivity, capturing the phase relationship between signals from different electrodes [8]. Incorporating PLVs allows for the inclusion of spatial information and also brings the computational model closer to the brain's actual functioning. Such computational model could offer a more genuine approach to recognizing emotions, providing a compelling rationale for their use. Furthermore, we have summarized some of the literature which make use of brain functional connectivity-based features for the recognition of human emotional states in Table 1.

Table 1. Summary of some research works on EEG-based emotion recognition which make use of brain functional connectivity-based features

Author(s)	Dataset	Features	Model	Accuracy (%)		
				Valence	Arousal	Dominance
Cui et al. (2023) [2]	DEAP	PLV	CNN	85	83	-
Wang et al. (2019) [15]	DEAP	PLV+DE	P-GCNN	73.31	77.03	79.20
Guo et al. (2018) [5]	DEAP	GC	CapsNet	88.09	87.37	-
Wang et al. (2019) [13]	DEAP	NMI	SVM	74.41	73.64	-
Song et al. (2019) [11]	DREAMER	-	DGCNN	86.23	84.54	85.02

In summary, existing methodologies either suffer from limited generalizability or overlook the importance of functional connectivity features like PLVs, which capture both spatial information and mirror neurobiological processes. To the best of our knowledge, no prior work has aimed to create a generalizable neural network model that integrates these facets. Our research aims to fill these gaps by developing a model applicable to multiple emotion recognition datasets, ensuring generalizability and neurobiological interpretability.

3   Proposed Method

3.1   Dataset Description

We have used three datasets in our work: DEAP, DREAMER and AMIGOS. They are described below.

The Database for Emotion Analysis using Physiological Signals (DEAP) [7] is a multimodal dataset, which contains EEG and peripheral physiological signals of 32 participants who were exposed to 40 one-minute long video clips intended to elicit various emotional states. Although the videos were of one minute, recording was started 3 s prior, resulting in recordings of 63 s with a sampling rate of 512 Hz, downsampled to 128 Hz). At the end of each trial, the participants self-reported ratings about their emotional levels in the form of valence, arousal, dominance, liking and familiarity.

The DREAMER dataset [6] is a multimodal emotional dataset consisting of EEG and eye-tracking data collected from 23 participants who were subjected to emotional video clips sourced from movies and music videos. The EEG data were recorded using 14 channels, with a sampling rate of 128 Hz. Eye-tracking data were captured simultaneously, providing additional modalities for emotion recognition. Participants viewed a total of 18 clips, and the emotional states were labeled along the dimensions of valence and arousal using Self-Assessment Manikins (SAM). The baseline signal recorded was of 61 s and the recorded signal was of varying duration of 65 s–393 s.

The Affective Multimodal Interfaces for Guiding Operator's Selection (AMIGOS) dataset [10] is unique in its focus on both individual and collaborative tasks to elicit emotional states. The dataset includes EEG, ECG, and peripheral physiological data from 40 participants engaged in activities such as watching video clips

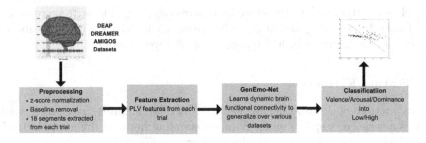

Fig. 1. Block diagram of our methodology

or playing games (we only take data of 34 participants in this work). The EEG data were acquired using 14 channels and sampled at a rate of 256 Hz. Given the variety of tasks and the presence or absence of interaction between participants, the dataset offers multiple configurations. The baseline signal recorded was of 5 s and the recorded signal was of varying duration of 54 s–150 s. Self-assessment was used to record participants' levels of valence, arousal and dominance.

To create binary-classification tasks, the emotion scales in each dataset are divided into two categories from the mid-value. Each of valence-arousal-dominance is categorized as low valence-arousal-dominance (ranging from start to mid) and high valence-arousal-dominance (ranging from mid to end) based on the respective scales. Block diagram of our methodology is shown in Fig. 1.

3.2 Data Preprocessing

As we have combined three datasets in our study, therefore we follow the same protocol for all three to make the dataset compatible. Initially for preprocessing, we have normalized our target data using baseline data which helps eliminate the individual variations and biases present in the recorded signals. The baseline data is recorded at resting state when no actual videos were shown to subjects. The mean (μ) and standard deviation (σ) of baseline data is obtained for each EEG channel and then z-score normalization is done as shown in Eq. 1.

$$x_{\mathrm{norm}}(t) = \frac{x(t) - \mu}{\sigma} \tag{1}$$

where, x is target signal and x_{norm} is normalized target signal of each EEG channel. In DEAP, EEG signals are recorded for 63 s with baseline signal of 3 s followed by 60 s target signal for 32 subjects and 40 trials. In case of DREAMER, target signals are recorded for varying duration of 65 s–393 s in addition to 61 s baseline signals for 23 subjects and 18 trials. Similarly, in case of AMIGOS also target signals are recorded for varying duration of 54 s–150 s in addition to 5 s baseline signals for 40 subjects (we only use 34 in this work) and 16 trials. Here, we have used last 54 s target signals for each trial recorded from 14 EEG channels, namely AF3, AF4, F3, F4, F7, F8, FC5, FC6, T7, T8, P7, P8, O1 and O2, in our study as depicted in Fig. 2. These normalized signals from respective channels are further segmented using 3 s non-overlapping window to increase the size of data for learning algorithm. Thus, from each channel, 18 segments are obtained for each trial, which is repeated for all three datasets.

3.3   Feature Extraction

The extraction of Phase-Locking-Value (PLV) [8] features from each trial, involves a series of steps to quantify phase coupling between different electrode pairs which includes the 14 electrodes as shown in Fig. 2.

The extraction process involves the following steps. Firstly, for each time segment, the EEG signals are processed using the Hilbert transform $(H(x(t)))$ to obtain the instantaneous phase information, as shown in Eq. 2.

$$\phi(t) = \arctan\left(\frac{H(x(t))}{x(t)}\right) \tag{2}$$

Once the instantaneous phase information is obtained, pairwise phase difference value is computed for each electrode pair. Finally, PLV for each electrode pair is calculated by averaging of the absolute value of the complex exponential of the pairwise phase differences $(\Delta\phi_n)$ over the entire segment, as shown in Eq. 3.

$$\text{PLV} = \left|\frac{1}{N}\sum_{n=1}^{N}\exp(i\Delta\phi_n)\right| \tag{3}$$

Since there are 14 EEG electrode nodes, we get PLV matrices of 14x14 dimensions for each of the 18 time segments, which reflect the connections in brain over time. Hence, finally we have combined data of dimensions 2238 x 14 x 14 x 18 ((subjects x trials) x PLV-matrix x time-segments).

3.4   Model Architecture

Graph Neural Networks (GNNs) are designed to apply neural network architecture in the realm of graph theory, facilitating data processing in graph-based domains, like EEG recordings. Here, each EEG channel represents a vertex node, while the edges are the connections between different vertex nodes. A specialized form of GNNs, called Graph Convolutional Neural Networks (GCNNs), expands upon conventional Convolutional Neural Networks (CNNs) by incorporating elements of spectral theory. GCNNs offer a distinct advantage when it comes to feature extraction from signals in discrete spatial domains, as compared to traditional CNNs. Their capacity to articulate the spatial relationships between varying nodes on a graph, is highly relevant for understanding correlations among multiple EEG channels in emotion recognition tasks. However, spatial and functional connections among EEG channels are not necessarily synonymous, which is a fundamental limitation of GCNNs.

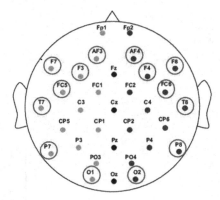

Fig. 2. 14-channel location map of EEG recording

To address the shortcomings of the conventional GCNN framework, Dynamic Graph Convolutional Neural Network (DGCNN) was proposed by Song et al. [11]. Unlike GCNNs, where the adjacency matrix has to be set in advance of training, this model introduces adjacency matrix as a learnable parameter allowing for dynamic updating of the adjacency matrix during model training and hence in this way learns to construct vertex node connections via an evolving adjacency matrix.

Since, brain functional connections for emotional states remain largely constant across population, we hypothesized that an adaptive learning of these connections over time should provide a path towards building a generalizable yet biologically interpretable model, by having a more genuine representation of emotion states in the brain. Therefore, taking inspiration from this idea, we build an architecture - GenEmo-Net (GENeralizable EMOtion NETwork), which can learn temporal functional connectivity patterns for each of the emotional states from Phase-Locking Values. To achieve this, our model architecture learns a weighted adjacency matrix from grid-like PLV matrices, as well as introduces a LSTM layer to capture temporal patterns in functional connectivity. These spatial functional connectivity and temporal patterns are fused together to make predictions about the emotional state. As the model is trained, the matrix is adaptively updated, enabling it to better capture the intrinsic functional relationships among EEG channels over time. In this way, GenEmo-Net is able to capture temporal functional connectivity patterns.

The proposed model, GenEmo-Net is trained with the following hyperparameters: RMSprop optimizer is used along with initial learning rate of 1e−2 which is updated by a factor of 0.1 for every 20 epochs in which the validation loss does not improve by a threshold of 1e−3. The minimum learning rate is set as 1e−5. The model is trained for 100 epochs with a batch size of 256. The architecture of the proposed model is shown is shown in Fig. 3.

Additionally, we experimented further by building another version of this architecture - GenEmo-X-Net (GENeralizable EMOtion eXperimental NET-

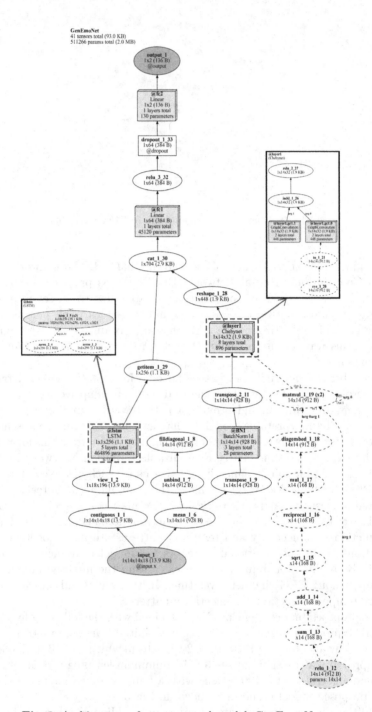

Fig. 3. Architecture of our proposed model: GenEmo-Net

Table 2. Comparison of our model against others

Model	Accuracy		
	Valence (%)	Arousal (%)	Dominance (%)
ViT	54.56±1.71	54.93±0.69	57.01±1.67
EEGNet	63.69±1.28	62.20±0.56	64.58±0.56
SVM	60.23±0.96	59.97±1.27	58.63±0.76
GenEmo-X-Net (Ours)	66.52±1.82	60.27±0.36	64.44±0.56
GenEmo-Net (Ours)	**70.98±0.73**	**65.47±0.56**	**70.09±0.37**

work), which is able to capture temporal variance in functional connections in the brain as well, unlike GenEmo-Net which rather captures a temporal 'summary'. Since the temporal variability in the functional connectivity encapsulates information about re-organization of functional networks over time during the task of emotion-recognition, we hypothesized that this would promote adaptability of the networks to dynamic contexts, as it does in the brain. For this experimental architecture, we introduce learnable connectivity matrices for each time segment. Therefore, as the model is trained, it learns functional connectivity at each time segment, and these learned dynamic temporal patterns of functional connectivity in the brain correspond to each of the emotional states, helping predict the same.

3.5 Model Evaluation

Since there is no such existing work based on generalizations across combinations of datasets, we trained on this generalizability task, some state-of-the-art model architectures which have been previously used for EEG-based emotion recognition tasks establishing competitive performance - ViT (Vision Transformer) [3], EEGNet (Compact Convolutional Neural Network) [9] and SVM (Support Vector Machines) [1] and compared our model against them. Additionally, we also compared the performance of our model, GenEmo-Net against the alternate experimental model architecture that we developed - GenEmo-X-Net. Accuracy of all the models are compared on binary classification tasks on each of the emotional dimensions of - valence, arousal, and dominance on 10% of the dataset, set aside as the testing set. Mean and standard deviation of accuracy of all the models is computed across three runs, which is used for the comparisons.

4 Results and Discussion

The GenEmo-Net model is compared against other models (as specified in Sect. 3.5), and the comparison results are shown in Table 2, as well as in the form of box-plots in Fig. 4.

We observe that the GenEmo-Net model out-performs all others in it's ability to generalize over diverse datasets obtained from different settings and subjects, achieving mean accuracy of 70.98%, 65.47%, and 70.09%, for each of valence, arousal and dominance emotion dimensions. Additionally note that our experimental model GenEmo-X-Net, performs below-par as compared to the former, which was a counter-intuitive result that we obtained.

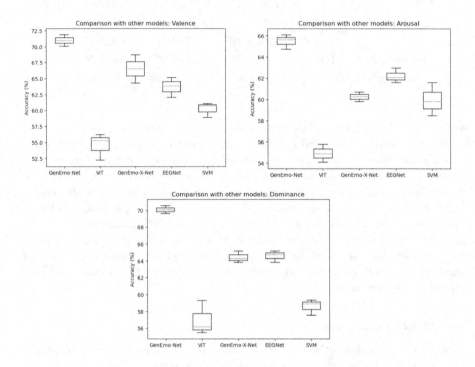

Fig. 4. Box plots of accuracy of our model versus others

In order to test the biological interpretablity of our model, we use the final weighted adjacency matrix learnt by the GenEmo-Net model to plot top 90 percentile connections in the human brain connectome. The MNI coordinates of the 14 EEG electrodes have been approximated and used for plotting these connections. These are shown in Fig. 5. Previous studies have suggested functional connectivity among ACC (Anterior Cingulate Cortex), DLPFC (Dorsolateral Prefrontal Cortex), Amygdala, MPFC (Medial Prefrontal Cortex), OFC (Orbitofrontal Cortex), IOG (Inferior Occipital Gyrus) and STS (Superior Temporal Sulcus), for emotional valence processing in the brain [12]. For arousal processing, connectivity among Amygdala, VLPFC (Ventrolateral Prefrontal Cortex), and ACC (Anterior Cingulate Cortex) has been suggested [4]. Similarly, for dominance processing, connectivity among VMPFC (Ventromedial Prefrontal Cortex), LPFC (Lateral Prefrontal Cortex), Amygdala and Striatum has been evidenced [16]. We observed from the plots in Fig. 5 that our model GenEmo-Net has been able to learn most of these functional connections corresponding to respective emotional states.

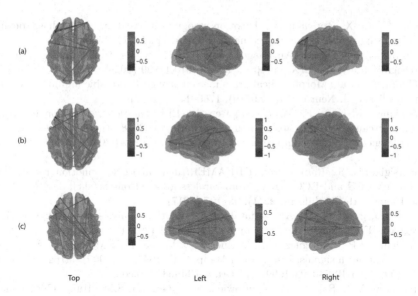

Fig. 5. Brain Functional Connectivity learned by GenEmo-Net model for (a) Valence (b) Arousal (c) Dominance, emotional states

5 Conclusion

In this paper, we proposed a model, GenEmo-Net, to address the generalization and biological interpretation limitations of existing work on emotion recognition from EEG signals. The results indicated that GenEmo-Net is able to generalize across diverse datasets, experimental settings and subjects, and performs better on the generalization task, as compared to conventional models. Additionally, we visualised the brain functional connections of emotion states learnt by the GenEmo-Net and observed that they are very close to what has been reported by existing cognitive neuroscience findings, which indicates that our model has learnt a genuine representation of emotion states in the brain. This marks a significant step forward towards models which are both generalizable and biologically interpretable in the EEG-based emotion recognition domain. For future prospects, we are more fascinated towards the applicability of our model for discrete emotions utilizing the transfer learning concepts.

Acknowledgements. This work is supported under Institute of Eminence (IoE) grant of Banaras Hindu University (B.H.U.), Varanasi, India.

References

1. Boser, B.E., Guyon, I.M., Vapnik, V.N.: A training algorithm for optimal margin classifiers. In: Proceedings of the Fifth Annual Workshop on Computational Learning Theory, pp. 144–152 (1992)

2. Cui, G., Li, X., Touyama, H.: Emotion recognition based on group phase locking value using convolutional neural network. Sci. Rep. **13**(1), 3769 (2023)
3. Dosovitskiy, A., et al.: An image is worth 16x16 words: transformers for image recognition at scale. arXiv preprint arXiv:2010.11929 (2020)
4. Gupta, A., et al.: Morphological brain measures of cortico-limbic inhibition related to resilience. J. Neurosci. Res. **95**(9), 1760–1775 (2017)
5. Jinliang, G., Fang, F., Wang, W., Ren, F.: EEG emotion recognition based on granger causality and capsnet neural network. In: 2018 5th IEEE International Conference on Cloud Computing and Intelligence Systems (CCIS), pp. 47–52. IEEE (2018)
6. Katsigiannis, S., Ramzan, N.: DREAMER: a database for emotion recognition through EEG and ECG signals from wireless low-cost off-the-shelf devices. IEEE J. Biomed. Health Inform. **22**(1), 98–107 (2017)
7. Koelstra, S., et al.: DEAP: a database for emotion analysis; using physiological signals. IEEE Trans. Affect. Comput. **3**(1), 18–31 (2011)
8. Lachaux, J.P., Rodriguez, E., Martinerie, J., Varela, F.J.: Measuring phase synchrony in brain signals. Hum. Brain Mapp. **8**(4), 194–208 (1999). https://doi.org/10.1002/(sici)1097-0193(1999)8:4<194::aid-hbm4>3.0.co;2-c
9. Lawhern, V.J., Solon, A.J., Waytowich, N.R., Gordon, S.M., Hung, C.P., Lance, B.J.: EEGNet: a compact convolutional neural network for EEG-based brain-computer interfaces. J. Neural Eng. **15**(5), 056013 (2018)
10. Miranda-Correa, J.A., Abadi, M.K., Sebe, N., Patras, I.: Amigos: a dataset for affect, personality and mood research on individuals and groups. IEEE Trans. Affect. Comput. **12**(2), 479–493 (2018)
11. Song, T., Zheng, W., Song, P., Cui, Z.: EEG emotion recognition using dynamical graph convolutional neural networks. IEEE Trans. Affect. Comput. **11**(3), 532–541 (2018)
12. Underwood, R., Tolmeijer, E., Wibroe, J., Peters, E., Mason, L.: Networks underpinning emotion: a systematic review and synthesis of functional and effective connectivity. Neuroimage **243**, 118486 (2021)
13. Wang, Z.M., Hu, S.Y., Song, H.: Channel selection method for EEG emotion recognition using normalized mutual information. IEEE Access **7**, 143303–143311 (2019)
14. Wang, Z., Liu, Y., Zhang, R., Zhang, J., Guo, X.: EEG-based emotion recognition using partial directed coherence dense graph propagation. In: 2022 14th International Conference on Measuring Technology and Mechatronics Automation (ICMTMA), pp. 610–617 (2022). https://doi.org/10.1109/ICMTMA54903.2022.00127
15. Wang, Z., Tong, Y., Heng, X.: Phase-locking value based graph convolutional neural networks for emotion recognition. IEEE Access **7**, 93711–93722 (2019)
16. Watanabe, N., Yamamoto, M.: Neural mechanisms of social dominance. Front. Neurosci. **9**, 154 (2015)

Ear-EEG Based-Driver Fatigue Detection System Augmented by Computer Vision

Ngoc-Dau Mai, Ha-Trung Nguyen, and Wan-Young Chung$^{(\boxtimes)}$

Department of AI Convergence, Pukyong National University, Busan, South Korea
wychung@pknu.ac.kr

Abstract. Driver fatigue is a significant danger to road safety, resulting in numerous accidents and fatalities on a global scale. To tackle this problem, researchers have been exploring innovative techniques for identifying and preventing driver drowsiness during real-world driving situations. This study proposes a novel approach for detecting driver fatigue by merging EEG (Electroencephalogram) data from sensors positioned behind the ear with computer vision-based analysis of facial characteristics. Behind-the-ear (BTE) EEG provides a more practical and user-friendly alternative than traditional scalp EEG methods. In addition to Ear-EEG signals, computer vision technology enhances fatigue detection accuracy by examining drivers' facial images while driving. The study introduces a custom-designed wearable device for gathering EEG data from four sensor electrodes behind the ear. Continuous wavelet transform (CWT) converts these EEG signals into scalograms. These scalograms and facial images captured by a camera focused on key facial areas such as the left eye, right eye, mouth, and entire face serve as inputs for a deep learning model developed for identifying driver fatigue. Subsequently, a comparative assessment is conducted to gauge the performance of the proposed system when using only Ear-EEG signals, only camera images, or a combination of both data sources. The test results validate the practicality and effectiveness of the proposed system in identifying driver fatigue. Additionally, a companion smartphone application has been developed to simplify and promptly monitor and alert drivers when they exhibit drowsiness while driving in traffic.

Keywords: BCI · Ear EEG · Driver Fatigue Detection · Deep Learning

1 Introduction

Drowsiness represents an intermediary state between full wakefulness and sleep, characterized by reduced cognitive functions and processing abilities [1]. This condition significantly impairs a driver's performance and increases the risk of traffic accidents. The emergence of driver fatigue detection (DFD) systems provides a potential solution for identifying drowsiness and improving driver vigilance. EEG-based DFD is a highly reliable and effective approach due to its direct connection to brain activity. Ear EEG electrodes are strategically placed in or near the outer ear, offering improved visibility, mobility, and skin-to-electrode contact, enhancing signal quality. Deep learning [2] has emerged as a potent paradigm in computer vision and biomedical signal processing, particularly in detecting drowsiness through facial cues and classifying EEG signals. In facial drowsiness detection, deep learning techniques have demonstrated remarkable capabilities in identifying subtle facial cues indicative of drowsiness.

Therefore, this study proposes developing a wearable DFD system that integrates behind-the-ear (BTE) EEG signals with an analysis of facial characteristics using a proposed deep-learning approach. The primary contributions of this study encompass (1) The introduction of a comprehensive and groundbreaking design for a BTE-EEG embedded device, promising to revolutionize the field of DFD systems. (2) A detailed exploration and rigorous analysis of the feasibility of combining BTE EEG-based features with computer vision-based analysis of facial characteristics for DFD. (3) Creating an advanced deep learning model meticulously tailored for DFD, utilizing BTE EEG and facial characteristics insights. (4) The development of a user-friendly smartphone application designed to offer simplified and immediate monitoring and alerting capabilities, aiding drivers in maintaining optimal alertness levels and ensuring their safety on the road.

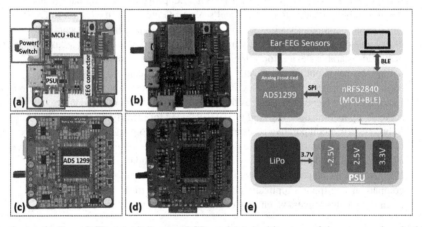

Fig. 1. (a) (b) Front PCB, (c) (d) Bottom PCB, and (e) Architecture of the proposed embedded system.

2 Proposed System

2.1 Ear-EEG Embedded Device

Figure 1 illustrates our embedded device's primary components and Printed Circuit Board (PCB) layout. This proposed device comprises three key units: the sensor collecting unit (SCU), the processor and wireless transceiver unit (PWTU), and the power supply unit (PSU). The SCU is responsible for acquiring EEG signals and utilizes the highly precise analog front-end integrated chip ADS1299. This chip features eight low-noise and eight 24-bit resolution analog-to-digital converters (ADC) for simultaneous sampling. The PWTU is equipped with the multi-protocol Bluetooth 5.3 system-on-chip, nRF52840, from Nordic Semiconductor. The PWTU employs a programmable peripheral interface (PPI) to manage the data-ready pin of the ADS1299. The PSU incorporates the MCP73831 battery charge management controller to ensure high efficiency.

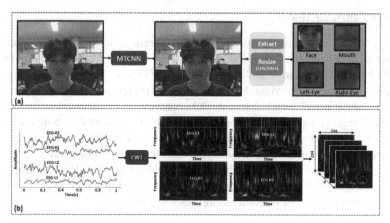

Fig. 2. (a) Computer vision-based face detection and alignment, (b) Ear-EEG-based CWT scalogram creation.

2.2 Computer Vision-Based on Face Features

Figure 2(a) illustrates a series of steps employed to acquire essential facial components, which play a pivotal role in driver fatigue detection through a front-facing camera setup. This data collection process encompasses two primary phases. The initial step involves joint face detection and alignment. At the same time, the subsequent phase entails extracting facial components such as the entire face, left eye, right eye, and mouth, followed by their resizing to meet specific dimensions. For face detection and alignment, the Multi-Task Cascaded Convolutional Networks (MTCNN) is deployed [3], chosen for its recognized speed and accuracy as a face detector. MTCNN, leveraging its cascaded architecture, excels in swiftly and accurately accomplishing the joint tasks of face detection and alignment.

2.3 Experiment

In our experimental investigation, we recruited 12 subjects aged 24 to 38. These individuals were carefully selected to meet specific criteria: they exhibited sound physical and mental health, possessed unimpaired vision, did not have sleep disturbances, and held valid driver's licenses. Our experimental design comprised two distinct measurement sessions: (1) The initial session involved recording subjects' EEG signals while they operated a driving simulator in a heightened alert state. (2) The subsequent session included EEG signal collection during a phase in which participants were predisposed to drowsiness while engaged in simulated driving scenarios. Participants were explicitly instructed to maintain a vehicle speed of approximately 50 mph. We incorporated a camera into our experimental setup to document participants' facial expressions, as depicted in Fig. 3(a–d).

2.4 Deep Learning Model

This study introduces an advanced deep-learning model designed for driver fatigue detection. This model employs a multi-input approach, utilizing facial component images and Ear-EEG-based Continuous Wavelet Transform (CWT) scalograms [4]. A comprehensive depiction of the proposed model's architecture is presented in Fig. 4. The structural framework of our novel deep learning model encompasses distinctive feature extractor blocks, each tailored to a specific type of input (comprising a total of 5 inputs). These feature extractor blocks consist of convolutional layers that extract and learn meaningful features from the input data. These layers capture essential information from facial component images and the Ear-EEG-based CWT scalograms. The deep learning model leverages the learned features to discern signs of fatigue, contributing to enhanced road safety by alerting drivers or relevant authorities when necessary.

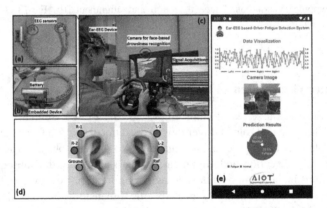

Fig. 3. (a–d) Experimental procedure, (e) Smart phone app for driver fatigue detection.

3 Results and Discussion

The Ear-EEG signals are acquired from four positions behind the ear using our newly designed embedded device. Subsequently, these signals undergo preprocessing through a band-pass filter to isolate the desired EEG signals within the 1Hz to 60Hz range while eliminating unwanted frequencies. The EEG data underwent a segmentation process in a trial to collect signals. This segmentation involved using a sliding window, lasting 1 s and consisting of 256 data points. Importantly, an overlap rate of 50% was applied during this segmentation. This research employs the Continuous Wavelet Transform (CWT) to analyze and represent the signals as scalogram images. Meanwhile, the images captured by the front-facing camera of the driver undergo two processing steps: joint face detection and alignment and the extraction of facial components, including the entire face, left eye, right eye, and mouth. The obtained results encompass Ear-EEG-based CWT Scalograms and images of facial components, which are used as inputs for the proposed deep-learning model to detect driver fatigue.

Figure 5 displays our deep learning model's loss curve and confusion matrix employed in driver fatigue detection. To assess the performance of this detection process, we utilized a comprehensive set of metrics encompassing accuracy, recall, precision, and the F1-score. In our research, we evaluated the efficacy of our proposed deep learning model utilizing multiple inputs, specifically combining Ear-EEG features with computer vision-based facial features compared to using either Ear-EEG features alone or solely relying on computer vision-based facial features. Our results, as elaborated in Table 1, demonstrate that the most superior performance in detecting driver fatigue states (fatigue state and normal state) was achieved when utilizing the combination of both Ear-EEG features and computer vision-based facial features, yielding respective values of 0.979, 0.983, 0.975, and 0.979 for accuracy, recall, precision, and F1-score. Additionally, we developed a companion smartphone application to facilitate real-time monitoring and timely alerts for drivers showing signs of drowsiness while navigating traffic, as depicted in Fig. 3(e).

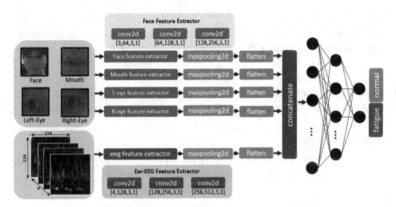

Fig. 4. The architecture of the proposed deep learning model.

4 Conclusion

This study introduced an innovative approach to identifying driver fatigue by combining EEG data collected from sensors behind the ear with computer vision analysis of facial features. BTE EEG offers a more practical and user-friendly alternative to traditional scalp EEG methods. In conjunction with Ear-EEG signals, computer vision technology enhances fatigue detection accuracy by analyzing drivers' facial images while driving. The study features a specially designed wearable device with four sensor electrodes behind the ear to gather EEG data. These EEG signals are then transformed into scalograms using CWT. In addition to facial images captured by a camera focusing on key facial areas such as the eyes and mouth, these scalograms serve as inputs for a deep learning model created to identify driver fatigue. Subsequently, a comparative evaluation assesses the system's performance when using only Ear-EEG signals, only camera images, or a combination of both data sources. The test results confirm the practicality and effectiveness of the proposed system in detecting driver fatigue. Furthermore, a companion smartphone app has been developed to simplify and promptly monitor drivers, alerting them when they display signs of drowsiness while driving in traffic.

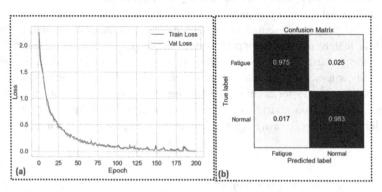

Fig. 5. The loss curves and the confusion matrix of the proposed model with the combined features of face images and ear-EEG-based CWT scalograms.

Table 1. Performance comparison of the proposed deep learning model with three types of features: Ear-EEG-based CWT scalograms, Face expression images, and Combined features

Evaluation Metrics	Features		
	Ear-EEG-based CWT scalograms	Face expression images (Left-Eye, Right-Eye, Mouth, Face)	Combined features
Accuracy	0.943	0.909	0.979
Precision	0.954	0.914	0.975
Recall	0.931	0.903	0.983
F1-score	0.942	0.908	0.979

Acknowledgements. This work was supported by the National Research Foundation of Korea (NRF) grant funded by the Korea government (MSIT) (NRF-2019R1A2C1089139).

References

1. Nguyen, H.T., Mai, N.D., Lee, B.G., Chung, W.Y.: Behind-the-ear EEG-based wearable driver drowsiness detection system using embedded tiny neural networks. IEEE Sens. J. (2023). https://doi.org/10.1109/JSEN.2023.3307766
2. LeCun, Y., Bengio, Y., Hinton, G.: Deep learning. Nature **521**(7553), 436–444 (2015)
3. Xiang, J., Gengming, Z.: Joint face detection and facial expression recognition with MTCNN. In: 2017 4th International Conference on Information Science and Control Engineering (ICISCE). IEEE (2017)
4. Sinha, S., et al.: Spectral decomposition of seismic data with continuous wavelet transform. Geophysics **70**(6), P19–P25 (2005)

Cross Cultural Comparison of Emotional Functional Networks

Mohammad Asif[1]($\boxtimes$) , Sudhakar Mishra[2] , Jerald Kannath[1],
Tarun Jayadevan[1], Divakar Singh[1], Gauttam Goyal[1], Aalok Bhuyar[1],
and Uma Shanker Tiwary[1]

[1] Indian Institute of Information Technology, Allahabad, India
pse2017001@iiita.ac.in
[2] Indian Institute of Technology, Kanpur, India
sudhakarm@iitk.ac.in

Abstract. Experience and expression of Emotions have cultural influences. In behavioural experiments, Western and Eastern cultures are shown to have differences in emotional experience and expression. Western culture promotes the expression of emotional experience, whereas, in Eastern culture, emotional expressions are not very explicit and are sometimes restricted by social norms. However, there is limited evidence for this observation in terms of brain activity. In this study, we analyzed two different datasets, the DEAP data for the Western population and the DENS data for the Eastern population, focusing specifically on happy and sad emotions. We calculated functional connectivity among EEG electrodes using phase locking value and found that activity in the frontal electrodes is more pronounced in Eastern culture compared to Western culture. Meanwhile, activity in the centro-parietal electrodes is more dominant in Western culture. Activity in the frontal brain regions is high during emotion regulation. On the other hand, activity in the centro-parietal regions is more associated with sensori-motor activity.

Keywords: Emotions · EEG · Functional Connectivity · DENS Dataset · DEAP Dataset · Cross-Cultural Study · Functional Networks

1 Introduction

Human experience is fundamentally shaped by emotions, which have an impact on our attitudes, actions, and social interactions. They combine physiological, cognitive, and subjective elements and are complex and multifaceted. Neuroscience research has a strong emphasis on figuring out the neural basis of emotions and how the brain processes them. The human brain is a complex network of neurons, and it's an emerging subject for neuroscience research. Thanks to modern brain-scanning techniques, we can explore how different parts of the brain work together, especially when it comes to handling our emotions. In this

M. Asif et al.—All authors have equal contributions.

study, we're focusing on the brain's emotional network and how it's influenced by culture.

Emotion experiences can vary between Eastern and Western cultures. These differences can be attributed to a combination of cultural, historical, and societal factors. Cultural factors can significantly influence how people perceive, express, and experience emotions [8,9]. It's important to note that the terms "Eastern" and "Western" are broad categories that encompass numerous diverse cultures, so there can be significant variations within each category [4].

Some general differences in emotional experiences between Eastern and Western cultures can be described in terms of emotional expression, cultural norms, emotional valence, coping strategies, and non-verbal cues. For example, Western cultures, particularly in North America and Europe, tend to encourage more open and direct expression of emotions. In contrast, many Eastern cultures, such as those in East Asia, often value emotional restraint and indirect communication [5,10]. Cultural norms in Eastern cultures may emphasize collectivism and harmony, which can lead to an emphasis on suppressing negative emotions in social settings to maintain group cohesion. In Western cultures, individualism is often more pronounced, leading to more individualized and expressive emotional responses [3,11,16].

Although, it is important to emphasize that these are general trends and that individual experiences and cultural variations within each category can be substantial. Also, there is an increasing globalization of culture and the exchange of emotional norms and expressions due to factors like globalization, increased international travel, and the spread of information through the internet and media [17]. Therefore, while there are differences, it's crucial to recognize the complexity and diversity of emotional experiences across cultures.

Most of the works emphasizing cultural differences are done behaviourally. To our knowledge, there is no research available which is looking for evidence of brain activity pertaining to different cultures. In this work, we are probing the functional connections during the emotional experience of Indian and European participants. We hypothesize that we might observe dominant activity in brain regions, which can support the cultural differences claim.

2   Methodology

2.1   Datasets

DEAP (Database for Emotion Analysis Using Psychological Signals). The DEAP dataset [6] is a collection of physiological and EEG signals taken while watching music videos by human subjects. The dataset was created by the Université catholique de Louvain (UCL) in Belgium as a component of the DEAP project, which aimed to understand how emotions and music interact with one another. The dataset contains data from 32 participants, each of whom watched a set of 40 music videos. Physiological signals recorded include measurements of skin conductance, heart rate, respiration, and blood volume pulse. EEG signals were recorded from 32 scalp electrodes. Emotions, namely - love, lovely, cheerful, melancholy, senti, sad, happy1, mellow, sentimental, terror, shock, hate, joy, excited, depressing, happy, fun were recorded. The dataset also includes ratings of the emotional valence and arousal of each music video by the participants, obtained using a self-assessment manikin (SAM) tool.

DENS (Dataset on Emotion with Naturalistic Stimuli). The DENS dataset [1,12] is a collection of physiological and EEG signals taken while watching naturalistic stimuli by the participants. The dataset was created at the Indian Institute of Information Technology, Allahabad (IIITA) in India. It aimed to target the precise temporal information of the emotions felt by the participants that are referred to as "emotional events". There are 465 emotional events available as preprocessed EEG data along with ECG and EMG data of the participants. EEG signals were recorded from 128 channels of high-density scalp electrodes. The targeted emotions of the dataset are adventurous, afraid, alarmed, amused, angry, aroused, calm, disgusted, enthusiastic, excited, happy, joyous, melancholic, miserable, sad, and triumphant. The dataset also includes ratings of the emotional valence and arousal of each music video by the participants, obtained using a self-assessment manikin (SAM) tool.

2.2 Complex Network Analysis of the Brain During Emotion

We have considered the preprocessed EEG signals for both datasets and calculated functional connectivity between phases of different electrodes using phase locking value (PLV). PLV is a statistical measure used in the field of neuroscience and signal processing to quantify the degree of phase synchronization between two or more oscillatory signals [14]. It is a valuable tool for understanding the functional connectivity and coordination of neural activity in the brain and is commonly used in studies involving electroencephalography (EEG), magnetoencephalography (MEG), and other neuroimaging techniques. PLV is particularly useful for investigating phase relationships between these oscillatory signals. It quantifies the consistency of phase differences between two signals over time. When the phase of one signal is consistently related to the phase of another signal, it suggests that there is a significant synchronization or coordination between the underlying neural processes.

PLV Calculation. PLV quantifies how well different parts of the brain or other oscillatory systems coordinate their activities by examining the consistency of phase relationships between their oscillations. It started with the instantaneous phase of two neural signals. Next, the difference between the phases of these two signals at a particular time is calculated. PLV can be calculated as follows for a pair of signals at each time point t:

$$PLV(t) = |\frac{\Sigma_{j=1}^{N} e^{i\Delta\phi}}{N}|$$

where N is the number of time points within an epoch, and $\Delta\phi$ is the phase difference at each time point.

Statistical Analysis and Visualization. To determine if the observed PLV values are significantly different from what would be expected by chance, we used the adjacency matrix of PLV values across the subjects for any experiment condition to calculate an average adjacency matrix for an experiment condition of interest. To rule out the possibility of a spurious connection, we calculated a threshold (average PLV value) from the average adjacency matrix.

$$Th_c = \frac{\Sigma_{i=1}^{31} \Sigma_{j=i+1}^{31} A\_PLV_{ij}}{465}$$

where A_PLV_{ij} is the average calculated across subject for any condition c. Th_c is the calculated threshold to filter out spurious connections. The above procedure has been applied to happy and sad emotion categories from both datasets.

The PLV values obtained from the analysis can be displayed as functional connectivity matrices. These matrices show the strength of the connections between all pairs of electrodes. Each electrode is represented by a heat map with the corresponding strength of the connection. To make the visualization even clearer, the data can be binarized with a threshold point as calculated above.

3   Results and Observations

We analyzed the functional connections among 32 electrodes for DEAP and DENS data. As mentioned in the method section, we selected robust connections by calculating a threshold and selecting connections above the threshold. We observed electrodes with more number of connections on the frontal electrodes for Indian participants in the DENS data. Whereas, for the western participants, electrodes with more number of connections are located in the central and centro-parietal regions, which are active in motor and sensory-motor networks (see Table 1 and Fig. 1). The functional connectivity matrices before and after binarization have been shown in Fig. 2 and Fig. 3.

Table 1. The table displays information about the electrodes in two datasets, DEAP and DENS, along with the number of connections each electrode has with other electrodes. The DENS dataset had a total of 128 electrodes. To make the two datasets comparable, we selected 32 electrodes from the DENS dataset that matched the electrodes in the DEAP dataset. The last column of the table is color-coded, with pink indicating that a particular electrode has more connections in the DENS dataset than in the DEAP dataset, and orange indicating that a particular electrode has more connections in the DEAP dataset than in the DENS dataset.

Electrodes		DEAP		DENS		Remarks
DEAP	DENS	Happy	Sad	Happy	Sad	
Fp1	22	4	5	16	17	Fp1
AF3	23	14	11	17	19	
F7	33	15	12	14	13	
F3	24	11	11	16	15	F3
FC1	13	11	8	8	12	
FC5	28	7	10	11	12	
T7	45	10	8	4	5	
C3	36	2	2	7	8	
CP1	37	6	6	6	6	
CP5	47	19	16	8	7	CP5
P7	58	6	5	12	12	P7
P3	52	15	13	10	10	P3
Pz	62	7	7	15	14	Pz
PO3	67	12	10	15	12	
O1	70	16	13	15	15	
Oz	75	16	13	16	16	
O2	83	21	14	15	15	
PO4	77	9	5	15	15	PO4
P4	92	21	17	11	12	
P8	96	20	17	12	13	
CP6	98	14	9	11	10	
CP2	87	13	13	6	7	CP2
C4	104	18	8	7	8	C4
T8	108	14	13	5	5	T8
FC6	117	4	2	10	9	
FC2	112	5	7	8	9	
F4	124	14	11	12	12	
F8	122	23	22	12	7	
AF4	3	17	16	17	18	
Fp2	9	5	7	16	17	Fp2
Fz	11	16	12	17	17	

4 Discussion

In this work, we assessed the impact of culture on functional connectivity of emotional experiences. The connectivity results confirms to our hypothesis that in Western culture brain regions which are associated with motor expression as well as sensory feeling are showing more connectivity. Whereas, in the brain connectivity in the Indian participants is more prominent is frontal and posterior sites.

Activity in the frontal sites could be due to emotion suppression and regulation as per the cultural norms [2,7]. Whereas, activity in the central and

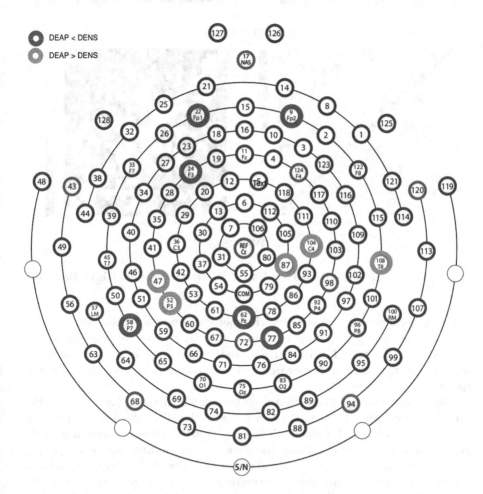

Fig. 1. Highlighted in this figure are the electrodes that display higher connection in one dataset compared to another. In the DENS dataset, the frontal and posterior sites exhibit greater connectivity, while participants in the DEAP dataset demonstrate dominant connectivity nodes in the sensory-motor regions.

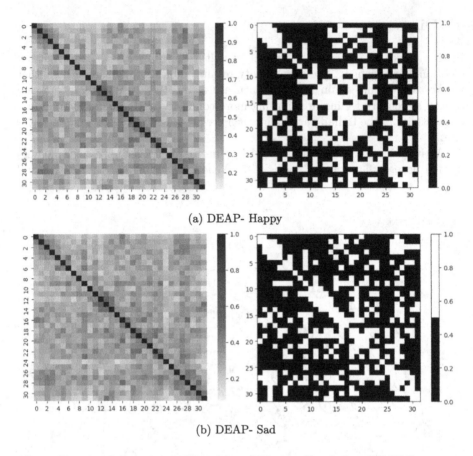

(a) DEAP- Happy

(b) DEAP- Sad

Fig. 2. Functional Connectivity Heat Map of Western Population- DEAP Dataset.

centro-parietal regions is associated with motor and sensory-motor activity [15]. Although, we acknowledge our limitation that our results are priliminary and based on the analysis of functional connectivity only. However, these exploratory results leads to confirmatory research with seed nodes which are reported in the results.

Our results also emphasize that even though globalization has increased the communication among different cultures, the subtler effect of culture on emotion may not had been much influenced. Since emotion is a learnt experience which evolves in a society and culture, the societal and cultural norms are deeply ingrained in emotional experience and expression. Even it is reported that emotion vocabulary sometimes differ from culture to culture. In our earlier work. We developed two datasets [12,13] to promote cross-cultural research on emotion. It is important to do more future research on cross-cultural aspects of different aspects of emotional experience and expression. It is more important to learn

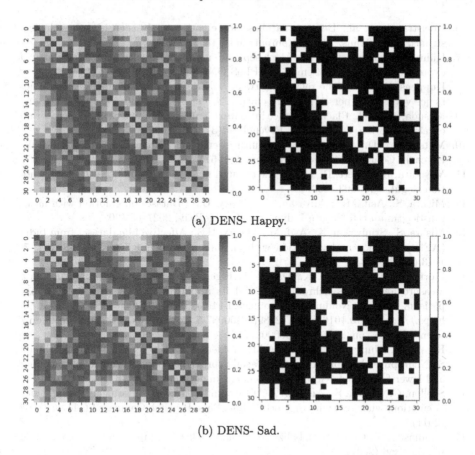

(a) DENS- Happy.

(b) DENS- Sad.

Fig. 3. Functional Connectivity Heat Map of Indian Population- DENS Dataset.

the cultural influence in emotional experience to develop behavioural therapies for different emotional disorders.

References

1. Asif, M., Mishra, S., Vinodbhai, M.T., Tiwary, U.S.: Emotion recognition using temporally localized emotional events in EEG with naturalistic context: DENS# dataset. IEEE Access (2023)
2. Choi, D., Sekiya, T., Minote, N., Watanuki, S.: Relative left frontal activity in reappraisal and suppression of negative emotion: evidence from frontal alpha asymmetry (FAA). Int. J. Psychophysiol. **109**, 37–44 (2016)
3. Eid, M., Diener, E.: Norms for experiencing emotions in different cultures: inter- and intranational differences. J. Pers. Soc. Psychol. **81**(5), 869 (2001)
4. Halder, M., Binder, J., Stiller, J., Gregson, M.: An overview of the challenges faced during cross-cultural research. Enquire **8**, 1–18 (2016)
5. Keltner, D., Tracy, J., Sauter, D.A., Cordaro, D.C., McNeil, G.: Expression of emotion. Handb. Emotions **4**, 467–482 (2016)

6. Koelstra, S., et al.: Deap: a database for emotion analysis; using physiological signals. IEEE Trans. Affect. Comput. **3**(1), 18–31 (2011)

7. Kühn, S., Gallinat, J., Brass, M.: Keep calm and carry on: structural correlates of expressive suppression of emotions. PLoS ONE **6**(1), e16569 (2011)

8. Kuppens, P., et al.: The relation between valence and arousal in subjective experience varies with personality and culture. J. Pers. **85**(4), 530–542 (2017)

9. Masuda, T., et al.: Placing the face in context: cultural differences in the perception of facial emotion. J. Pers. Soc. Psychol. **94**(3), 365 (2008)

10. Matsumoto, D., Hwang, H.S.C.: Culture, emotion, and expression. Cross-Cult. Psychol.: Contemp. Themes Perspect., 501–515 (2019)

11. Mesquita, B., Albert, D., et al.: The cultural regulation of emotions. Handb. Emotion Regul. **486**, 503 (2007)

12. Mishra, S., Asif, M., Srinivasan, N., Tiwary, U.S.: Dataset on emotion with naturalistic stimuli (DENS) on Indian samples. bioRxiv, 2021–08 (2021)

13. Mishra, S., Srinivasan, N., Asif, M., Tiwary, U.S.: Affective film dataset from India (AFDI): creation and validation with an Indian sample. J. Cult. Cognit. Sci. **7**, 1–13 (2023)

14. Mormann, F., Lehnertz, K., David, P., Elger, C.E.: Mean phase coherence as a measure for phase synchronization and its application to the EEG of epilepsy patients. Phys. D: Nonlinear Phenomena **144**(3), 358–369 (2000). ISSN 0167–2789. https://doi.org/10.1016/S0167-2789(00)00087-7. https://www.sciencedirect.com/science/article/pii/S0167278900000877

15. Ross, E.D., Gupta, S.S., Adnan, A.M., Holden, T.L., Havlicek, J., Radhakrishnan, S.: Neurophysiology of spontaneous facial expressions: I. motor control of the upper and lower face is behaviorally independent in adults. Cortex 76, 28–42 (2016)

16. Scollon, C.N., Koh, S., Au, E.W.: Cultural differences in the subjective experience of emotion: when and why they occur. Soc. Pers. Psychol. Compass **5**(11), 853–864 (2011)

17. Tomlinson, J.: Cultural globalization. The Blackwell companion to globalization, pp. 352–366 (2007)

Emotion Recognition Using Phase-Locking-Value Based Functional Brain Connections Within-Hemisphere and Cross-Hemisphere

Ruchilekha[1], Varad Srivastava[2(✉)], and Manoj Kumar Singh[3]

[1] DST- Centre for Interdisciplinary Mathematical Sciences,
Banaras Hindu University, Varanasi, India
ruchilekha.cims@bhu.ac.in

[2] Quantitative Analytics, Barclays, London, UK
varad.srivastava@barclays.com

[3] Department of Computer Science, Banaras Hindu University, Varanasi, India
manoj.dstcims@bhu.ac.in

Abstract. Research in cognitive neuroscience has found emotion-induced distinct cognitive variances between the left and right hemispheres of the brain. In this work, we follow up on this idea by using Phase-Locking Value (PLV) to investigate the EEG based hemispherical brain connections for emotion recognition task. Here, PLV features are extracted for two scenarios: Within-hemisphere and Cross-hemisphere, which are further selected using maximum relevance-minimum redundancy (mRmR) and chi-square test mechanisms. By making use of machine learning (ML) classifiers, we have evaluated the results for dimensional model of emotions through making binary classification on valence, arousal and dominance scales, across four frequency bands (theta, alpha, beta and gamma). We achieved the highest accuracies for gamma band when assessed with mRmR feature selection. KNN classifier is most effective among other ML classifiers at this task, and achieves the best accuracy of 79.4%, 79.6%, and 79.1% in case of cross-hemisphere PLVs for valence, arousal, and dominance respectively. Additionally, we find that cross-hemispherical connections are better at predictions on emotion recognition than within-hemispherical ones, albeit only slightly.

Keywords: Phase-Locking Value (PLV) · brain functional connectivity · mRmR · chi-square test

1 Introduction

Emotions play a fundamental role in human communication and interaction, significantly influencing our behavior, decision-making, and overall well-being. Accurately detecting and interpreting human emotions has far-reaching applications, ranging from affective computing to mental health monitoring and human-robot interaction. Among the various modalities utilized for emotion analysis, electroencephalography (EEG) stands out as a promising and non-invasive technique for capturing the neural correlates of emotions. The use of EEG for emotion recognition presents numerous advantages; one being, EEG is non-invasive and relatively affordable, along with a high temporal resolution, allowing for the precise examination of rapid changes in emotional states. Moreover, EEG is capable of capturing brain activity associated with both conscious and unconscious emotional processes, offering a comprehensive perspective on emotional experiences. Unlike emotions captured using physiological signals from Autonomous Nervous System (like heart rate, galvanic skin response) which are vulnerable to noise, those captured directly from the Central Nervous System (like the EEG) capture the expression of emotional experience from its origin. This has sparked extensive research in the field of EEG-based emotion recognition, aiming to harness the power of EEG signals to advance our understanding of emotions and pave the way for practical applications [10]. Leveraging machine learning and pattern recognition techniques has made it possible to translate complex EEG data into meaningful emotional states, bridging the gap between neuroscience and technology.

There are two widely accepted emotion models around which such research has centered around - discrete and dimension. Based on the discrete basic emotion description approach, emotions can be categorized into six fundamental emotions: sadness, joy, surprise, anger, disgust, and fear. Alternatively, the dimension approach enables emotions to be classified based on multiple dimensions (valence, arousal, and dominance). Valence pertains to the level of positivity or negativity experienced by an individual, while arousal reflects the degree of emotional excitement or indifference. The dominance dimension encompasses a spectrum ranging from submissive (lack of control) to dominance (assertiveness). In practice, emotion recognition predominantly relies on the dimension approach due to its simplicity in comparison to the detailed description of discrete basic emotions [13], which also allows for a quantitative analysis. In this work, our investigations explore the latter. Emotion recognition from EEG involves extracting relevant time or frequency domain feature components in response to stimuli evoking different emotions. However, a common limitation of existing methods is the lack of spatial correlation between EEG electrodes in univariate feature extraction. EEG brain network is a highly valuable approach for examining EEG signals, wherein each EEG channel serves as a node and the connections between nodes are referred to as edges. The concept of brain connectivity encompasses functional connectivity and effective connectivity [1, 2]. Moreover, findings in cognitive neuroscience have provided evidence for the structural and functional dissimilarities between the brain hemispheres [5].

To address this limitation and leverage hemispherical functional brain connections for emotion recognition, the phase locking value (PLV) method [9] has been utilized in our work which enables the investigation of task-induced changes in long-range neural activity synchronization in EEG data.

Based on this, we investigate the connections both within-hemisphere, and cross-hemisphere. Therefore, this paper proposes an EEG emotion recognition scheme based on significant Phase Locking Value (PLV) features extracted from

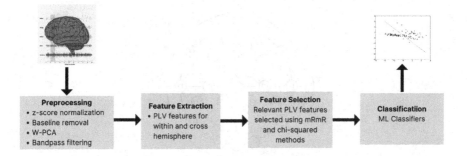

Fig. 1. Block diagram of our proposed method

hemispherical brain regions in the EEG data acquired as part of DEAP dataset [7], to understand the functional connections underlying within same hemisphere and cross hemisphere. By investigating performance of various machine learning models in being able to recognize the human emotions from EEG signals, this work throws light on which rhythmic EEG bands (alpha, beta, theta, gamma, all), and hemispherical brain connections (within or cross) are most efficient and responsive to emotions to measure the emotional state.

2   Related Work

There have been many studies conducted on using DEAP dataset for emotion recognition. Wang et al. (2018) [14] used an EEG specific 3-D CNN architecture to extract spatio-temporal emotional features, which are used for classification. Chen et al. (2015) [4] used connectivity features representation for valence and arousal classification.

Current findings in cognitive neuroscience have provided evidence for the structural and functional dissimilarities between the brain hemispheres [5]. Zheng et al. (2015) [17] conducted an investigation on emotional cognitive characteristics induced by emotional stimuli, revealing distinct cognitive variances between the left and right hemispheres. The study indicated that the right hemisphere exhibits enhanced sensitivity towards negative emotions. Similarly, Li et al. (2021) [11] employed the calculation of differential entropy between pairs of EEG channels positioned symmetrically in the two hemispheres, and used bi-hemisphere domain adversarial neural network to learn emotional features distinctively from each hemisphere.

Consequently, the analysis of EEG signals in both the left and right hemispheres holds immense significance in advancing emotional recognition techniques. Following this, Zhang et al. (2022) [16] focused on the asymmetry of the brain's hemispheres and employed cross-frequency Granger causality analysis to extract relevant features from both the left and right hemispheres, highlighting the significance of considering functional connectivity between hemispheres and leveraging cross-frequency interactions to improve the performance of EEG-based emotion recognition systems.

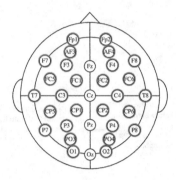

Fig. 2. 32 electrode positions in the international 10-20 system; Selected electrodes from each hemisphere for PLV features: left (in green color) and right (in blue color) (Color figure online)

Wang et al. (2019) [15] used Phase-Locking Value (PLV), to extract information about functional connections along with the spatial information of electrodes and brain regions.

3   Proposed Method

3.1   Dataset Description

The DEAP (Database for Emotion Analysis using Physiological Signals) dataset [7] consists of data from 32 participants, who were exposed to 40 one-minute video clips with varying emotional content. These video clips were carefully selected to elicit different emotional states.

While the EEG signal is recorded at 512 Hz, it is down-sampled to 128 Hz sampling frequency in this work. Although the videos were of one minute, recording was started 3 s prior, resulting in recordings of 63 s. Therefore, data of dimension 32 × 40 × 32 × 8064 (participants × videos × channels × EEG sampling points) was recorded.

At the end of each trial, the participants self-reported ratings about their emotional levels in the form of valence, arousal, dominance and liking and familiarity. Each of these scales spans from a low value of one to a high value of nine. To create binary-classification tasks, the scales are divided into two categories. Each of valence-arousal-dominance is categorized as high valence-arousal-dominance (ranging from five to nine) and low valence-arousal-dominance (ranging from one to five) based on the respective scales.

A block diagram depicting our proposed method is shown in Fig. 1, while Algorithms 1 and 2 depict our proposed approach.

3.2   Data Preprocessing

Preprocessing involved several steps to enhance the quality and extract relevant information from the recorded signals, which are described as follows.

Algorithm 1. Emotion Recognition: Preprocessing

1: **procedure** PREPROCESSING(Data)
2:     **for** each subject in Subjects **do**
3:         **for** each video in Videos **do**
4:             **for** each signal X from EEG electrode **do**
5:                 Perform z-score normalization based baseline removal
6:                 Perform wavelet-based multiscale PCA
7:                 **for** each band in $\alpha, \beta, \theta, \gamma$, all **do**
8:                     Apply bandpass filtering to decompose signals into band

---

Algorithm 2. Emotion Recognition: Feature Extraction and Selection

1: **procedure** FEATUREEXTRACTIONANDSELECTION(Preprocessed_Data)
2:     **for** each subject in Subjects **do**
3:         **for** each video in Videos **do**
4:             **for** each band in $\alpha, \beta, \theta, \gamma$, all **do**
5:                 **for** each video segment 1:10 **do**
6:                     Compute PLV features for within-hemisphere
7:                     Compute PLV features for cross-hemisphere

$$\text{PLV} = \left| \frac{1}{N} \sum_{n=1}^{N} \exp(i\Delta\phi_n) \right|$$

8:     **for** each band in $\alpha, \beta, \theta, \gamma$, all **do**
9:         Select 50 most significant features using mRmR and chi-square
10:        **for** each hemisphere in within hemisphere, cross hemisphere **do**
11:            **for** each set of 50 significant features **do**
12:                **for** each dim in Valence, Arousal, Dominance **do**
13:                    Apply machine learning classifiers on the data

---

Firstly, we used Z-score normalization, which helps eliminate the individual variations and biases present in the recorded signals. Since the recording was started 3 s before the actual video, the signal in the first three seconds is used for z-score based baseline removal. Assuming $x(i)$ is the input signal, it is performed by first calculating the mean (μ) and standard deviation (σ) of the signal upto the first three seconds (N = 3 × 128), and then normalizing as shown in Eq. 1.

$$x_{\text{normalized}}(i) = \frac{x(i) - \mu}{\sigma} \tag{1}$$

We then used Wavelet-based multiscale PCA [3] to remove noise from the normalized signal. Firstly, a covariance matrix C_j of the wavelet coefficients is computed as shown in Eq. 2. Top eigenvectors with highest eigenvalues are chosen after eigenvalue decomposition. However, in this work, we selected the whole set of the principle components, instead of taking a subset.

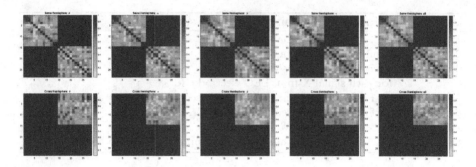

Fig. 3. PLV features extracted for within-hemisphere (top) and cross-hemisphere (below) for first subject and first trial

$$C_j = \frac{1}{N} \sum_{n=1}^{N} W_{\text{coeff}}(n, j) \cdot W_{\text{coeff}}(n, j)^T \tag{2}$$

Finally, the normalized EEG signal is projected onto the selected eigenvectors V_{ij} (ith eigenvalue at scale j) at each scale to obtain the wavelet-based multiscale PCA features, as shown in Eq. 3.

$$\text{PCA}_{ij}(n) = x_{\text{normalized}}(n) \cdot V_{ij} \tag{3}$$

After applying these steps, bandpass filtering is performed on the EEG signals by decomposing them into α (8–15 Hz), β (16–30 Hz), θ (4–7 Hz), and γ (30–45 Hz) bands.

3.3 Feature Extraction

The extraction of Phase-Locking-Value (PLV) features from each EEG band - α (alpha), β (beta), θ (theta), and γ (gamma) bands, involves a series of steps to quantify phase coupling between different electrode pairs within each frequency band. The electrodes selected from each hemisphere for extraction of these features are shown in Fig. 2, which include 14 electrodes from each of the left and right hemispheres.

The extraction process involves the following steps. Firstly, we segment the preprocessed EEG signals into 10 time windows of 6 s each. Next, for each segment, the EEG signals within the selected frequency band are processed using the Hilbert transform $H(x(t))$, to obtain the instantaneous phase information, as shown in Eq. 4.

$$\phi(t) = \arctan\left(\frac{H(x(t))}{x(t)}\right) \tag{4}$$

Table 1. Classifier Details

Classifier	Algorithm Type	Parameters and Details
K-Nearest Neighbors	Supervised learning (Instance-based)	No. of neighbors (k) = 1, Distance Metric: Euclidean
Decision Tree	Supervised learning (Tree-based)	Max category levels: 10, Max no. of decision splits: size(X, 1)–1, Min Leaf Size: 1, Min Parent Size: 10
Support Vector Machine	Supervised learning (Kernel-based)	Kernel Function: linear, Kernel Scale: 1
Random Forest	Ensemble learning (Bagging)	Base Estimator: DT, K-fold: 10
Adaboost	Ensemble learning (Boosting)	Base Estimator: DT, K-fold: 10

Once the instantaneous phase information is obtained, pairwise phase difference value is computed for each electrode pair within the frequency band of interest. Finally, PLV for each electrode pair within the frequency band is calculated by averaging of the absolute value of the complex exponential of the pairwise phase differences ($\Delta\phi_n$) over the entire segment, as shown in Eq. 5. An example of PLV features extracted for within-hemisphere and cross-hemisphere is shown in Fig. 3.

$$\text{PLV} = \left| \frac{1}{N} \sum_{n=1}^{N} \exp(i\Delta\phi_n) \right| \qquad (5)$$

In our proposed approach, we calculate PLVs within each frequency band, for both within-hemisphere and cross-hemisphere, which is described as follows. The 28 electrodes left after removal of the four middle electrodes (Fz, Cz, Pz, and Oz) are symmetrical. To investigate the role of hemispherical functional brain connections, we compute PLVs on these electrode pairs through two kinds of combinations: (1) within-hemisphere (wherein, electrodes in each hemisphere form a pair with every other electrode in the same hemisphere), and (2) cross-hemisphere (wherein, electrodes in one hemisphere form a pair with each electrode from the other hemisphere). While the former reflects the connections in each hemisphere, the latter reflects the connections across hemispheres and between the left and right hemispheres.

Since there are 14 EEG electrode nodes in each hemisphere, the number of effective PLV values in the case of cross-hemisphere is $14 * 14 = 196$, while in the case of within-hemisphere is $14 * 14 * 2 = 392$.

Table 2. Accuracy on the DEAP dataset for Valence Classification using features from Within and Cross Hemisphere

Within-hemisphere

Model	Theta			Alpha			Beta			Gamma			All		
	all	mRmR	chi	all	mRmR	chi	all	mRmR	chi	all	mRmR	chi	all	mRmR	chi
KNN	44.7	56.2	56.4	44.7	59.3	57.7	44.7	70.7	69.5	44.7	**78.1**	74.7	44.7	66.4	63.4
	±0.0	±1.0	±0.7	±0.0	±0.8	±1.2	±0.0	±1.4	±1.2	±0.0	±1.5	±0.5	±0.0	±1.3	±1.9
DT	55.3	54.9	54.7	55.3	56.7	55.0	55.3	63.8	63.1	55.3	66.7	66.1	55.3	60.9	60.6
	±0.0	±1.4	±0.9	±0.0	±1.2	±1.6	±0.0	±0.8	±1.8	±0.0	±1.4	±1.2	±0.0	±1.6	±1.1
SVM	55.3	55.2	55.3	55.3	56.4	55.2	55.3	56.7	57.4	55.3	58.9	58.4	55.3	57.9	55.2
	±0.0	±0.2	±0.0	±0.0	±1.1	±0.5	±0.0	±1.1	±0.9	±0.0	±1.0	±0.9	±0.0	±1.0	±0.7
RF	55.3	61.6	61.5	55.3	64.0	62.6	55.3	72.1	71.5	55.3	77.0	76.2	55.3	69.6	68.1
	±0.0	±1.1	±1.0	±0.0	±1.0	±0.8	±0.0	±1.2	±0.9	±0.0	±1.5	±1.3	±0.0	±1.2	±1.3
AB	55.3	59.0	58.7	55.3	59.8	59.3	55.3	65.7	66.4	55.3	67.8	67.0	55.3	64.1	63.3
	±0.0	±0.8	±1.5	±0.0	±1.1	±1.0	±0.0	±0.9	±1.0	±0.0	±1.2	±1.0	±0.0	±1.3	±1.0

Cross-hemisphere

Model	Theta			Alpha			Beta			Gamma			All		
	all	mRmR	chi	all	mRmR	chi	all	mRmR	chi	all	mRmR	chi	all	mRmR	chi
KNN	50.9	56.1	56.9	51.4	60.2	59.8	51.3	73.2	72.9	50.4	**79.4**	78.6	50.6	66.8	66.6
	±1.6	±1.1	±1.4	±1.3	±0.8	±1.1	±1.8	±1.1	±0.9	±1.5	±1.0	±1.1	±1.5	±1.1	±1.0
DT	51.0	54.6	55.0	52.0	56.6	55.3	51.5	64.7	63.7	51.0	67.6	68.1	51.4	61.0	61.2
	±1.0	±1.1	±1.1	±1.8	±1.6	±1.6	±1.8	±1.6	±1.8	±1.1	±1.1	±1.2	±1.4	±1.9	±2.0
SVM	55.3	57.4	57.6	55.3	57.4	57.7	55.3	60.7	59.5	55.3	61.0	60.6	55.3	58.2	58.3
	±0.0	±1.2	±1.0	±0.0	±1.2	±0.8	±0.0	±0.9	±1.2	±0.0	±0.8	±1.2	±0.0	±1.0	±1.1
RF	50.9	61.8	62.2	51.3	64.3	64.1	51.3	73.9	73.1	50.4	77.7	77.8	50.6	70.3	69.9
	±1.6	±0.7	±0.8	±1.3	±0.6	±1.2	±1.8	±0.8	±0.9	±1.5	±1.3	±1.1	±1.5	±1.2	±0.6
AB	54.7	59.5	59.6	55.0	61.1	61.1	55.1	67.7	67.1	55.7	68.9	68.8	55.7	64.4	65.0
	±1.0	±1.3	±1.2	±0.4	±0.6	±1.0	±0.9	±1.2	±1.9	±0.9	±1.5	±1.1	±1.1	±1.1	±0.7

3.4   Feature Selection

We select relevant PLV features for classification using two feature selection methods - mRmR (maximum Relevance-minimum Redundancy) and chi-square, explained as follows. The mRmR algorithm [12] aims to select features that have a high relevance to the target variable (e.g., emotion classification) while minimizing redundancy among selected features, as shown for a specific feature F_i in Eq. 6.

$$\text{mRmR}(F_i) = \text{Relevance}(F_i) - \alpha \times \text{Redundancy}(F_i) \tag{6}$$

We also use chi-square statistic [6] to select the features with the highest statistical significance, with respect to the target variable. It's calculation for a specific feature F_i with c classes in the target variable and observed O_{ij} and expected frequencies E_{ij} is shown in Eq. 7. We use these methods to identify the fifty most relevant and discriminative PLV features for our task.

$$\chi^2(F_i) = \sum_{j=1}^{c} \frac{(O_{ij} - E_{ij})^2}{E_{ij}} \tag{7}$$

Table 3. Accuracy on the DEAP dataset for Arousal Classification using features from Within and Cross Hemisphere

Within-hemisphere

Model	Theta all	mRmR	chi	Alpha all	mRmR	chi	Beta all	mRmR	chi	Gamma all	mRmR	chi	All all	mRmR	chi
KNN	42.4	60.3	60.0	42.4	63.1	61.4	42.4	72.9	71.2	42.4	**79.0**	77.2	42.4	68.4	67.1
	±0.0	±1.0	±1.7	±0.0	±1.5	±1.7	±0.0	±1.2	±1.3	±0.0	**±1.4**	±1.2	±0.0	±1.0	±1.0
DT	57.6	57.9	57.9	57.6	59.8	59.0	57.6	64.1	64.6	57.6	67.5	66.5	57.6	62.8	62.5
	±0.0	±1.2	±1.4	±0.0	±1.8	±1.2	±0.0	±1.0	±1.6	±0.0	±1.2	±1.5	±0.0	±1.1	±1.4
SVM	57.6	62.5	62.6	57.6	64.0	63.0	57.6	64.2	62.0	57.6	63.9	62.7	57.6	64.1	63.1
	±0.0	±1.5	±1.0	±0.0	±1.5	±0.6	±0.0	±1.0	±1.1	±0.0	±1.2	±2.0	±0.0	±1.0	±1.1
RF	57.6	65.3	65.0	57.6	67.6	67.3	57.6	74.6	73.8	57.6	78.0	76.6	57.6	71.7	70.9
	±0.0	±1.0	±1.1	±0.0	±1.2	±1.5	±0.0	±0.8	±1.4	±0.0	±1.6	±1.4	±0.0	±1.0	±0.6
AB	57.6	63.7	63.1	57.6	64.4	64.0	57.6	67.3	66.7	57.6	69.1	68.2	57.6	65.9	65.7
	±0.0	±1.7	±1.5	±0.0	±1.5	±1.1	±0.0	±1.2	±1.2	±0.0	±1.3	±1.0	±0.0	±1.5	±1.1

Cross-hemisphere

Model	Theta all	mRmR	chi	Alpha all	mRmR	chi	Beta all	mRmR	chi	Gamma all	mRmR	chi	All all	mRmR	chi
KNN	51.0	59.7	60.2	51.9	63.0	62.6	51.9	75.0	73.8	51.5	79.6	**79.6**	51.1	68.7	68.8
	±1.2	±1.4	±1.2	±1.3	±2.0	±1.1	±1.4	±0.8	±1.1	±1.1	±1.2	**±1.5**	±1.4	±1.0	±0.8
DT	52.5	57.3	58.2	52.1	59.1	59.2	52.8	64.3	65.3	52.0	67.9	67.5	51.7	63.2	63.2
	±1.4	±1.3	±1.4	±1.5	±1.7	±1.4	±0.8	±1.9	±1.2	±0.9	±1.8	±1.2	±1.1	±1.4	±1.1
SVM	57.6	63.1	63.2	57.6	64.2	64.2	57.6	64.5	64.6	57.6	64.8	64.7	57.6	63.8	64.1
	±0.0	±1.2	±0.8	±0.0	±1.0	±1.1	±0.0	±1.4	±1.2	±0.0	±1.2	±1.0	±0.0	±1.4	±1.1
RF	51.0	65.5	66.1	51.9	67.3	67.2	51.9	75.4	75.1	51.6	78.4	78.2	51.0	72.2	71.7
	±1.1	±0.9	±1.4	±1.3	±1.3	±1.6	±1.3	±1.5	±1.5	±1.2	±1.3	±1.3	±1.4	±0.8	±0.9
AB	58.2	63.3	63.2	57.9	64.6	64.1	58.0	68.2	67.9	57.9	69.2	69.3	58.1	66.2	66.7
	±0.4	±1.2	±1.5	±0.8	±1.5	±1.2	±0.8	±1.5	±1.2	±0.4	±1.2	±1.1	±0.9	±1.2	±1.3

3.5   Classification

PLV features are extracted from each band, for each of within and cross-hemisphere. Fifty most significant features are selected through mRmR and Chi-squared methods, on which classification is performed. We employed several popular machine learning classifiers to learn the underlying patterns and relationship, namely K-Nearest Neighbors (KNN), Decision Tree, Support Vector Machines (SVM), Random Forest, and Adaboost. The technical specifications of these models are shown in Table 1. The trained models were then evaluated using cross-validation for 10 folds. The mean of accuracies obtained and their standard deviation are used as an evaluation metric to assess the performance of the models and corresponding approaches on emotion recognition task.

4   Results and Discussion

Table 2 shows the accuracy for valence classification from within and cross hemispheres. We observe that the approach involving features from the Gamma band, selected through the mRmR method, using the KNN classifier perform best at valence classification. Additionally, PLV features from cross-hemisphere seem to be performing better (accuracy of 79.4%) than those from within-hemisphere

Table 4. Accuracy on the DEAP dataset for Dominance Classification using features from Within and Cross Hemisphere

Within-hemisphere

Model	Theta			Alpha			Beta			Gamma			All		
	all	mRmR	chi	all	mRmR	chi	all	mRmR	chi	all	mRmR	chi	all	mRmR	chi
KNN	39.1	61.5	60.0	39.1	63.0	62.8	39.1	72.9	71.5	39.1	**77.1**	76.2	39.1	69.1	67.0
	±0.0	±1.5	±1.3	±0.0	±1.2	±1.0	±0.0	±0.8	±1.2	±0.0	**±0.9**	±1.0	±0.0	±1.4	±1.3
DT	60.9	59.3	59.3	60.9	60.4	59.6	60.9	65.0	65.7	60.9	67.5	67.4	60.9	63.6	62.8
	±0.0	±0.8	±1.5	±0.0	±1.6	±1.3	±0.0	±1.1	±1.4	±0.0	±1.0	±1.4	±0.0	±1.7	±0.8
SVM	60.9	64.0	61.2	60.9	64.5	62.9	60.9	65.3	62.9	60.9	64.7	61.6	60.9	64.4	64.6
	±0.0	±1.1	±0.3	±0.0	±1.1	±1.2	±0.0	±1.1	±0.9	±0.0	±0.8	±0.8	±0.0	±0.9	±1.1
RF	60.9	66.0	65.8	60.9	67.6	67.4	60.9	73.8	74.3	60.9	76.6	75.8	60.9	71.4	70.5
	±0.0	±1.7	±1.1	±0.0	±1.9	±1.6	±0.0	±1.4	±0.9	±0.0	±0.9	±1.1	±0.0	±1.1	±0.9
AB	60.9	63.3	64.0	60.9	64.7	65.0	60.9	67.9	68.3	60.9	69.5	68.2	60.9	67.3	67.0
	±0.0	±1.0	±1.5	±0.0	±0.7	±1.1	±0.0	±1.5	±1.4	±0.0	±1.2	±1.0	±0.0	±1.5	±1.2

Cross-hemisphere

Model	Theta			Alpha			Beta			Gamma			All		
KNN	52.5	61.2	61.1	52.4	63.1	63.4	53.8	75.6	74.2	53.1	**79.1**	**78.8**	53.1	69.3	68.6
	±1.4	±1.2	±1.1	±1.7	±1.1	±1.0	±1.5	±1.1	±1.2	±1.5	**±0.6**	**±0.9**	±1.8	±0.9	±1.3
DT	52.9	60.1	59.8	53.2	60.7	61.1	54.4	66.6	65.6	54.4	69.1	69.1	53.5	64.8	64.4
	±1.3	±2.1	±1.2	±1.4	±1.4	±1.4	±1.0	±1.4	±1.0	±1.2	±1.7	±1.7	±1.5	±0.9	±0.9
SVM	60.9	65.0	64.9	60.9	65.4	65.2	60.9	66.0	65.5	60.9	65.6	66.0	60.9	65.4	65.2
	±0.0	±0.8	±1.2	±0.0	±1.2	±1.2	±0.0	±0.9	±1.1	±0.0	±1.1	±1.3	±0.0	±0.7	±1.0
RF	52.5	66.4	66.7	52.4	68.8	68.9	53.8	75.2	74.9	53.1	78.1	78.0	53.1	72.5	72.2
	±1.4	±1.2	±1.3	±1.7	±1.4	±1.8	±1.5	±1.4	±1.3	±1.5	±0.9	±1.2	±1.8	±1.1	±0.5
AB	60.9	65.2	65.2	60.7	64.8	65.5	60.9	68.8	69.0	60.7	71.2	70.6	60.8	68.1	67.8
	±0.2	±1.4	±1.1	±0.6	±1.1	±1.3	±0.3	±1.1	±1.6	±0.5	±1.1	±1.3	±0.3	±1.2	±1.3

(accuracy of 78.1%). Table 3 shows the accuracy for arousal classification from within and cross hemispheres. We observe that the approach involving features from the Gamma band, selected through the mRmR/Chi-square method, using the KNN classifier perform best at arousal classification. Additionally, PLV features from cross-hemisphere seem to be performing slightly better (accuracy of 79.6%) than those from within-hemisphere (accuracy of 79.0%). Table 4 shows the accuracy for dominance classification from within and cross hemispheres. We observe that the approach involving features from the Gamma band, selected through the mRmR method, using the KNN classifier perform best at dominance classification. Additionally, PLV features from cross-hemisphere seem to be performing better (accuracy of 79.1%) than those from the within-hemisphere (accuracy of 77.1%). On comparing our results with the state of arts (Table 5), we find that our approach performs better with state-of-the-art accuracy.

Overall, the experimental results demonstrate that gamma EEG band is most relevant for emotion recognition and among machine learning classifiers, KNN achieves the best performance across all three ratings. Additionally, there is a minor increment in accuracy when PLV features are acquired from cross-hemisphere as compared to within-hemisphere.

Table 5. Comparison with the state of arts

Research Work	Accuracy (%)		
	Valence	Arousal	Dominance
Chen et al. (2015) [4]	76.2	73.6	–
Wang et al. (2018) [14]	73.3	72.1	–
Wang et al. (2019) [15]	73.3	77.0	79.2
Kumari et al. (2022) [8]	77.5	78.4	79.4
Ours	**79.4**	**79.6**	**79.1**

5   Conclusion

In this paper, we have performed an emotion recognition task based on brain functional connectivity. Firstly, EEG signals are processed and denoised using wavelet based multiscale PCA. Then, PLV features are extracted from these processed signals and further mRmR feature section is done to examine the performance of brain connections demonstrated for within-hemisphere and cross-hemisphere. The obtained results manifest that gamma band is most effective and relevant for the evaluation of emotion recognition task. We achieved the best performance with KNN classifier across three rating dimensions of emotions (valence, arousal and dominance) for cross-hemisphere connections. Although there is a very slight difference between both the scenarios, we concluded that phase information obtained across cross-hemisphere connections is more reliable in comparison to same hemisphere one. Besides, as we know the information extracted via brain connections requires more and more numbers of EEG electrodes to enhance the performance of emotion recognition, simultaneously increases complexity for data acquisition. Thus, we are interested in multivariate phase synchronisation which improves the estimation of region-to-region source space connectivity with lesser number of EEG channels while eliminating useless electrodes. We leave this interesting topic as our future work.

References

1. Cao, J., et al.: Brain functional and effective connectivity based on electroencephalography recordings: a review. Hum. Brain Mapp. **43**(2), 860–879 (2021). https://doi.org/10.1002/hbm.25683
2. Cao, R., et al.: EEG functional connectivity underlying emotional valance and arousal using minimum spanning trees. Front. Neurosci. **14**, 355 (2020). https://doi.org/10.3389/fnins.2020.00355
3. Chavan, A., Kolte, M.: Improved EEG signal processing with wavelet based multiscale PCA algorithm. In: 2015 International Conference on Industrial Instrumentation and Control (ICIC), pp. 1056–1059 (2015). https://doi.org/10.1109/IIC.2015.7150902

4. Chen, M., Han, J., Guo, L., Wang, J., Patras, I.: Identifying valence and arousal levels via connectivity between EEG channels. In: 2015 International Conference on Affective Computing and Intelligent Interaction (ACII), pp. 63–69 (2015). https://doi.org/10.1109/ACII.2015.7344552

5. Dimond, S.J., Farrington, L., Johnson, P.: Differing emotional response from right and left hemispheres. Nature **261**(5562), 690–692 (1976). https://doi.org/10.1038/261690a0

6. Pearson, K.: X. on the criterion that a given system of deviations from the probable in the case of a correlated system of variables is such that it can be reasonably supposed to have arisen from random sampling. Lond. Edinburgh Dublin Phil. Maga. J. Sci. **50**(302), 157–175 (1900). https://doi.org/10.1080/14786440009463897

7. Koelstra, S., et al.: Deap: a database for emotion analysis; using physiological signals. IEEE Trans. Affect. Comput. **3**(1), 18–31 (2012). https://doi.org/10.1109/T-AFFC.2011.15

8. Kumari, N., Anwar, S., Bhattacharjee, V.: Time series-dependent feature of EEG signals for improved visually evoked emotion classification using EmotionCapsNet. Neural Comput. Appl. **34**(16), 13291–13303 (2022). https://doi.org/10.1007/s00521-022-06942-x

9. Lachaux, J.P., Rodriguez, E., Martinerie, J., Varela, F.J.: Measuring phase synchrony in brain signals. Hum. Brain Mapp. **8**(4), 194–208 (1999). https://doi.org/10.1002/(sici)1097-0193(1999)8:4⟨194::aid-hbm4⟩3.0.co;2-c

10. Li, X., Hu, B., Sun, S., Cai, H.: EEG-based mild depressive detection using feature selection methods and classifiers. Comput. Methods Programs Biomed. **136**, 151–161 (2016). https://doi.org/10.1016/j.cmpb.2016.08.010

11. Li, Y., Zheng, W., Zong, Y., Cui, Z., Zhang, T., Zhou, X.: A bi-hemisphere domain adversarial neural network model for EEG emotion recognition. IEEE Trans. Affect. Comput. **12**(2), 494–504 (2021). https://doi.org/10.1109/taffc.2018.2885474

12. Peng, H., Long, F., Ding, C.: Feature selection based on mutual information criteria of max-dependency, max-relevance, and min-redundancy. IEEE Trans. Pattern Anal. Mach. Intell. **27**(8), 1226–1238 (2005). https://doi.org/10.1109/TPAMI.2005.159

13. Singh, M.K., Singh, M.: A deep learning approach for subject-dependent and subject-independent emotion recognition using brain signals with dimensional emotion model. Biomed. Signal Process. Control **84**, 104928 (2023). https://doi.org/10.1016/j.bspc.2023.104928

14. Wang, Y., Huang, Z., McCane, B., Neo, P.: Emotionet: a 3-d convolutional neural network for EEG-based emotion recognition. In: 2018 IJCNN, pp. 1–7 (2018). https://doi.org/10.1109/IJCNN.2018.8489715

15. Wang, Z., Tong, Y., Heng, X.: Phase-locking value based graph convolutional neural networks for emotion recognition. IEEE Access **7**, 93711–93722 (2019). https://doi.org/10.1109/access.2019.2927768

16. Zhang, J., Zhang, X., Chen, G., Huang, L., Sun, Y.: EEG emotion recognition based on cross-frequency granger causality feature extraction and fusion in the left and right hemispheres. Front. Neurosci. **16**, 974673 (2022). https://doi.org/10.3389/fnins.2022.974673

17. Zheng, W.L., Lu, B.L.: Investigating critical frequency bands and channels for EEG-based emotion recognition with deep neural networks. IEEE Trans. Auton. Ment. Dev. **7**(3), 162–175 (2015). https://doi.org/10.1109/tamd.2015.2431497

Context-Aware Facial Expression Recognition Using Deep Convolutional Neural Network Architecture

Abha Jain[1], Swati Nigam[1,2(✉)], and Rajiv Singh[1,2]

[1] Department of Computer Science, Banasthali Vidyapith, Radha Kishnpura, Rajasthan 304022, India
swatinigam.au@gmail.com
[2] Centre for Artificial Intelligence, Banasthali Vidyapith, Radha Kishnpura, Rajasthan 304022, India

Abstract. A frame of reference, which includes additional contextual information, can provide a more accurate and comprehensive understanding of the individual's emotional state. This context might encompass factors such as the person's surroundings, body language, gestures, tone of voice, and the specific situation or events taking place. Previous research in this field has often struggled to recognize emotions within a contextual framework. However, by considering contextual elements in addition to facial expressions, we can gain a more nuanced and precise picture of the individual's emotions. In this paper, we used both context-aware datasets (Emotic, CAER, and CAER-S) and only the facial emotion datasets (Affectnet and AEFW) to signify the context. In this Emotic dataset images are labeled with 26 emotional categories. We utilized these datasets to build a convolutional neural network model that effectively examines both the individual and the overall scenario to accurately identify a wide range of information pertaining to emotional states. The features obtained from these two modules are combined using a specialized fusion network. Through this approach, we demonstrate the significance of emotion recognition within a visual context.

Keywords: Context-aware · convolution neural network · emotion

1 Introduction

Context-aware facial emotion recognition technology is the key to unlocking a new era of emotional intelligence. Context-aware facial emotion recognition refers to the ability of a system or technology to accurately detect and interpret facial expressions in real-time, taking into consideration the surrounding context in which the expressions occur. This includes factors such as the individual's environment, social cues, and other relevant contextual information.

Previous research in the subject of computer vision has primarily focused on the examination of facial expressions, often involving the categorization of these expressions into the six (or seven) basic emotions [1–3]. By incorporating context awareness,

B. J. Choi et al. (Eds.): IHCI 2023, LNCS 14531, pp. 127–139, 2024.
https://doi.org/10.1007/978-3-031-53827-8_13

facial emotion recognition systems become more robust and reliable in accurately identifying and understanding emotions displayed on a person's face. Contextual factors, such as social settings, cultural norms, and individual experiences, can influence emotions, making it particularly important to incorporate context awareness into facial emotion recognition systems. [4, 6].

In many instances, when we broaden our perspective beyond an individual and consider the surrounding environment, we can discern additional emotional nuances that would otherwise remain hidden without context. For example, we can observe from the scenario depicted in Figure 1(a) that this individual is experiencing feelings of worry and pressure. But if you consider the contextual boundaries in Figure 1(b), it appears that he is ready to launch an attack on his opponent in a game and is prepared to counter any offensive moves made by the opponent. Moreover, we can infer that his overall emotional state is alarmed, as he appears confident in the actions he is about to undertake. So he is in disquiet about the situation.

Traditional facial emotion recognition systems primarily focus on analyzing facial features and patterns to determine emotions. The way emotions are expressed varies, including through facial expressions [3, 4], speech [6], and body language [7]. However, these systems often overlook the influence of context, which can significantly impact the interpretation and understanding of emotions. For example, a smile at a social gathering may indicate happiness, while the same smile at a business meeting may indicate politeness or agreement rather than genuine joy. Indeed, when taking the context into account, it becomes possible to make reasonable conjectures regarding emotional states even in cases where the person's face is not visible.

In this paper, we address the problem of recognizing emotion states in context. We used two popular datasets, Emotic [4] and CAER [5]. The EMOTIC and CAER (Context-Aware Emotion Recognition Networks) databases comprise images featuring people within their respective contexts, each annotated to reflect the emotional states that an observer can deduct from the overall situation. We structured the networks using a two-stream architecture, which consists of two feature encoding streams: one for facial encoding and the other for context encoding. Our primary concept revolves around the search for pertinent contexts, a factor that aids the model in mitigating ambiguity and enhancing accuracy in emotion recognition. Our study focused on evaluating the efficacy of a convolutional neural network (CNN) model in accurately identifying emotions within a contextual framework.

This research presents a technique that utilizes contextual information along with facial expression to demonstrate the practicality of accurately recognizing the suitable emotion within a given environment. In order to achieve this objective, we have established the concept that a model's emotions and context convey connections and limitations among various elements. This study represents the first known instance of employing deep learning to comprehensively investigate the integration of contextual information and facial information in order to achieve emotion recognition.

Section 2 provides an overview of the proposed context-aware emotion recognition system. Section 3 demonstrates the methodology of integrating contextual information with face expression identification. Section 4 showcases the findings of the experiments conducted. Section 5 provides the concluding remarks of the study.

(a)   (b)

Fig. 1. Facial Expression and (b) Facial Expression with contextual information

2 Related Work

A comprehensive literature survey on context-aware facial emotion recognition (FER) reveals a significant body of research and advancements in this area. The following is an overview of some key studies and contributions in the field:

Li et al. (2019) [8] introduced a dynamic attention-based convolutional neural network that effectively captures both local and global context information for the purpose of facial emotion recognition. The model dynamically attends to different facial regions based on their relevance to the emotional context, improving the accuracy of emotion recognition.

Zhang et al. (2020) [9] concentrated on integrating many modalities, including facial expressions, speech, and body motions, in order to enhance context-aware FER. The study develops a deep learning-based framework that effectively combines these modalities to enhance emotion recognition accuracy.

Li et al. (2017) [10] introduced a sophisticated adaptive attention network for accurately identifying face emotions in real-world situations. The model dynamically adjusts attention to different facial regions based on their discriminative power, taking into account contextual information to improve emotion recognition accuracy.

Caon et al. (2013) [11] provided a comprehensive overview of context-aware affective computing, including context-aware FER. It explores different contextual factors, such as social context, environmental context, and temporal context, and their influence on emotion recognition. The study also discusses various approaches and challenges in context-aware affective computing.

Zhao et al. (2019) [12] proposed a context-aware FER framework based on deep neural networks. The study considers both facial expressions and contextual information, such as scene context and temporal dynamics, to improve emotion recognition accuracy.

The model effectively integrates contextual information with facial features for enhanced performance.

These studies highlight the importance of considering contextual factors in facial emotion recognition to improve accuracy and understand emotions in a more comprehensive manner. They demonstrate the effectiveness of various techniques, such as attention mechanisms, multi-modal fusion, and deep learning approaches, in achieving context-aware FER.

The area of context-aware facial expression recognition (FER) is continuously progressing, with ongoing research and improvements. This literature study offers a brief overview of the current corpus of research and establishes a basis for future investigation and advancement in this captivating academic field.

3 Proposed Method

Within this section, we introduce a simple yet powerful structure for the detection of emotions in photos and movies that takes into account the surrounding environment. This paradigm utilizes both facial expressions and environmental information in a complementary and cooperative manner to improve recognition accuracy.

A straightforward approach involves utilizing the holistic visual features, as demonstrated in prior work [13, 14, 28]. However, such a model may not effectively capture important contextual regions. Recognizing that emotions can be better understood by considering both the contextual elements of a scene and facial expressions [15, 16], we introduce an attention inference module designed to estimate contextual information in both images and videos. By temporarily concealing facial regions in the input data and focusing on attention regions, our networks are capable of identifying more discriminative contextual regions. This, in turn, enhances the accuracy of emotion recognition in a context-aware manner.

To establish the proposed set of emotional categories as outlined in Table 1, we conducted a comprehensive collection of affective state vocabulary and concluded 26 groups of words to represent the exact human emotion state [4].

To simplify, let consider an image denoted as "I" and a video $V = \{I_1, \ldots, I_T\}$ comprised of a sequence of "T" images. Our primary objective is to determine the emotion label "y" from a set of "K" emotion labels, $\{y1, \ldots, yK\}$, assigned to either the image "I" or the video clip "V" using deep Convolutional Neural Networks (CNNs). To address this challenge, we introduce a network architecture composed of two distinct sub-networks: a two-stream encoding network and an adaptive fusion network, as depicted in Figure 2. These two-stream encoding networks encompass a face stream and a context stream, each responsible for encoding facial expressions and context information separately. By merging these two sets of features within the adaptive fusion network, our approach achieves optimal performance in the context-aware recognition of emotions.

3.1 Model Architectures

We present a comprehensive model, as illustrated in Figure 6, that can simultaneously predict both emotion and contextual characteristics. Our networks incorporate a facial expression encoding module, which is comparable to existing approaches used for determining facial expressions [9, 10, 17]. In order to create the input for the face stream, we first detect and separate the facial areas using easily accessible face detectors [10]. Moreover, supplementary feature extraction modules have been created as a condensed iteration of the low-rank filter convolutional neural network first shown in [5]. The main benefit of this network lies in its ability to offer great precision while simultaneously reducing the number of parameters and computational complexity. The initial network comprises 16 convolutional layers with 1-dimensional kernels, which effectively simulate 8 layers by employing 2-dimensional kernels. Afterwards, a fully connected layer is added, creating a direct link to the SoftMax layer. In our revised version, we remove the fully connected layer and instead transmit the features obtained from the activation map of the final convolutional layer. The selection is predicated upon the objective of preserving the crucial geographical data necessary for the work.

The attributes obtained from these two modules are then merged using a specialised fusion network. The fusion module initiates the process by implementing a global average pooling layer on each feature map, thereby reducing the dimensionality of the data greatly. Subsequently, a primary fully connected layer functions as a dimensionality reduction layer for the pooled features, yielding a vector with 256 dimensions. Subsequently, a bigger fully linked layer is added, allowing the training process to acquire distinct representations for each task, in accordance with the concepts described in [5]. This layer is utilized for the identification of emotion categories, encompassing a total of 26 distinct emotional states. Each convolutional layer is thereafter followed by Batch normalization and rectifier linear activation.

The three modules' parameters are simultaneously learned using stochastic gradient descent with momentum. The batch size has been adjusted to 52, which is twice the number of unique categories in the dataset. Our method employs uniform sampling per category to ensure that each discrete category is represented by at least one instance in every batch. Based on empirical evidence, we have determined that this strategy produces better outcomes in comparison to randomly rearranging the training set.

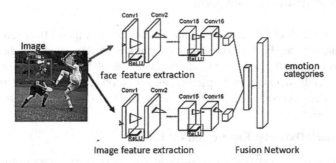

Fig. 2. Propose Model Architecture for Context aware Facial Emotion Recognition

The overall loss function used for model training is defined as a weighted combination of two distinct losses: $L\_comb = \lambda\_disc * L\_disc + \lambda\_cont * L\_cont$. Here, λ_disc, cont represents the weight that determines the importance of each loss component, while L_disc and L_cont denote the losses associated with the tasks of learning discrete categories and learning continuous dimensions, respectively.

We approach this multiclass-multilabel problem by framing it as a regression task. To address the class imbalance inherent in the dataset, we employ a weighted Euclidean loss function. Through empirical analysis, we have determined that this particular loss function outperforms alternatives such as Kullback-Leibler divergence or a multi-class multi-classification hinge loss. To be precise, the loss is defined as follows:

$$L_{disc} = \frac{1}{N} \sum_{i=1}^{N} w_i \left(\hat{y}_i^{disc} - y_i^{disc} \right)^2 \tag{1}$$

where N represent the number of categories (N = 26 as per case), $\hat{y}$ disc i is the caculated estimated result for the i-th category and y_i ^{disc} is the original-truth label. The parameter w_i is the weight assigned to each category. Weight values are defined as $w_i = 1/(\ln(c+pi))$, where pi is the probability of the i-th category and c is a parameter to control the range of valid values for w_i. Using this weighting scheme, the values of w_i are bounded as the number of instances of a category approach to 0. This is particularly relevant in our case as we set the weights based on the occurrence of each category in every batch.

It is essential to merge the derived characteristics from two modules in order to effectively identify the emotion by utilizing both facial and contextual information simultaneously. The feature extraction modules are initiated by utilizing pre-trained models from two distinct extensive classification datasets, specifically ImageNet [18] and Places [19]. ImageNet contains a diverse collection of photos that represent common items, including people. This makes it a helpful tool for understanding the visual content of the image area that includes the person of interest. Conversely, Places is a deliberately designed dataset for advanced visual comprehension tasks, specifically for recognizing scene categories. Therefore, by pretraining the image feature extraction model using the Places dataset, it guarantees the inclusion of global (high-level) contextual information.

4   Experiments and Discussion

In this section, we discuss the two benchmark datasets and their effectiveness in the proposed context-aware Facial Expression Recognition (FER) system [20]. Initially, we provide an overview of the benchmark datasets rather than the details of the experimental setup. Subsequently, we compare the performance of the other model on these benchmark datasets with their approach and their efficiency and effectiveness on the same dataset.

4.1   Benchmark Datasets: Emotic and CAER

The EMOTIC database [4] consists of images sourced from MSCOCO [20], Ade20k [21], and the Google search engine. The collection consists of 18,316 pictures, each containing 23,788 individuals with annotations. Figure 1 exhibits instances of images contained in the database, accompanied by their corresponding comments. The "EMOTIC" framework has 26 distinct emotional categories, which cover a wide range of emotional states. The categories are elaborated and delineated in Table 1.

The table's definitive list of categories includes the six fundamental emotions (category 7, 10, 18, 20, 22, and 23) [22]. Category 18, designated as "Aversion," functions as a more comprehensive category that includes the basic feeling of disgust.

CAER is a compilation of extensive video snippets extracted from television programmes, which are then annotated to facilitate the recognition of emotions in a context-aware manner. Every video clip underwent manual annotation, categorizing them into six distinct emotions: "anger", "disgust", "fear", "happy", "sad", and "surprise", in addition to a category labelled as "neutral". The collection comprises 3,201 video segments, totaling around 1.1 million frames.

Furthermore, Lee and Kim [5] have derived approximately 70,000 static images from CAER, resulting in the formation of a static image subset referred to as CAER-S. Figure 1(b)illustrate the images from CAER-S. This dataset considers only images with one emotional label and ignores images with more than two annotations.

Table 2 conducts a comparison and gives a description of context-aware datasets CAER CAER [5] and Emotic [4] datasets and several other widely used datasets, including CAER-S [5], Affect-Net [23], AFEW [24], and Video Emotion datasets [25] (Fig. 3).

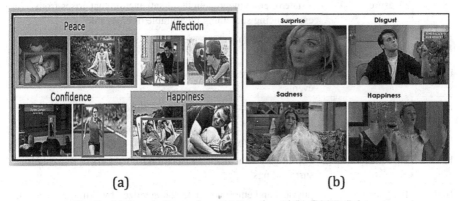

(a)   (b)

Fig. 3. (a) Sample Image from (a) Emotic and (b) CAER-S dataset

Table 1. Emotion Categories as per EMOTIC Dataset

Sr. No.	Emotion Labels	Feelings
1	Peace	a state of well-being and calmness by the absence of worry, the presence of positive thoughts or sensations, and a feeling of satisfaction
2	Affection	"Fond feelings" encompass with love, care, and affection
3	Esteem	refers to the positive opinion and a sense of high regard, respect, or admiration one holds for someone
4	Anticipation	involves a sense of expectation and hope for something positive or significant to occur
5	Engagement	Act of showing the genuine interest or focusing one's attention and energy on the task at hand, often with enthusiasm and dedication
6	Confidence	It involves having a robust belief in one's capacity to tackle challenges, make informed decisions, and accomplish goals
7	Happiness	feelings of joy, contentment, and satisfaction
8	Pleasure	a state of enjoyment, delight, or satisfaction that arises from experiencing something pleasurable or enjoyable
9	Excitement	state of enthusiasm, eagerness, or a heightened sense of anticipation
10	Surprise	astonishment or disbelief in response to unexpected or startling events
11	Sympathy	someone who is experiencing pain, suffering, or hardship. It involves a sense of care
12	Doubt/Confusion	states marked by uncertainty or lack of clarity
13	Disconnection	a feeling of detachment or isolation from others or from one's surroundings
14	Fatigue	physical and emotional state characterized by extreme tiredness, weakness
15	Embracement	state characterized by open acceptance, warmth, and a willingness to embrace someone or something
16	Yearning	a deep and intense desire or longing for something
17	Disapproval	indicating a negative judgment or lack of acceptance
18	Aversion	is a strong feeling of dislike or avoidance
19	Annoyance	is a mild form of irritation or frustration caused by something that disrupts or disturbs one's peace
20	Anger	feelings of displeasure, hostility, and a desire to react to a perceived injustice, frustration, or provocation

(*continued*)

Table 1. (*continued*)

Sr. No.	Emotion Labels	Feelings
21	Sensitivity	to the capacity to perceive and react to stimuli, emotions, or external influences with awareness
22	Sadness	marked by feelings of sorrow, unhappiness, and a sense of loss or disappointment
23	Fear	triggered by the perception of a threat, danger, or harm
24	Pain	sensation characterized by discomfort, distress, or suffering, often resulting from injury, illness, or emotional distress
25	Suffering	Is state of experiencing physical or emotional pain, distress, or hardship
26	Disquietment	state marked by restlessness, unease, or a lack of tranquility

Table 2. Description of the different datasets

Dataset name	Image/video	Size of data	Setting	Annotation type	Context
Emotic [4]	Image	18316	Web	26 categories	Yes
Affecnet [23]		450,000	Web	8 categories	No
CAERS [5]	Image	70,000	TV show	7 Categories	Yes
AFEW [24]	Video	1,809 Clip	Movie	7 categories	No
CAER [5]	Video	13,201Clip	TV Show	7 categories	Yes

4.2 Experimental Setup

In this implementation, OpenCV was employed to crop the face images. We implemented this fusion model using the Pytorch library. We used the pretrained model Resnet 18 with the Places dataset. We conducted training on three variations of the CNN model: one exclusively for facial data, another solely for contextual information, and a third that combined both. These configurations are illustrated in Figure 6, utilising different input types and utilising distinct loss functions. Afterwards, we evaluated the performance of these models using the testing set. For every case, we determined the training parameters by considering the validation set. Table 2 displays the average precision (AP), which indicates the extent of accuracy obtained by the test set across different categories, as represented by the area under the precision-recall curves. The results in the first three columns are obtained by employing a unified loss function (Lcomb) with CNN architectures that only process the face (F, first column), solely the image (C, second column), and both the body and the image simultaneously (F + C, third column).

Incorporating information from both the body and image inputs yields the best results for all categories except "esteem." This underscores the effectiveness of incorporating information from both sources for discrete category recognition. Notably, the results obtained using only the image context (C) generally perform less favorably when compared to the other two inputs (F, C, and F+C). This observation aligns with the understanding that within the same scene, different individuals may exhibit varying emotions, even though they share most of the context.

This paper focuses on the issue of identifying emotional states within a given setting. The EMOTIC database is a collection of images featuring people in various real-life settings, rather than controlled conditions. The images are labelled based on the individuals' discernible emotional states, utilising a combination of two distinct types of annotations: the 26 emotional categories suggested and elucidated in this study, together with a CNN model designed for the purpose of estimating emotions within a given environment. The model utilises cutting-edge methods for visual recognition and serves as a standard for the task of measuring emotional states in a given scenario. A technology capable of perceiving emotions in a manner like to humans has a multitude of possible applications in fields such as human-computer interaction, human-assistive technologies, and online education, among others.

The primary objective of this study is to precisely determine emotional states in a particular context. The EMOTIC database is a collection of photos captured in unregulated environments, featuring persons in their personal surroundings. The photographs are annotated to portray the perceived emotional states of the individuals portrayed. This task involves the use of two distinct forms of annotations: the 26 emotional categories, which are elucidated and delineated in this investigation, and the three customary continuous emotional dimensions (valence, arousal, and dominance). Moreover, a Convolutional Neural Network (CNN) model is shown to precisely forecast emotions in particular contextual settings. The model incorporates cutting-edge techniques in visual recognition and establishes a benchmark for predicting contextual emotional states.

The utilisation of an advanced technology that can precisely discern emotions, akin to human perception, holds significant potential in various domains, including human-computer interaction, assistive technologies, and online education, among others (Fig. 4 and Table 3).

Fig. 4. Implemented model show the annotated emotion for given images

Table 3. Precision value for Emotic dataset

Sr. No.	Category	Face	Context	Face + Context
1	Peace	21.65	20.46	20.4
2	Affection	19.17	16.67	21.65
3	Esteem	9.34	17.56	18.45
4	Anticipation	56.34	49.28	52.49
5	Engagement	82.16	78.92	80
6	Confidence	73.59	65.6	69
7	Happiness	54.9	49.45	52.81
8	Pleasure	46.33	44.56	49
9	Excitement	74.78	68.45	70
10	Surprise	21.89	19.55	20
11	Sympathy	11.56	11.67	15.43
12	Doubt/ Confusion	33.49	32.1	33.36
13	Disconnection	14.47	12.49	16.25
14	Fatigue	9.53	8.34	10.76
15	Embracement	2.59	3.05	3.24
16	Yearning	8.66	8.01	8.92
17	Disapproval	11.97	6.32	10.04
18	Aversion	7.89	3.43	9.81
19	Annoyance	11.64	6.09	16.93
20	Anger	7.76	5.69	11.29
21	Sensitivity	5.14	4.77	5.08
22	Sadness	9.06	6.11	19.34
23	Fear	15.55	14.68	16.77
24	Pain		2.09	
25	Suffering	10.35	5.77	8.96
26	Disquietment	18.24	15.78	19.88

5 Conclusions

The primary objective of this study is to precisely discern emotional states in a particular context. The EMOTIC database is a collection of photographs captured in unregulated environments, displaying persons in their personal surroundings. The photographs are annotated to represent the perceived emotional states of the individuals represented. This is done using two types of annotations: the 26 emotional categories, which are introduced and explained in this study, and the three classic continuous emotional dimensions (valence, arousal, and dominance). Moreover, this research presents a convolutional neural network (CNN) model that can precisely forecast emotions in different contextual settings. This model sets a benchmark for assessing contextual emotional states by using advanced techniques in visual recognition. The applicability of a system capable of discerning emotions in a manner akin to human perception is significant in various domains, including human-computer interaction, assistive technology, and online education.

References

1. Ekman, P.: Cross-cultural Studies of Facial Expression. Darwin and Facial Expression, pp. 169–220. Malor Books, Los Altos (2006)
2. Ekman, P., Friesen, W.V.: Constants across cultures in the face and emotion. J. Pers. Soc. Psychol. **17**, 124–129 (1971)
3. Fridlund, A.J.: Human facial expression: an evolutionary view. Nature **373**, 569 (1995)
4. Kosti, R., Alvarez, J.M., Recasens, A., Lapedriza, A.: Emotion recognition in context. In: CVPR (2017)
5. Lee, J., Kim, S., Kim, S., Park, J., Sohn, K.: Context-aware emotion recognition networks. IEEE explore (2019)
6. Soleymani, M Pantic, M.; Pun, T. Multimodal Emotion Recognition in Response to Videos. IEEE Trans. Affect. Comput. 2012,3, 211–223
7. Noroozi, F., Marjanovic, M., Njegus, A., Escalera, S., Anbarjafari, G.: Audio-visual emotion recognition in video clips. IEEE Trans. Affect. Comput. **10**, 60–75 (2019)
8. Li, X., Peng, X., Ding, C.: Sequential interactive biased network for context-aware emotion recognition. In: 2021 IEEE International Joint Conference on Biometrics (IJCB), pp. 1–6. IEEE (2021)
9. Zhang, D., et al.: Multi-modal multi-label emotion recognition with heterogeneous hierarchical message passing. In: Proceedings of the AAAI Conference on Artificial Intelligence, vol. 35, no. 16, pp. 14338–14346 (2021)
10. Li, Y., Lu, G., Li, J., Zhang, Z., Zhang, D.: Facial expression recognition in the wild using multi-level features and attention mechanisms. IEEE Trans. Affect. Comput. (2020)
11. Caon, M., Angelini, L., Yue, Y., Khaled, O.A., Mugellini, E.: Context-aware multimodal sharing of emotions. In: Kurosu, M. (ed.) HCI 2013. LNCS, vol. 8008, pp. 19–28. Springer, Heidelberg (2013). https://doi.org/10.1007/978-3-642-39342-6_3
12. Wu, S., Zhou, L., Hu, Z., Liu, J.: Hierarchical context-based emotion recognition with scene graphs. IEEE Trans. Neural Netw. Learn. Syst. (2022)
13. Mehendale, N.: Facial emotion recognition using convolutional neural networks (FERC). SN Appl. Sci. **2**(3), 1–8 (2020). https://doi.org/10.1007/s42452-020-2234-1
14. Mohan, K., Seal, A., Krejcar, O., Yazidi, A.: FER-net: facial expression recognition using deep neural net. Neural Comput. Appl. **33**, 9125–9136 (2021)
15. Johannßen, D., Biemann, C.: Neural classification with attention assessment of the implicit-association test OMT and prediction of subsequent academic success. In: KONVENS (2019)
16. Shenoy, A., Sardana, A.: Multilogue-net: a context aware RNN for multi-modal emotion detection and sentiment analysis in conversation (2020). arXiv preprint arXiv:2002.08267
17. Bendjillali, R.I., Beladgham, M., Merit, K., Taleb-Ahmed, A.: Improved facial expression recognition based on DWT feature for deep CNN. Electronics **8**, 324 (2019)
18. Deng, J., Dong, W., Socher, R., Li, L., Li, K., Fei-Fei, L.: Imagenet: a large-scale hierarchical image database. In CVPR (2009)
19. B. Zhou, A. Khosla, A. Lapedriza, A. Torralba, and A. Oliva. Places: An image database for deep scene understanding. CoRR, abs/1610.02055, 2015

20. Ismatov, A., Enriquez, V.G., Singh, M.: FaceHub: facial recognition data management in blockchain. In: Lee, S.-W., Singh, I., Mohammadian, M. (eds.) Blockchain Technology for IoT Applications. BT, pp. 135–153. Springer, Singapore (2021). https://doi.org/10.1007/978-981-33-4122-7_7

21. Lin, T., etal.: Microsoft COCO: common objects in context. CoRR abs/1405.0312 (2014)

22. Zhou, B., Zhao, H., Puig, X., Fidler, S., Barriuso, A., Torralba, A.: Semantic understanding of scenes through ade20k dataset (2016)

23. Prinz, J.: Which emotions are basic. Emot. Evol. Rational. **69**, 88 (2004)

24. Mollahosseini, A., Hasani, B., Mahoor, M.H.: Affectnet: a database for facial expression, valence, and arousal computing in the wild. IEEE Trans. Affect. Comput. **10**(1), 18–31 (2017)

25. Dhall, A., Goecke, R., Lucey, S., Gedeon, T., et al.: Collecting large, richly annotated facial-expression databases from movies. IEEE Multimedia (2012)

26. Jiang, Y.G., Xu, B., Xue, X.: Predicting emotions in user-generated videos. In: AAAI (2014)

27. Tran, D., Bourdev, L., Fergus, R., Torresani, L., Paluri, M.: Learning spatiotemporal features with 3d convolutional networks. In: ICCV (2015)

28. You, Q., Jin, H., Luo, J.: Visual sentiment analysis by attending on local image regions. In: AAAI (2017)

29. Ko, B.C.: A brief review of facial emotion recognition based on visual information. Sensors **18**, 401 (2018)

Human Centred AI

Exploring Multimodal Features to Understand Cultural Context for Spontaneous Humor Prediction

Ankit Kumar Singh[1]([✉]), Shankhanil Ghosh[2], Ajit Kumar[1],
and Bong Jun Choi[1]

[1] Computer Science and Engineering, Soongsil University, Seoul, South Korea
aks.bihta@gmail.com, davidchoi@soongsil.ac.kr
[2] University of Hyderabad, Hyderabad, Telangana, India

Abstract. This study aims to predict humor in a binary label, i.e., the presence or absence of humor in video recordings. The challenge here is to predict the variable in a cross-cultural manner, where the training data is in German, and the testing is done on the recordings of English-language-speaking football coaches. The novelty of this paper lies in exploring audio and textual features to predict humor in a cross-cultural setting. It is interesting to study audio and text-based features due to the cross-cultural nature of the problem, which remains largely unexplored when studying pose and facial features. The paper explores several audio (mms-lid, wav2vec 2.0) and textual (LaBSE, multilingual-e5-base) features and then uses them to train both RNN and Transformer encoder architectures. Experiments have been performed on the transformer encoder architectures with and without position encoding to study the effects of the absence of positional encoding in those features for humor detection. Late fusion has also been studied with combinations of all three modalities. We achieved our best AUC Score of 0.9251 and 0.8245 for the development and test set, respectively, out of five given submissions.

Keywords: Humor Detection · Spontaneous Humor Prediction · Multimodal data training · Multimodal Fusion · Late Fusion · Understand Cultural Context

This research was supported by the Ministry of Science and ICT (MSIT) Korea under the National Research Foundation (NRF) Korea (NRF-2022R1A2C4001270), by the MSIT Korea Korea under the India-Korea Joint Programme of Cooperation in Science & Technology (NRF-2020K1A3A1A68093469), and by the ITRC (Information Technology Research Center) support program (IITP-2022-2020-0-01602) supervised by the IITP (Institute for Information & Communications Technology Planning & Evaluation).

B. J. Choi et al. (Eds.): IHCI 2023, LNCS 14531, pp. 143–152, 2024.
https://doi.org/10.1007/978-3-031-53827-8_14

1   Introduction

Understanding human humor is a challenging task for AI models. The problem becomes more complex when we add a multicultural element to it. The difference in language and vocal expression while expressing humor makes humor analysis a very interesting study in multimodal deep learning. In this context, we study the challenge of predicting humor in a cross-cultural multimodal setting. We refer to the problem statement described in The Multimodal Sentiment Analysis Challenge (MuSe) 2023 [3] paper, which describes the MuSe-Humor problem. We studied the MuSe-Humor problem, which aims to predict spontaneous humor in a binary label, i.e., the presence or absence of humor in the video recordings. The challenge is set up cross-culturally, using press conference recordings of 10 German and 6 English football coaches. The training and validation data is in German-speaking football coaches, while the testing (evaluation) is done on the recordings of 6 English-language-speaking football coaches. The main goal is to find novel multimodal solutions to predict humor using audio, visual (face image), and textual modalities.

For model training and evaluating purposes, we referenced the code as the baseline to train two types of deep learning models, GRU-based and Transformer encoder-based. For feature fusion, we studied early fusion (concatenation of feature set before model training) and late fusion, emphasizing late fusion. We also performed an extensive hyperparameter search on the trained models to ensure no overfitting.

We studied the results from the baseline [2], and the result provided for MuSe-humor by Christ et al. [4]. We observed that extracted features had experimented with different setups from both the baseline papers. With the RNN network, the best result obtained has an Area Under the Curve (AUC) value of 0.8310. We focused on engineering and extracting new features and explored the possibilities to improve the performance of spontaneous humor prediction. We mainly explored verbal clues, i.e., audio and text modality, in-depth and kept the baseline features for video modality. Christ et al. have stated that facial expressions are most promising for spontaneous satire detection. However, the main focus of this paper is to predict humor by learning cultural cues through deep learning methods. Hence, the paper focuses only on audio and text modalities where cultural cues could be understood through multilingual deep learning models. We explored and experimented with various pre-trained audio and text large models to extract features.

For text modality, we experimented with sentiment extraction features fine-tuned on such datasets on BERT Base (multilingual) model (`bert-base-sent`) [6]. For text features, `multilingual-e5-base` [13] and `LaBSE` [9] were the better performers with maximum AUC scores of 0.8390 and 0.8106, respectively. The performance has significantly improved against feature fusion.

We acknowledge the challenges posed by the cross-cultural nature of the problem, which has made it difficult to achieve optimal results using text and audio modalities. However, the experiments conducted on these modalities have yielded intriguing findings and results. These studies provide valuable insights that contribute to our understanding of the problem.

The rest of the paper provides a detailed explanation of our approach and result. Section 2 provides the features and model used in our approach. Section 3 provides the experiments and results. Section 4 concludes the paper.

2 Features and Model

2.1 Audio

Audio data plays a vital role in detecting humor in videos as it contains essential features such as timing, intonation, emphasis, and vocal cues. These features enable the identification of humorous elements within speech patterns and delivery styles. We have used two pre-trained models for audio feature extraction. These pre-trained models (Wav2Vec2.0) are based on the Massively Multilingual Speech (MMS) dataset, a new dataset by Facebook based on readings of publicly available religious texts. The following subsection explains both feature extraction processes in brief.

mms-lid-256: New studies indicate that employing transformer architecture-based speech models trained using self-supervised methods on extensive datasets yields generalized representations, which can be finetuned for diverse speech-related tasks. Multilingual speech data is important for more generalized speech representation in cross-cultural speech tasks. For the muse-humor sub-challenge, we utilized Wav2Vec2 architecture, which consists of 1 billion parameters and is finetuned on 256 languages of the MMS (Massively Multilingual Speech) [10] dataset for language identification tasks. We extracted a 1280-dimensional representation by taking the mean over the last hidden layer representations with a sliding window of 3 s with a step size of 500ms over each audio file.

From Table 1, we can observe that mms-lid-256 has the best AUC 0.9013 and 0.9006 for RNN and transformer models, respectively. So, it is the best-performing feature for audio modality, and the performance is consistent in both RNN and transformer models.

mms-300m: The model undergoes pretraining with Wav2Vec2's self-supervised training objective, utilizing approximately 500,000 h of speech data encompassing over 1,400 languages. By employing a sliding window of 3 s and a step size of 500 ms, we computed a 1024-dimensional representation by averaging the last hidden layer representations across each audio file.

2.2 Video

In MuSe 2023, we have been given three pre-extracted features for the video modality: FAU (faus) [8], FaceNet512 (facenet) [12], and Vision Transformer (vit) [1]. In the proposed work, we used these three feature sets with different combinations and other given and new audio and text features. We have not

Table 1. The performance of RNN and Transformer on various audio features.

Features	Model	DEval
egemaps	RNN	0.6864
	TF	
ds	RNN	0.6938
	TF	
w2v-msp	RNN	0.8431
	TF	0.8423
mms-lid-256	RNN	**0.9013**
	TF	**0.9006**
mms-300m	RNN	0.8151
	TF	0.8444

experimented with any visual features because the main focus of the study was to build methods to understand affective variables in a cross-cultural setting. Visual features have little variation when cultures are exchanged. The argument is that visual features such as pose and facial key points do not vary much over different cultures.

2.3 Text

Understanding text embeddings to extract feature embeddings for cross-cultural humor understanding is a very interesting problem that constitutes a separate domain of study. This paper has analyzed how pre-trained multilingual large language models could be used to understand humor in an uni-modal setting. Furthermore, the impact of those unimodal embeddings with late-fusion with other modalities has also been studied. Interestingly, text modalities have contributed to a significant improvement in the AUC scores for late fusion, as has been discussed in further sections. Pretrained models that extract feature vectors are chosen to get close vector embeddings for the same text piece in different languages, including German and English. The dataset [5] has only English and German languages. The intuition is that if a good feature vector is extracted from text, the proposed model architecture could learn the weights to predict humor in a multi-modal and multicultural setting.

Our research involves studying large models that are trained on multilingual benchmark datasets. LaBSE and multilingual-e5-base are the top performers when understanding cross-contextual humor. Table 3 shows that multilingual-e5-base has the best AUC of 0.8394 and 0.8357 for RNN and transformer models, respectively. So, it is the best-performing feature for the text modality, and the performance is consistent in both RNN and transformer models.

Other features that have been analyzed are variations of sentence transformers [11] and variations of the bert-base-sent, which is finetuned `bert-base-uncased` model [7] on a sentiment analysis dataset, finetuned on sentiment analysis datasets (which includes data from English and German, among other languages). For unimodal analysis, LaBSE, `multilingual-e5-base`, and `bert-base-sent`.

LaBSE: LaBSE is a language-agnostic BERT sentence embedding that generates high dimensional vectors while being language-independent to a certain extent. It is capable of capturing the context and encoding text into high-dimensional vectors. The model is trained and optimized to produce similar representations exclusively for bilingual sentence pairs that are translations of each other. So, it can be used for mining for translations of a sentence in a larger corpus.

multilingual-e5-base: `multilingual-e5-base` belongs to a family of text embeddings called E5 [13] that is trained with weak-supervision signals from the CCPairs dataset. This embedding is a very good choice for this kind of task because it is a general-purpose embedding model that generates a single-vector representation that can be used for training in a downstream task such as the one in focus in this paper.

The similarity of the corpus has been analyzed by taking a sentence pair in German and English, then generating the LaBSE and `multilingual-e5-base` embeddings. We calculated the cosine similarity for those embeddings. One sample result has been shown below in Table 2. The sample text used here is Article 1 of the Universal Declaration of Human Rights[1] and its official German translation. The cosine similarities suggest that the vector embedding generated by these models is very close to each other despite the difference in languages. Hence, this indicates that these models can be used to train a downstream model for humor detection.

Table 2. Exploring cosine similarity between `multilingual-e5-base` and LaBSE to demonstrate the language-independent embedding generation in German and English. The difference in vector embedding generated by LaBSE and `multilingual-e5-base` prediction is ≈0.016.

Language	Text
English	All human beings are born free and equal in dignity and rights. They are endowed with reason and conscience and should act towards one another in a spirit of brotherhood
German	Alle Menschen sind frei und gleich an Würde und Rechten geboren. Sie sind mit Vernunft und Gewissen begabt und sollen einander im Geist der Brüderlichkeit begegnen
LaBSE	0.9072
mult-e5-base	0.8912

---

[1] https://www.un.org/en/about-us/universal-declaration-of-human-rights.

Table 3. The performance of RNN and Transformer on various text features.

Features	Model	DEvel
bert-multilingual	RNN	0.8154
	TF	0.8175
LaBSE	RNN	0.8219
	TF	0.8099
sentence-transformer	RNN	0.8035
	TF	0.8044
bert-base-sent	RNN	0.8032
	TF	0.8139
multilingual-e5-base	RNN	**0.8394**
	TF	**0.8357**

2.4 Fusion

Experiments with late fusion have been conducted using the method proposed in the baseline paper [3]. Late fusion experiments have been categorized into 4 types.

- *Late fusion with 2 modalities:* Two modalities, preferably one audio feature and one text fusion, are fused to generate the final prediction values.
- *Late fusion with 3 modalities:* Three modalities, which include one from audio, one from text, and another from visual (as described in the baseline paper) features, have been fused to generate final prediction values.
- *Late fusion with multiple modalities:* In this scenario, two best-performing unimodal features from audio and two best-performing unimodal features are usually used to generate the final prediction values.

3 Experiments and Results

3.1 Experimental Setup

For running the experiments designed in this paper, we used a workstation with 2x NVIDIA GeForce RTX 3090 GPUs, with each GPU having a memory capacity of 24 GB. The workstation has 36x CPUs Intel(R) Core(TM) i9-10980XE CPU @ 3.00 GHz.

3.2 Dataset and Data Preparation

The paper uses the Passau Spontaneous Football Coach Humor (Passau-SFCH) dataset [5] to analyze and train the models. The dataset contains videos of football coaches, where some coaches make humourous comments during an interview. The 10 German coaches' videos are used for training, and 6 English coaches' videos are used for testing.

The raw dataset contains video files, which are then processed to extract facial images of the coaches, along with the audio and the transcription of the conversation from the audio. Since the authors already did the preprocessing in the baseline paper [3], we have not contributed to any further processing of this nature. However, we have used the raw audio and the text to generate and align the embeddings according to the alignment information extracted from the metadata. A simple algorithm has been used to align the raw data to a 500 milliseconds timeframe. The alignment method used here has been described in Eqs. 1 and 2.

$$t_a = \left\lfloor \frac{t_\alpha}{500} \right\rfloor \times 500 \tag{1}$$

$$t_b = \left\lfloor \frac{t_\beta}{500} + 1 \right\rfloor \times 500 \tag{2}$$

where t_α is the initial starting timestamp for the audio/text segment, and t_β is the final ending timestamp for the same. t_a and t_b are the aligned timestamps, quantized on 500 ms duration.

3.3 Hyperparameters

To find the ideal combination of various hyperparameter settings, we explored input features, model type, hidden state dimension (h), number of RNN layers or number of TF attention heads (r), and learning rate (lr), a grid search was conducted. Table 4 presents the specific values tested for each hyperparameter during the search.

Table 4. Various hyperparameters and their values.

hyperparameters	Values
features	egemaps, ds, vit, facenet, faus, w2v-msp, mms-lid-256, mms-300m, bert-multilingual, LaBSE, sentence-transformer, bert-base-sent, multilingual-e5-base
model type	RNN, TF
model_dims (h)	32, 64, 128, 256
rnn_n_layers/n_head (r)	2, 4
lr (lr)	0.001, 0.005, 0.0001

3.4 Results

Table 6 shows the results obtained for our five submissions. All other submissions employed late fusion except for the first submission, which utilized the predictions from the best audio mms-lid-256. The A (audio), V (video), and T (text) values represent the best development scores obtained for each modality, for which the feature and hyperparameter combinations used are summarized in Table 5.

Table 5. Trained model of three modularity selected for late fusion.

Name	Feature	Model	h	r	lr	AUC
Audio						
A	mms-lid-256	RNN	128	2	0.001	**0.9013**
A1	mms-lid-256	TF	256	2	0.001	0.9007
A2	w2v-msp	RNN	128	2	0.005	0.8445
A3	mms-lid-256	TF	128	2	0.005	0.8736
A4	w2v-msp	TF	128	2	0.0001	0.8374
Text						
T	multilingual-e5-base	RNN	128	2	0.005	**0.8394**
T1	multilingual-e5-base	TF	128	2	0.0001	0.8357
T2	bert-multilingual	RNN	256	2	0.001	0.8155
T3	bert-base-sent	TF	128	4	0.0001	0.8139
T4	bert-multilingual	TF	128	2	0.001	0.8107
Video						
V	vit	RNN	64	2	0.0001	0.8272

Table 6. Performance of each Submission on the Development (DEval) and the Test(TEval) Testing Dataset. Feature and hyperparameter combinations used are summarized in Table 5.

SubNo.#	Fusion	DEval	TEval	Diff
1	**A**	0.9013	0.6324	0.27
2	**A + T**	0.9206	0.7357	0.19
3	**A + T + V**	0.9208	0.7381	0.19
4	**A1 + A2 + T1 + T2 + V**	0.9083	0.7985	0.11
5	**A3 + A4 + T3 + T4**	0.8859	**0.8247**	0.06

3.5 Discussion

We have made all five submissions using late fusion on different modalities except the first, for which we submitted the best result obtained through audio modality. As mentioned, we have experimented with RNN and transformer architecture, so the submissions combine these architectures and three data modalities. In addition, we have also repeated the modality; for example, in submission 4, we have used two audio and text features along with the video.

We presented our best-performing outcome on the development set for the initial submissions. However, to our surprise, the evaluation test (TEval) showed a significant decline in performance despite the high development AUC score. To investigate this discrepancy, we determined that the model was primarily trained on German language data, leading to overfitting when evaluating English data.

To address this issue, we implemented two straightforward approaches for subsequent submissions. In the first approach, referred to as submission 4, we employed architecture fusion. This involved the late fusion of the best-performing text features (T1, T2) and audio features (A1, A2) using RNN and transformer models. As a result, the performance gap between development and evaluation was reduced to a margin of only 0.11.

In the second approach, used for submission 5, we focused on experimenting with the transformer architecture and two top-performing audio and text modality features. By training fifty models with different parameters for all four combinations (A3, A4, T3, T4), we selected the lowest-performing model from each and performed late fusion. This was based on the premise that a model with lower development performance would be less prone to overfitting and more adaptable to cross-lingual test datasets. As anticipated, submission 5 exhibited the best performance among all our submissions when evaluated on the test set, thus validating our hypothesis.

While our best performance is nearly on par with the baseline, considering our understanding of the problem and the extensive experimentation conducted, we firmly believe that further improvements can be made to enhance overall performance.

4 Conclusions and Future Work

In the proposed work, we explored the multimodal features to improve spontaneous humor detection for the MuSe-humor sub-challenge. Out of three modalities, we experimented mainly with audio and text features by extracting new and existing features in the baseline. Specifically, we extended the baseline RNN-based experiments by performing various experiments combining all three modalities (with new and baseline features) with a transformer-based model. We also tested late fusion approaches like majority voting, average, and weighted average. We achieved the AUC value of 0.8859 and 0.8247 for the development and test set (our best result from five given submissions), respectively. Furthermore, through experiments, we investigated the degradation in effectiveness obtained for the test dataset, and we suggested different strategies to improve further the effectiveness of late fusion, such as data preprocessing and data augmentation.

References

1. Caron, M., et al.: Emerging properties in self-supervised vision transformers. In: Proceedings of the IEEE/CVF International Conference on Computer Vision, pp. 9650–9660 (2021)
2. Christ, L., et al.: The muse 2023 multimodal sentiment analysis challenge: mimicked emotions, cross-cultural humour, and personalisation. arXiv preprint arXiv:2305.03369 (2023)
3. Christ, L., et al.: The muse 2022 multimodal sentiment analysis challenge: humor, emotional reactions, and stress. In: Proceedings of the 3rd International on Multimodal Sentiment Analysis Workshop and Challenge, pp. 5–14 (2022)

4. Christ, L., Amiriparian, S., Kathan, A., Müller, N., König, A., Schuller, B.W.: Multimodal prediction of spontaneous humour: a novel dataset and first results. arXiv preprint arXiv:2209.14272 (2022)
5. Christ, L., Amiriparian, S., Kathan, A., Müller, N., König, A., Schuller, B.W.: Multimodal prediction of spontaneous humour: a novel dataset and first results (2022)
6. Devlin, J., Chang, M.W., Lee, K., Toutanova, K.: Bert: pre-training of deep bidirectional transformers for language understanding. arXiv preprint arXiv:1810.04805 (2018)
7. Devlin, J., Chang, M., Lee, K., Toutanova, K.: BERT: pre-training of deep bidirectional transformers for language understanding. CoRR abs/1810.04805 (2018). http://arxiv.org/abs/1810.04805
8. Ekman, P., Friesen, W.V.: Facial action coding system. Environ. Psychol. Nonverbal Behav. (1978)
9. Feng, F., Yang, Y., Cer, D., Arivazhagan, N., Wang, W.: Language-agnostic bert sentence embedding (2022)
10. Pratap, V., et al.: Scaling speech technology to 1,000+ languages. arXiv preprint arXiv:2305.13516 (2023)
11. Reimers, N., Gurevych, I.: Sentence-bert: sentence embeddings using Siamese bert-networks (2019)
12. Schroff, F., Kalenichenko, D., Philbin, J.: Facenet: a unified embedding for face recognition and clustering. In: Proceedings of the IEEE Conference on Computer Vision and Pattern Recognition, pp. 815–823 (2015)
13. Wang, L., et al.: Text embeddings by weakly-supervised contrastive pre-training. arXiv preprint arXiv:2212.03533 (2022)

Development of Pneumonia Patient Classification Model Using Fair Federated Learning

Do-hyoung Kim[✉], Kyoungsu Oh, Seok-hwan Kang, and Youngho Lee

Gachon University, 1342, Seongnam-daero, Sujeong-gu, Seongnam, Gyeonggi-do 13120,
Republic of Korea
rlaehgud9604@gachon.ac.kr

Abstract. Worldwide, pneumonia has been a major problem for the past few centuries. Currently, medical staff are having a hard time due to the increase in COVID-19 in many countries. Chest X-ray is the most common method for screening and diagnosing chest diseases. However, there are difficulties in building the model due to data confidentiality between patients and hospitals and problems with collecting large amounts of data within hospitals. As a solution to this, we propose FFLFCN, which uses federated learning and deep learning to diagnose pneumonia. FFLFCN built a central model by training local models in multiple hospitals with their own data while maintaining the privacy of patient data. In this paper we aim to inspire federated learning research for pneumonia patients using FFLFCN, and achieve improved AUC and shorter learning time than before.

Keywords: Fair Federated Learning · Protected Health Information · Deep Learning

1 Introduction

Coronavirus disease 2019(COVID-19) has caused a serious health crisis and caused unprecedented social disruption worldwide [1–3]. By the end 2021, more than 43% of the world's population has been infected with COVID-19 at least once, and the COVID-19 pandemic has killed more than 5.94 million people worldwide [4, 5]. After the initial outbreak, COVID-19 is on the rise again in many countries, especially southern Asian countries such as India and Nepal. Therefore, research on detecting and diagnosing COVID-19 patients is very meaningful [6, 7]. Clinical symptoms of patients infected with COVID-19 mainly include fever, chills, dry cough, and body pain. A key step in determining and treating COVID-19 is effectively screening infected patients. One of the main screening methods is using a radiology chest X-ray. Therefore, in this study, we plan to develop a deep learning model that uploads chest X-ray images and predicts pneumonia.

Pneumonia is dangerous disease that can cause fatal injuries in elderly people suffering from diabetes and asthma [8]. However, the progress of medical research is hampered by a lack of publicly available datasets due to privacy concerns. To alleviate this problem, federated learning can be used to address data privacy and data silo issues.

Federated Learning (FL) is machine learning method that processes data within a user's individual device or each distributed server and then shares learned parameters with a central server [9]. FL is being used to solve data privacy issues and mitigate the risk of clinical information leakage because it does not require data transmission and centralization required in existing centralized methods.

In this study, we optimize the sum of unweighted binary cross entropy loss using adversarial loss and optimized parameters, and show higher performance than previous models in classifying pneumonia patients. Additionally, issues of transparency, accuracy, and bias were addressed using Minimum Information for Medical AI Reporting (MIN-IMAR). Finally, we proposed the Fair Federated Learning Loss Function Chest X-ray Dense Convolutional Network (FFLFCN), a global model that uses federated learning to learn all datasets without sharing the data.

2 Related Work

2.1 Federated Learning

FL is used in the energy field, healthcare field, and many other fields. Each local stores a local model, and the central server stores a global model with the same architecture for the model. The local model is trained and its parameters are uploaded to the global model, which aggregates the parameters and broadcasts them back to the local model. After joint optimization, the central server returns to the global state of each device and continues to accept updated data computed by each client in the new global state.

FL can solve the problem of large-scale unprotected personal data and complete update learning of devices without exchanging large amounts of data. This decentralized education model approach offers personal health information, security, regulatory, and economic advantages. FL presents new statistical and systematic challenges when training machine models on distributed device networks [10]. FL, which relies on distributed data, serves several aspects of research.

Fi Chen et al. identified the combination of FL and meta-learning as a major advancement in FL [11]. Konstantin Sozinov et al. have made some progress in applying FL to human activity identification [12]. Khan et al. used a deep learning model with a FL approach to solve the problem of collecting large amounts of data [13]. Rieke et al. highlighted considerations of how FL could provide solutions for the future of digital health [14]. NG, Dianwen et al. described joint efforts to alleviate the problem of insufficient training of AI models with small datasets and achieve FL for medical imaging models [15].

However, FL involves training statistical models on large heterogeneous networks. Naively minimizing the total loss function in such networks may disproportionately advantage or disadvantage some devices. Therefore, fairness, flexibility, and efficiency are improved by fair resource allocation in wireless networks that promotes fair and accurate distribution [16]. Additionally, LI, Tian, et al., show that more accurate and fairer models are possible compared to state-of-the-art processes and strong baselines in federated datasets [17].

2.2 DenseNet (Dense Convolutional Network) [18]

DenseNet is a network architecture in which each layer is directly connected to all other layers in a feedforward manner. For each layer, the feature maps of all previous layers are treated as separate inputs, while its own feature maps are passed as input to all previous layers. This concatenation pattern provides state-of-the-art accuracy on CIFAR 10/100 and SVHN with and without data augmentation. On the large-scale ILSVRC 2012 (ImageNet) dataset [19], DenseNet achieves similar accuracy as ResNet [20], but uses less than half the amount of parameters and roughly half the number of FLOPs.

2.3 CheXNet [21]

The pneumonia detection task is a binary classification problem, where the input is a frontal chest X-ray image and the output is a binary label indicating the presence or absence of pneumonia, respectively. CheXNet is a 121-layer DenseNet trained on the Chest X-ray 14 dataset. DenseNet improves the flow of information and gradients through the network, making optimization of very deep networks tractable. The final fully connected layer is replaced with a layer with a single output and then sigmoid nonlinearity is applied. The weights of the network are initialized with the weights of the pre-trained model in ImageNet. The network uses standard parameters and is trained end-to-end using Adam to select the model with the lowest validation loss.

2.4 Loss Function

Adversarial Loss is one of the loss functions used Generative Adversarial Networks (GAN), which models the competition between generators and discriminators. This loss helps the generator produce more realistic fake data and the discriminator to distinguish more accurately. A ConvNet trained on MRI images via adversarial loss was validated by applying it to unpaired CT data for cardiac structure segmentation and obtained promising results [22]. Contextual Loss is a loss function used in tasks such as image inpainting and autoencoders. This loss measures the contextual similarity between the original image and the generated image, helping the generated image look more natural and replacing missing parts appropriately. Using Contextual Loss, an alternative loss function that doesn't require alignment was proposed, providing an effective and simple solution to a new problem space [23]. Encoder Loss is used in models such as autoencoders. It is mainly used with reconstruction loss and adversarial loss, and the model effectively represents and restores the input data. These loss functions play an important role in model learning and performance improvement in applications such as various computer vision tasks and medical image processing.

2.5 IID, Non-IID Condition

McMahan et al. [24] argued that FedAvg can process Non-Independent and Identically Distributed (Non-IID) data to some extent. However, many studies point out that a decrease in FL accuracy is almost inevitable in non-IID and heterogeneous data [25]. The basic assumption of independent and identical distributions in FL, which can usually be due to temporal or spatial correlation of the data or non-normal distribution of the training and testing data sets, is no longer met.

The ultimate goal of FL is to create a global model that learns all data sets without sharing data. Therefore, in this study, we conducted various studies on various non-IID data sets to develop a personalized approach for processing non-IID data in FL using FFLFCN.

3 Improve Transparency

3.1 MINIMAR (MINImum Information for Medical AI Reporting): Reporting Standards for Artificial Intelligence in Health Care [26]

The medical industry is becoming familiar with AI-based solutions that are rapidly emerging in clinical settings. However, the nature of the data needed to evaluate the performance of these prediction models has not been adequately reported, leaving uncertainty and doubt about their application in the broader healthcare environment. An empirical evaluation of 81 studies comparing clinicians and AI models revealed major issues including lack of transparency, bias, and unjustified claims.

MINIMAR used in this study meets four requirements. The first requirement includes information about the data source, the population providing the training data in terms of cohort selection. The second requirement includes training data demographics in a way that can be compared to the population to which the model is applied. The third requirement provides detailed information about model architecture and development so that the intent of the model can be interpreted, compared to similar models, and allowed for replication. Finally, transparent reporting of model evaluation, optimization, and validation clarifies how to achieve local model optimization and enable replication and resource sharing.

Therefore, MINIMAR was used to solve the issues of standards, transparency, and bias regarding information for clinical AI research. Figure 1 and 2 was created by applying the data and requirements used in this study to the MINIMAR Standard Report.

MINIMAR Standard Report

Features	Checklist	Description	Chest X-ray8	Example	비고
1. Study population and setting population	■ Yes ☐ No	Population from which study sample was drawn	Patients undergoing elective surgery(Chest X-ray)	All patients	
Study setting	■ Yes ☐ No	The setting in which the study was conducted(e.g., academic medical left, community healthcare system, rural healthcare clinic)	Department of Radiology and Imaging Sciences, Clinical Center, National Center for Biotechnology Information, National Library of Medicine, National Institutes of Health, MD 20892	2 U.S. academic medical lefts	
Data source	■ Yes ☐ No	The source from which data were collected	University of California San Diego, Guangzhou Women and Children's Medical Center, PACS(Picture Archiving and Communication System)	EHRs	
Cohort selection	■ Yes ☐ No	Exclusion / inclusion criteria	108,948 frontal x-ray images of 32,717 unique patients (1992-2015) with common disease labels	All admissions for adult patients. Hospitalizations of 24 h or longer.	

Features	Checklist	Description	Chest X-ray8	Example	비고
2. Patient demographic characteristics	■ Yes ☐ No				
Age	■ Yes ☐ No	Age of patients included in the study	Mean 46.6 y	Median 56 y	
Sex	■ Yes ☐ No	Sex breakdown of study cohort	Female: 43.5% Male: 56.5%	Female 55.0%	
Race	■ Yes ☐ No	Race characteristics of patients included in the study	All patients	-	
Ethnicity	■ Yes ☐ No	Ethnicity breakdown of patients included in the study	All patients	-	
Socioeconomic status	■ Yes ☐ No	A measure or proxy measure of the socioeconomics status of patients included in the study	All patients	-	

Fig. 1. MINIMAR Standard Report 1, 2

Features	Checklist	Description	Chest X-ray8	Example	비고
3. Model architecture	■ Yes ☐ No				
Model output	■ Yes ☐ No	The computed result of the model	AUC(Area Under ROC Curve)	In-hospital deaths, 30-day unplanned readmission, length of stay, discharge status	
Target user	■ Yes ☐ No	The indented user of the model output (eg. clinician, hospital management team, insurance company)	Risk scores produced by the model will be used by the hospital team for pain management	Predictions produced by the model will be used by hospitals for care management	
Data splitting	■ Yes ☐ No	How data were split for training, testing, and validation	train : val : test = 7 : 1 : 2	80%/10%10% (train/validation/test)	
Gold standard	■ Yes ☐ No	Labeled data used to train and test the model	Use optimized parameters for each loss function	Death, readmission and ICD codes in EHRs	
Model task	■ Yes ☐ No	Classification or prediction	Classification and Prediction	Prediction	
Model architecture	■ Yes ☐ No	Algorithm type (eg. machine learning, deep learning, etc.)	FFLACN, CheXNet	Recurrent neural networks, attention-based time-aware neural network model and neural network	
Features	■ Yes ☐ No	List of variables used in the model and how they were used in the model in terms of categories or transformation	atelectasis, cardiomegaly, infiltration, mass, nodule, pneumonia, pneumothorax	Provided in detail for all models	
Missingness	■ Yes ☐ No	How missingness was addressed: reported, imputed, or corrected	Missing data were imputed using median of the variable distribution	-	

Features	Checklist	Description	Chest X-ray8	Example
4. Model evaluation	■ Yes ☐ No	AUC(Area Under ROC Curve)		
Optimization	■ Yes ☐ No	Model or parameter tuning applied	Use optimized parameters for each loss function	Documented and provided for all models in detail
Internal model validation	■ Yes ☐ No	Study internal validation	Internal 10-fold cross-validation	Hold-out validation set
External model validation	■ Yes ☐ No	External validation using data from another setting	Not performed	Not performed
Transparency	■ Yes ☐ No	How code and data are shared with the community.	Code and sample data available via GitHub	Data is not available; code is available via GitHub

Fig. 2. MINIMAR Standard Report 3, 4

4 Method

In this study, modified MNIST, CIFAR-10 MVTec Anomaly Detection, and Chest X-ray8 data sets were used. At this time, FL was verified in an environment that simulated MNIST, CIFAR-10 MVTec Anomaly Detection, and Chest X-ray8 data sets with actual data distribution.

4.1 Dataset

The first dataset, MNIST, used to build the model, used 60,000 training images and 10,000 test images consisting of handwritten digit images. The second dataset, CIFAR-10, was trained with 60,000 color images of 32 × 32 pixels with labels, and the test data was split 7:3 without overlap. The third data image resolution was MVTec Anomaly Detection [27], which consists of 700 × 700–1024 × 1024 pixels, had 15 labels, and used 3,629 high-resolution images for training data and 1,725 images for test data. For model training and testing, the training data in each data set consisted only of normal images, and the test data included both defective and non-defect images.

The fourth dataset is Chest X-ray8 [28], collected from 1992 to 2015. It consists of 108,948 frontal x-ray images with eight common disease labels text-mined, extracted from text radiology reports through NLP techniques from 32,717 unique patients. This is a chest X-ray image mined from the Picture Archiving and Communication Systems (PACS) system of the Department of Radiology and Imaging Sciences. The columns consisted of atelectasis, cardiomegaly, effusion, infiltration, mass, nodule, pneumonia, and pneumothorax, and each image was labeled with one or more pathology keywords or "normal." In Chest X-ray8, X-ray images were extracted directly from DICOM files. It is resized to a 1024 × 1024 bitmap image without significant loss of detail compared to the 512 × 512 image size of the OpenI dataset. The intensity range was rescaled using the default window settings stored in the DICOM header file. Of the 108,948 frontal X-ray images in the fourth dataset, 24,636 images contained one or more pathologies, and the remaining 84,312 images were normal. For pathology classification and localization tasks, the entire dataset was randomly shuffled into three subgroups, namely train, val, and test, in a 7:1:2 ratio for CNN fine-tuning via Stochastic Gradient Descent (SGD).

4.2 Data Pre-processing, Augmentation

The third dataset consists of texture images and object images. The black-and-white images were converted to gray-scale to apply SSIM (Structural Similarity Index Measure), and the color images were converted to gray-scale to apply SSIM-Autoencoder. The texture image in the training image was cut to a size of 128 × 128 using SSIM-based loss and expanded to a size of 256 × 256 using a convolution layer for input to the model. Objects were passed to Autoencoder as images and anomaly maps were generated using per-pixel SSIM. Afterwards, the training images were augmented to 10,000 by randomly rotating or applying mirroring.

For the fourth dataset, before feeding the images into the network, the images were scaled down to 224 × 224 and normalized based on the mean and standard deviation of the images in the ImageNet [29] training set. We also augmented the training set with random horizontal flipping.

4.3 FFLFCN (Fair Federated Learning Loss Function Chest X-ray Dense Convolutional Network) Model Construction

This study was inspired by the CheXNet model [21] and added three loss functions: Adversarial Loss [30], Contextual Loss [31], Encoder Loss [32] to compensate for the shortcomings of each loss function. Additionally, the sum of unweighted binary cross-entropy losses was optimized using the optimized parameters.

The Adversarial Loss used first is suitable for fooling the discriminator by taking advantage of realistic data generation, mode collapse mitigation, and generative diversity. Instability was reduced by using feature matching loss for adversarial learning [33]. However, if only the adversarial loss function is used, the generator is not optimized for learning context information about the input data due to training instability, hyperparameter sensitivity, and simple quality control. The second used Contextual Loss is strong in perceptual similarity, content preservation, and geometric transformation, so the distance between each layer shows more distinct results than before [31]. However, shortcomings such as complexity, hyperparameter tuning, and loss environment may lead to convergence problems during training. To compensate for these shortcomings, latent space control and normalization were performed using Encoder Loss. This loss of redundancy penalizes the similarity between encoded features, reducing irrelevant and repetitive data. Therefore, Adversarial Loss, Contextual Loss, and Encoder Loss functions were added to the CheXNet model to reduce optimization, distinct results, and repetitive data. (1) shows the objective function.

$$F(object) =$$
$$W1 * F(adversarial\ loss) + W2 * F(contextual\ loss) + W3 * F(encoder\ loss)$$

$$(1)$$

The relationship between the optimized FFLFCN model and the text (X-ray Report) is learned jointly. Figure 3 shows the FFLFCN Model Construction. And the fourth data set was randomly divided into train, val, and test 7:1:2. We also compared the class-specific AUROC of the model with the previous state of the art, held by Yao et al. (2017) for 13 classes and Wang et al. (2017) for the remaining one class.

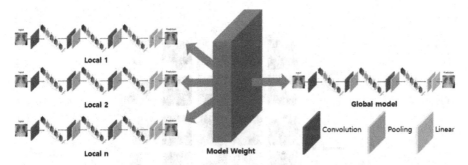

Fig. 3. Fair Federated Learning Loss Function Chest X-ray Dense Convolutional Network Model Construction

5 Results

In this study, the model was trained and evaluated for 20 epochs for each data set. It shows that the FFLFCN model achieved high AUC (Area Under the Curve) in the fourth data set. Table 1 shows that AUC is improved over existing performance in the

FL environment as a result of applying FL based on FFLFCN and CheXNet and using different seeds for the Digit and Class specified above.

In this study, we evaluated whether a universal data set can be created even when each local device has data of a specific class in the data set, which is a problem with FL. Because this is the problem facing federated learning. Figure 4 shows that during model testing, CAM (Class Activation Map) was used to provide interpretation and visual explanation of the deep learning model to improve model reliability and to improve and optimize the model. Figure 5 is an image divided into anomaly-free and anomalous generated by applying the FFLFCN model to the MVTec AD dataset. Also, in Fig. 6 shows that the model cannot generated abnormal objects. Table 2 shows the results of applying Federated Learning based on CNN and MLP. Performance evaluation used IID and Non-IID. As a result, it was confirmed that performance was improved over existing performance in the FL environment.

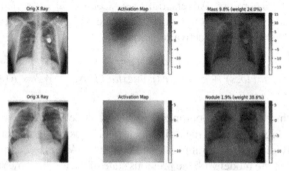

Fig. 4. Class Activation Map: Use the CAM to determine which regions of an image are associated with a class

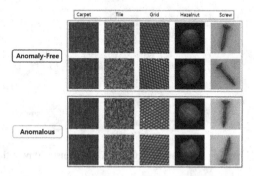

Fig. 5. Example images for some of all categories of the MVTec Anomaly Detection Dataset. For each category, anomaly-free as well as an anomalous example is shown.

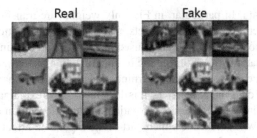

Fig. 6. CIFAR-10 Exemplar real and generated samples.

Table 1. Below is the result showing the AUC score

Dataset	Type	Model	AUC
MNIST	2	FFLFCN	0.926
		CheXNet	0.932
CIFAR-10	Plane	FFLFCN	0.868
		CheXNet	0.891
MVTec AD	Tile	FFLFCN	0.873
		CheXNet	0.911
Chest X-ray8	Pneumonia	FFLFCN	**0.9153**
		CheXNet	0.7680

Table 2. Comparsion of prediction results

Dataset	Model	IID	Non-IID
MNIST	MLP	94.17	90.17
	CNN	98.60	96.40
Chest X-ray8	MLP	94.10	90.91
	CNN	95.47	91.10
CIFAR-10	MLP	42.16	-
	CNN	59.47	-
F-MNIST	MLP	94.10	83.42
	CNN	9.80	9.80

6 Conclusion

In this study, we conducted an experiment on pneumonia identification using CXR images based on a federated learning framework. Additionally, it demonstrated the potential of FL in terms of performance and data protection, which is important for handling sensitive medical data. In particular, in FL, only weights are transmitted, participants are not aware of each other's local data sets, and personal information leakage can be prevented. The proposed approach can be used to complement existing approaches and prevent problems that may arise during the de-identification process.

However, the success of a federated learning model depends on many factors, including the quality of the data and parameters. It is difficult to prove explainability for medical FL that cannot currently survey or collect patient health information data. However, in future research, we will determine which part of the input data affects the output value of the deep neural network and solve the problem using XAI (eXplainable Artificial Intelligence), which removes the existing black-box characteristics.

Acknowledgment. "This paper is supported by Korean Agency for Technology and Standards under Ministry of Trade, Industry and Energy in 2023" (project title: Establishment of standardization basis for BCI and AI Interoperability, 20022362).

References

1. Clark, A., et al.: Global, regional, and national estimates of the population at increased risk of severe COVID-19 due to underlying health conditions in 2020: a modelling study. Lancet Glob. Health **8**, e1003–e1017 (2020)
2. Zhu, N., et al.: A novel coronavirus from patients with pneumonia in China, 2019. N. Engl. J. Med. **382**, 727–733 (2020)
3. Brooks, S.K., et al.: The psychological impact of quarantine and how to reduce it: rapid review of the evidence. The Lancet **395**, 912–920 (2020)
4. Barber, R.M., et al.: Estimating global, regional, and national daily and cumulative infections with SARS-CoV-2 through Nov 14, 2021: a statistical analysis. The Lancet **399**, 2351–2380 (2022)
5. Wang, H., et al.: Estimating excess mortality due to the COVID-19 pandemic: a systematic analysis of COVID-19-related mortality, 2020–21. The Lancet **399**, 1513–1536 (2022)
6. Yan, B., et al.: An improved method of COVID-19 case fitting and prediction based on LSTM (2020)
7. Zhang, Y., et al.: Covid-19 public opinion and emotion monitoring system based on time series thermal new word mining. arXiv preprint arXiv:2005.11458 (2020)
8. Chebib, N., et al.: Pneumonia prevention in the elderly patients: the other sides. Aging Clin. Exp. Res. **33**, 1091–1100 (2021)
9. AbdulRahman, S., Tout, H., Ould-Slimane, H., Mourad, A., Talhi, C., Guizani, M.: A survey on federated learning: the journey from centralized to distributed on-site learning and beyond. IEEE Internet Things J. **8**, 5476–5497 (2020)
10. Smith, V., Chiang, C.-K., Sanjabi, M., Talwalkar, A.S.: Federated multi-task learning. Adv. Neural Inf. Process. Syst. **30** (2017)
11. Chen, F., Luo, M., Dong, Z., Li, Z., He, X.: Federated meta-learning with fast convergence and efficient communication. arXiv preprint arXiv:1802.07876 (2018)

12. Sozinov, K., Vlassov, V., Girdzijauskas, S.: Human activity recognition using federated learning. In: 2018 IEEE International Conference on Parallel & Distributed Processing with Applications, Ubiquitous Computing & Communications, Big Data & Cloud Computing, Social Computing & Networking, Sustainable Computing & Communications (ISPA/IUCC/BDCloud/SocialCom/SustainCom), pp. 1103–1111. IEEE (2018)
13. Khan, S.H., Alam, M.G.R.: A federated learning approach to pneumonia detection. In: 2021 International Conference on Engineering and Emerging Technologies (ICEET), pp. 1–6. IEEE (2021)
14. Rieke, N., et al.: The future of digital health with federated learning. NPJ Dig. Med. **3**, 119 (2020)
15. Ng, D., Lan, X., Yao, M.M.-S., Chan, W.P., Feng, M.: Federated learning: a collaborative effort to achieve better medical imaging models for individual sites that have small labelled datasets. Quant. Imaging Med. Surg. **11**, 852 (2021)
16. Li, T., Sanjabi, M., Beirami, A., Smith, V.: Fair resource allocation in federated learning. arXiv preprint arXiv:1905.10497 (2019)
17. Li, T., Hu, S., Beirami, A., Smith, V.: Ditto: fair and robust federated learning through personalization. In: International Conference on Machine Learning, pp. 6357–6368. PMLR (2021)
18. Huang, G., Liu, Z., Van Der Maaten, L., Weinberger, K.Q.: Densely connected convolutional networks. In: Proceedings of the IEEE Conference on Computer Vision and Pattern Recognition, pp. 4700–4708 (2017)
19. Russakovsky, O., et al.: Imagenet large scale visual recognition challenge. Int. J. Comput. Vision **115**, 211–252 (2015)
20. Targ, S., Almeida, D., Lyman, K.: Resnet in resnet: generalizing residual architectures. arXiv preprint arXiv:1603.08029 (2016)
21. Rajpurkar, P., et al.: Chexnet: radiologist-level pneumonia detection on chest x-rays with deep learning. arXiv preprint arXiv:1711.05225 (2017)
22. Dou, Q., Ouyang, C., Chen, C., Chen, H., Heng, P.-A.: Unsupervised cross-modality domain adaptation of convnets for biomedical image segmentations with adversarial loss. arXiv preprint arXiv:1804.10916 (2018)
23. Mechrez, R., Talmi, I., Zelnik-Manor, L.: The contextual loss for image transformation with non-aligned data. In: Proceedings of the European conference on computer vision (ECCV), pp. 768–783 (2018)
24. McMahan, B., Moore, E., Ramage, D., Hampson, S., y Arcas, B.A.: Communication-efficient learning of deep networks from decentralized data. In: Artificial intelligence and statistics, pp. 1273–1282. PMLR (2017)
25. Zhao, Y., Li, M., Lai, L., Suda, N., Civin, D., Chandra, V.: Federated learning with non-iid data. arXiv preprint arXiv:1806.00582 (2018)
26. Hernandez-Boussard, T., Bozkurt, S., Ioannidis, J.P., Shah, N.H.: MINIMAR (MINimum Information for Medical AI Reporting): developing reporting standards for artificial intelligence in health care. J. Am. Med. Inf. Assoc. **27**, 2011–2015 (2020)
27. Bergmann, P., Fauser, M., Sattlegger, D., Steger, C.: MVTec AD–a comprehensive real-world dataset for unsupervised anomaly detection. In: Proceedings of the IEEE/CVF Conference on Computer Vision and Pattern Recognition, pp. 9592–9600 (2019)
28. Wang, X., Peng, Y., Lu, L., Lu, Z., Bagheri, M., Summers, R.M.: Chestx-ray8: hospital-scale chest x-ray database and benchmarks on weakly-supervised classification and localization of common thorax diseases. In: Proceedings of the IEEE Conference on Computer Vision and Pattern Recognition, pp. 2097–2106 (2017)
29. Deng, J., Dong, W., Socher, R., Li, L.-J., Li, K., Fei-Fei, L.: Imagenet: a large-scale hierarchical image database. In: 2009 IEEE Conference on Computer Vision and Pattern Recognition, pp. 248–255. IEEE (2009)

30. Oikarinen, T., Zhang, W., Megretski, A., Daniel, L., Weng, T.-W.: Robust deep reinforcement learning through adversarial loss. Adv. Neural. Inf. Process. Syst. **34**, 26156–26167 (2021)
31. Isola, P., Zhu, J.-Y., Zhou, T., Efros, A.A.: Image-to-image translation with conditional adversarial networks. In: Proceedings of the IEEE Conference on Computer Vision and Pattern Recognition, pp. 1125–1134 (2017)
32. Pillai, S., Vadakkepat, P.: Two stage deep learning for prognostics using multi-loss encoder and convolutional composite features. Expert Syst. Appl. **171**, 114569 (2021)
33. Salimans, T., Goodfellow, I., Zaremba, W., Cheung, V., Radford, A., Chen, X.: Improved techniques for training gans. Adv. Neural Inf. Process. Syst. **29** (2016)

Artificial Intelligence in Medicine: Enhancing Pneumonia Detection Using Wavelet Transform

Mekhriddin Rakhimov, Jakhongir Karimberdiyev[✉], and Shakhzod Javliev

Tashkent University of Information Technologies, Tashkent, Uzbekistan
jahongirkarimberdiyev618@gmail.com

Abstract. This research introduces a deep learning technique for classifying lung images using wavelet transform-based feature extraction in TensorFlow. The accurate and automated analysis of lung images is vital for diagnosing and treating lung diseases. In this study, we present a new approach that combines wavelet transform-based feature extraction with a convolutional neural network (CNN) to achieve precise lung image classification. The proposed approach involves several steps. First, the lung images undergo preprocessing to eliminate noise and enhance contrast. Next, the images are decomposed into various frequency subbands using wavelet transform. The resulting wavelet coefficients serve as features for the classification process. Additionally, we utilize our custom CNN architecture as a classifier in TensorFlow to categorize lung images as either normal or pneumonia. To assess the effectiveness of the proposed method, we utilized a dataset containing 5216 lung images. Experimental results demonstrate that the proposed approach achieves an impressive classification accuracy of 96.9% for lung images. Furthermore, our method outperforms other state-of-the-art techniques in the field of lung image classification.

Keywords: Artificial Intelligence · Medicine · Deep Learning · CNN classifier · Pneumonia classification · Wavelet transform

1 Introduction

AI can be used to analyze vast amounts of medical data, including electronic health records, medical images, genomic data, and clinical trial results, to uncover patterns and insights that would be difficult or impossible for humans to identify. This can lead to earlier and more accurate diagnoses, more personalized treatments, and better patient outcomes [1]. In medical imaging, AI can be used to improve the accuracy and speed of diagnosis. The speed of diagnosis in various fields, such as medical diagnosis or computer diagnostics, can be influenced by several factors, including the hardware being used [2]. The processing power of the hardware, typically measured in terms of the CPU (Central Processing Unit) or GPU (Graphics Processing Unit), can significantly affect diagnosis speed [3–5].

Judging by the numbers of recent years, we can see that the market value of the medical system is growing year by year. We can see this in Fig. 1 [6]. Despite the

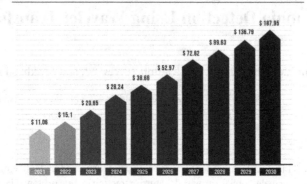

Fig. 1. Artificial intelligence in healthcare market size, 2021–2030

many potential benefits of AI in medicine, there are also challenges and concerns to be addressed. These include issues related to data privacy and security, the need for transparency and interpretability in AI systems, and the potential for AI to perpetuate biases and inequalities in healthcare. With careful attention to these issues, however, AI has the potential to transform medicine and improve the lives of patients around the world [7]. Several studies have demonstrated the effectiveness of deep learning technologies for the detection of pneumonia [8]. For example, a recent study used a CNN to analyze chest X-rays and accurately classify them as normal or indicative of pneumonia with an accuracy of over 95%. Another study used a deep learning algorithm to analyze CT (computed tomography) scans and accurately identify pneumonia with a sensitivity of over 95%.

Medical image classification is a critical task in the diagnosis and treatment of various diseases [9]. Wavelet transform is a powerful tool for feature extraction in image analysis [10]. It involves decomposing an image into different frequency sub-bands, which can capture both the spatial and frequency information of the image. Wavelet transform has been used in various medical image analysis applications, including image denoising, segmentation, and classification. Wavelet transform is a powerful tool for feature extraction in image analysis, which can capture both the spatial and frequency information of an image. By applying wavelet transform to chest X-rays, features can be extracted that are indicative of pneumonia [11]. These features can then be used to train a machine-learning model to classify new chest X-rays as either normal or indicative of pneumonia. The main purpose of this research is to help radiologists in accurate and efficient diagnosis of lung diseases by feature extraction based on wavelet transform combined with deep learning methods in TensorFlow using CNN architecture.

2 Related Works

A. Classification of medical images for lung cancer using a hybrid CNN-SVM approach.
    The use of computer-aided detection (CAD) to detect lung cancer early is a promising area of research. A recent study proposed a hybrid CNN-SVM algorithm that can

classify CT images of the lung and accurately detect cancer cells. CNNs are effective in automatically extracting relevant features from images, while SVMs are good at handling high-dimensional data [12].

B. *Detecting pneumonia in chest X-ray images using convolutional neural networks (CNNs) and transfer learning.*

Convolutional Neural Networks (CNNs) are neural network architectures inspired by the visual cortex of the brain. They excel at solving complex pattern recognition tasks, particularly in image classification, by recognizing both linear and non-linear patterns. CNNs are advantageous for image classification due to their ability to achieve high performance while requiring fewer parameters and connections compared to other neural networks. This characteristic makes training CNNs relatively easier.

C. *A segmentation and classification system for bone fracture detection using wavelet transform and neural networks.*

This paper focuses on detecting fractures in human bones using X-ray images and image processing techniques. The authors describe a method that combines wavelet-based segmentation and neural network-based classification to locate and identify fractures in medical images, specifically X-ray images.

3   Dataset Description/Methodology

In this paper, we have used a dataset consisting of lung re-inigration images. Some images from this dataset are shown in Fig. 2. We prepared the model for training by dividing the used data set into training, test and validation sets.[15]. In the examination of chest X-ray images, an initial screening was conducted to ensure the quality of the scans. This involved the removal of all scans that were deemed to be of low quality or unreadable. After this quality control process, a total of 5,126 high-quality images were retained in the dataset, which were then categorized into two groups: Pneumonia and Normal.

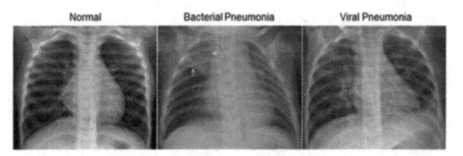

Fig. 2. Examples of Chest X-Rays depicting patients with Pneumonia.

4 Experiments and Results

To assess the effectiveness of the pneumonia classification system that utilizes wavelet transform in deep learning, a series of experiments were conducted using a publicly accessible dataset of chest X-ray images. The dataset comprises a total of 5,216 chest X-ray images, consisting of 1,341 normal images and 3,875 images showing signs of pneumonia.

Training and Testing
In our experimentation, we employed a convolutional neural network (CNN) based on a deep learning framework to perform classification of the chest X-ray images, distinguishing between normal cases and those indicating the presence of pneumonia [16]. To extract relevant features from the images, we utilized wavelet transform, which was subsequently utilized as input for the CNN during the classification process.

The CNN was trained using a training set comprising 4,500 images. The performance of the trained model was assessed by validating it on a separate validation set consisting of 500 images. Furthermore, the evaluation of the classification system was conducted on a testing set comprising 216 images.

Figure 3 illustrates the architecture of the CNN model utilized in this study. The model employed in this study was adapted from the ResNet-18 architecture, with slight modifications to accommodate input images of dimensions $460 \times 460 \times 3$. It incorporates four convolutional layers that function as feature extractors, utilizing kernel sizes of 3×3. Subsequently, each convolutional layer is followed by a 2×2 subsampling layer. The resulting output image dimensions from these layers are $13 \times 13 \times 16$, which are then converted into a 1×1024 dimension for feedforwarding to the subsequent hidden layer.

Wavelet transform was employed to extract features from the chest X-ray images in our study. More specifically, we utilized the discrete wavelet transform (DWT) on each image, which generated four sub-bands representing different frequency scales. These sub-bands were subsequently used as input for training the CNN.

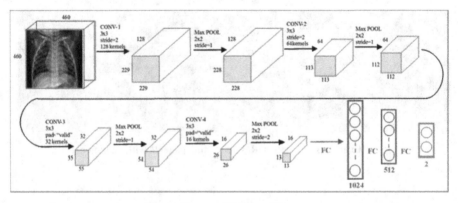

Fig. 3. Proposed CNN architecture.

During the training process, we utilized the Adam optimizer with a learning rate of 0.001 and a batch size of 32. The training was executed for a total of 50 epochs, and early stopping was implemented based on the validation loss to prevent overfitting.

The CNN architecture proposed in this study comprised four convolution- maxpool layers. After running the model for 50 epochs, it achieved an accuracy of 97.3% on the training set and 96.9% on the test set. Figure 4 illustrates the accuracy and loss curves for the training epochs, specifically focusing on the 50th epoch throughout the training process.

In addition, we conducted an ablation study to assess the impact of wavelet transform on the performance of the CNN architecture. We compared the performance of the model with and without wavelet transform using the same CNN architecture. The results of this comparison are presented in Fig. 5.

Testing:
To gain insights into the most significant regions of the chest X-ray images for pneumonia classification, we visualized the activation maps for the final convolutional layer of the CNN. This visualization provided a better understanding of the specific areas that played a crucial role in the classification process. Further details regarding this visualization can be found in reference [17].

Based on our experimental findings, our proposed lung classification system achieved impressive results on the testing set, with an accuracy of 96.9%. The precision, recall, and F1 score were also high, with values of 98%, 98%, and 98.5% respectively. These outcomes emphasize the efficacy of incorporating wavelet transform within deep learning for accurately classifying normal and pneumonia lungs in chest X-ray images.

The next step involves constructing a confusion matrix, which is a tabular summary of the number of correct and incorrect predictions made by a classifier. It provides a useful tool for evaluating the performance of a classification model by calculating performance metrics such as accuracy, precision, recall, and F1-score (Table 1).

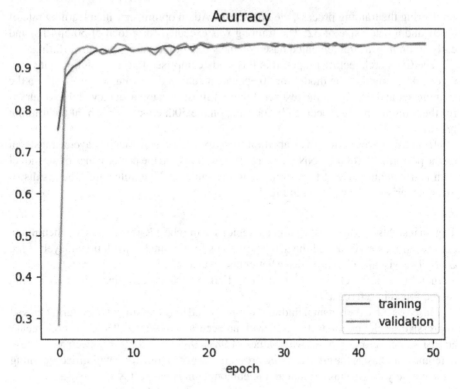

Fig. 4. Model accuracy curves with 50 epochs with wavelet transform.

Table 1. Classification report.

Classes	Precision	Recall	F1-score	Accuracy
Normal/Pneumonia	0.96	0.93	0.94	0.97

Accuracy is one of the most commonly used metrics for classification, and it is calculated using a formula based on the confusion matrix (1):

$$\text{Accuracy} = (TN + TP)/(TN + FP + FN + TP) \tag{1}$$

The formula for calculating F1-score is given below (2):

$$\text{F1-score} = (2 * \text{Precision} * \text{Recall})/(\text{Precision} + \text{Recall}) \tag{2}$$

It is crucial to note that the F1 score is computed based on the predicted classes rather than prediction scores.

The proposed method has the potential to assist radiologists in the accurate and efficient diagnosis of lung diseases. The results of this study suggest that wavelet transform-based feature extraction combined with deep learning techniques in TensorFlow can be a powerful tool for medical image analysis and classification.

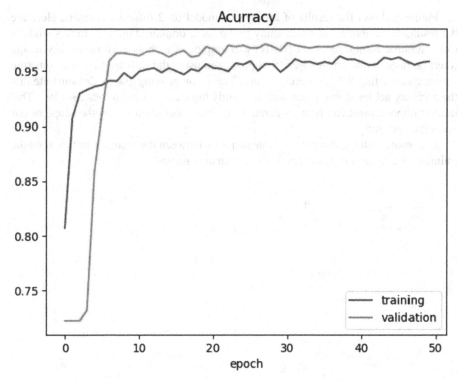

Fig. 5. Model accuracy curves with 50 epochs without wavelet transform.

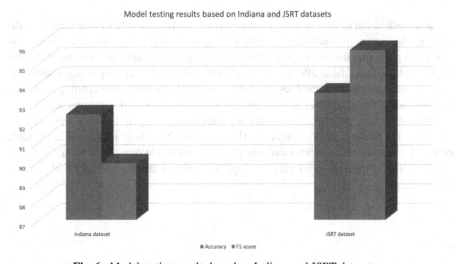

Fig. 6. Model testing results based on Indiana and JSRT datasets.

Figure 6 shows the results of testing our model on 2 different datasets. Here are the results from another research study in the same domain. Figure 7. In this study, a Convolutional Neural Network (CNN) was employed without performing any image filtering. Instead, the images were first processed using the wavelet transform function before constructing the CNN architecture. The accompanying graphs demonstrate that the accuracy achieved in our research is slightly higher compared to other studies. This modest improvement can be considered as a minor innovation within the scope of our scientific research.

In a separate study, there is a notable disparity between the accuracy achieved on the training set and the validation set. This discrepancy serves.

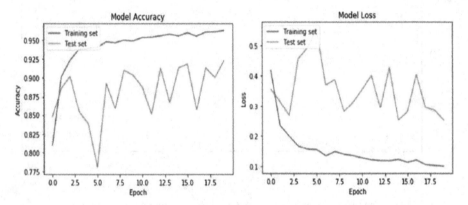

Fig. 7. Accuracy graph of another research paper.

as an indication of the presence of overfitting. Specifically, when the accuracy on the training data is exceptionally high while the accuracy on the validation data is notably low, it suggests that the model is overfitting. In our own research, we have taken careful measures to address the challenges of overfitting and underfitting. By being attentive to these issues, we have obtained results that mitigate the problem of overfitting, ensuring that the model's performance is not excessively biased towards the training data.

In this section, we have conducted a comparative analysis between our research work and the findings of other studies. The outcomes have been presented in the table below. The results demonstrate that our proposed methodology for pneumonia image classification is effective and yields a high F1 score, indicating strong performance (Table 2).

Table 2. Comparison table with the results of other similar studies

Method	Precision	Recall	F1-score	Accuracy
CNN with transfer learning [18]	0.90	0.91	NA	0.91
Deep learning-based classification [19]	0.93	0.92	0.94	0.93
Pneumonia diagnosis using multi-scale convolutional neural networks [20]	0.96	0.96	0.92	0.95
Automated detection of pneumonia using an ensemble of deep learning models [21]	0.94	0.95	0.89	0.90

5 Future Research Work

While our proposed lung classification system using wavelet transform in deep learning achieved promising results, there is still room for further research and improvement. The following are some potential areas for future research: Multimodal data integration: Our current classification system only uses chest X-ray images as input. However, the integration of other types of medical imaging data, such as CT scans or MRI, may improve the accuracy and reliability of lung disease diagnosis.

Real-time diagnosis: Our proposed classification system is designed for offline diagnosis of lung diseases. However, the development of real-time diagnosis systems could enable faster and more efficient diagnosis of lung diseases, which could be particularly useful in emergency situations.

6 Conclusion

In our study, we introduced a lung classification system that utilizes wavelet transform in deep learning to diagnose pneumonia and normal lungs in chest X-ray images. The experimental results we obtained validated the effectiveness of employing wavelet transform-based feature extraction, as it significantly enhanced the accuracy and efficiency of lung classification through deep learning techniques.

The proposed classification system has the potential to assist clinicians in the diagnosis and management of lung diseases, leading to better patient outcomes and reduced healthcare costs. The system could be integrated with existing medical imaging systems to provide automated and accurate diagnosis of lung diseases, reducing the workload of healthcare professionals and improving the efficiency of the diagnostic process.

In conclusion, the proposed lung classification system using wavelet transform in deep learning represents a significant step towards more accurate and efficient diagnosis of lung diseases. Continued research in this area has the potential to have a significant impact on healthcare and improve patient outcomes.

References

1. Yeasmin, S.: Benefits of artificial intelligence in medicine. In: 2019 2nd International Conference on Computer Applications & Information Security (ICCAIS), Riyadh, Saudi Arabia, pp. 1–6 (2019). https://doi.org/10.1109/CAIS.2019.8769557
2. Rakhimov, M., Akhmadjonov, R., Javliev, S.: Artificial intelligence in medicine for chronic disease classification using machine learning. In: 2022 IEEE 16th International Conference on Application of Information and Communication Technologies (AICT), Washington DC, DC, USA, pp. 1–6 (2022). https://doi.org/10.1109/AICT55583.2022.10013587
3. Abdusalomov, A.B., Safarov, F., Rakhimov, M., Turaev, B., Whangbo, T.K.: Improved feature parameter extraction from speech signals using machine learning algorithm. Sensors. **22**(21), 8122 (2022)
4. Rakhimov, M., Mamadjanov, D., Mukhiddinov, A.: A high-performance parallel approach to image processing in distributed computing. In: 2020 IEEE 14th International Conference on Application of Information and Communication Technologies (AICT), pp. 1–5 (2020)
5. Musaev, M., Rakhimov, M.: A method of mapping a block of main memory to cache in parallel processing of the speech signal. In: 2019 International Conference on Information Science and Communications Technologies (ICISCT), pp. 1–4 (2019)
6. Artificial intelligence-market. https://www.precedenceresearch.com/artificial-intelligence-market
7. Montalico, B., Herrera, J.C.: Classification and detection of pneumonia in X-Ray Images using deep learning techniques. In: 2022 IEEE Sixth Ecuador Technical Chapters Meeting (ETCM), Quito, Ecuador, pp. 01–05 (2022). https://doi.org/10.1109/ETCM56276.2022.9935757
8. Abdullaeva, M.I., Juraev, D.B., Ochilov, M.M., Rakhimov, M.F: Uzbek Speech synthesis using deep learning algorithms. In: Zaynidinov. H., Singh, M., Tiwary, U.S., Singh, D. (eds.) Intelligent Human Computer Interaction. IHCI 2022. Lecture Notes in Computer Science, vol. 13741. Springer, Cham (2022). https://doi.org/10.1007/978-3-031-27199-1_5
9. Ayyannan, M., Mohanarathinam, A., Sathya, D., Nithya, T., Tamilnidhi, M., Kumar, N.S.: Medicalimage classification using deep learning techniques: a review. In: 2023 Second International Conference on Electronics and Renewable Systems (ICEARS), Tuticorin, India, pp. 1327–1332 (2023). https://doi.org/10.1109/ICEARS56392.2023.10084948
10. Cui, Z., Zhang, G., Wu, J.: Medical image fusion based on wavelet transform and independent component analysis. In: 2009 International Joint Conference on Artificial Intelligence, Hainan, China, pp. 480–483 (2009). https://doi.org/10.1109/JCAI.2009.169
11. Kosasih, K., Abeyratne, U.R., Swarnkar, V., Triasih, R.: Wavelet augmented cough analysis for rapid childhood pneumonia diagnosis. IEEE Trans. Biomed. Eng. **62**(4), 1185–1194 (2015). https://doi.org/10.1109/TBME.2014.2381214
12. Saleh, A.Y., Chin, C.K., Penshie, V., Al-Absi, H.R.H.: Lung cancer medical images classification using hybrid CNN-SVM pneumonia detection in chest X-ray images using convolutional neural networks and transfer learning (2020). https://doi.org/10.1016/j.measurement.2020.108046
13. Bagaria, R., Wadhwani, S., Wadhwani, A.K.: A wavelet transform and neural network based segmentation & classification system for bone fracture detection. Optik **236**, 166687 (2021). https://doi.org/10.1016/j.ijleo.2021.166687
14. Singh, D., Singh, M., Hakimjon, Z.: Homogeneous Polynomial Splines for Digital Signal Systems. Springer, Heidelberg (2019). https://doi.org/10.1007/978-981-13-2239-6
15. Kermany, D.S., Goldbaum, M., Chai, W.: Identifying medical diagnoses and treatable diseases by image-based deep learning (2018). https://doi.org/10.1016/j.cell.2018.02.010

16. Musaev, M., Rakhimov, M.: Accelerated training for convolutional neural networks. In: 2020 International Conference on Information Science and Communications Technologies (ICISCT), pp. 1–5 (2020). https://doi.org/10.1109/ICISCT50599.2020.9351371

17. Urinbayev, K., Orazbek, Y., Nurambek, Y., Mirzakhmetov, A., Varol, H.A.: End-to-end deep diagnosis of X-ray images. In: 2020 42nd Annual International Conference of the IEEE Engineering in Medicine & Biology Society (EMBC), Montreal, QC, Canada, pp. 2182–2185 (2020). https://doi.org/10.1109/EMBC44109.2020.9175208

18. Salehi, A.W., et al.: A study of CNN and transfer learning in medical imaging: advantages, challenges, future scope. Sustainability **15**, 5930 (2023). https://doi.org/10.3390/su15075930

19. Ibrahim, A.U., Ozsoz, M., Serte, S., et al.: Pneumonia classification using deep learning from chest X-ray images during COVID-19. Cogn. Comput. (2021). https://doi.org/10.1007/s12559-020-09787-5

20. Yan, T., Wong, P.K., Ren, H., Wang, H., Wang, J., Li, Y.: Automatic distinction between COVID-19 and common pneumonia using multi-scale convolutional neural network on chest CT scans. Chaos Solitons Fractals **140**, 110153 (2020). https://doi.org/10.1016/j.chaos.2020.110153

21. Salehi, M., Mohammadi, R., Ghaffari, H., Sadighi, N., Reiazi, R.: Automated detection of pneumonia cases using deep transfer learning with paediatric chest X-ray images. Br. J. Radiol. **94**(1121), 20201263 (2021). https://doi.org/10.1259/bjr.20201263

Maximizing Accuracy in AI-Driven Pattern Detection in Cardiac Care

Ritu Chauhan[1] and Dhananjay Singh[2(✉)]

[1] Artificial Intelligence and IoT Lab, Centre for Computational Biology and Bioinformatics, Amity University, Noida, India
[2] ReSENSE Lab, School of Professional Studies, Saint Louis University, St. Louis, MO, USA
dan.usn@ieee.org

Abstract. Artificial Intelligence (AI) has laid down the platform where all the emerging fields can benefit for detection of patterns. In similar, evolving AI technology has revolutionized the traditional healthcare technology to other level of technological advancement. Hence, Machine learning paradigms are designed to transform the era of learning and retrieval of hidden patterns from the healthcare databases. The current scope of the work focuses on detecting patterns from car diac care database using machine learning. The database comprise of several features such as age, gender, smoking variable, blood pressure, diabetes, alcohol consumption, sleep variables and other features which can be the potential cause for the prognosis of the disease Further, the databases applicability is measured with different classifiers such K nearest neighbours (KNN), Adaboost, XGboost, Gradient Boost, Decision Tree Logistic Regression, and Random Forest Classifiers to determine the most relevant classifier for the prognosis of the disease. The results suggest that Xgboost works efficiently with higher accuracy rate as compared to other classifiers.

Keywords: Heart disease · Factors of heart diseases · Prediction · AI · Machine learning · Cardiovascular disease

1 Introduction

Artificial Intelligence (AI) is a cutting-edge technology which has overwhelmed researchers and scientists around the globe to obtain hidden and informative patterns from large scale databases. In similar apprehension, past studies have been indicative to ex ploit AI in numerous application domain to extract meaningful patterns and knowledge from the data [1]. Moreover, we can say that AI is a broad umbrella under which Machine Learning, Deep Learning, soft computing and other research area are broadly classified as per the end user and the domain requirement. So, the current technological experts suggest to apply Machine learning (ML) and other expertise to retrieve informatic patterns from the databases. In similar, the current scope of study is subjective on the implementation of machine learning algorithms and models which are capable of learning patterns, generating predictions and conclusions for healthcare databases.

B. J. Choi et al. (Eds.): IHCI 2023, LNCS 14531, pp. 176–187, 2024.
https://doi.org/10.1007/978-3-031-53827-8_17

In addition, Machine learning models can integrate with complex healthcare databases such as demographics, medical history, lifestyle factors, genetic information, Electronic Health records (EHR), laboratory databases and other which can be identified factors for personalized risk scores. These models are capable of identifying patients which can be at high risk for suffering from heart disease and facilitate targeted interventions and lifestyle modifications [2, 3]. Similarly, ML based algorithms can also analyze longitudinal data and detect subtle changes in physiological parameters that precede the onset of heart disease. Hence, the datasets play a critical role to ensure the significant pattern detection. Hence, we can say pre-processing is a perilous step that ensures data quality, handles missing values, normalizes variables, reduces dimensionality. By performing these preprocessing steps, it can enhance the reliability and valid ity of their findings, enabling them to draw meaningful conclusions and contribute to the scientific body of knowledge. In the current study of approach, pre-processing of data was conducted, which includes essential data transformations such as label encoding to convert categorical variables into numerical form. Subsequently, the dataset was divided into training and testing subsets using an 80:20 ratio, ensuring a representative distribution of data for model training and evaluation. Further, the datasets were modeled on six different classifiers for trained and the training data. The performance of each classifier was then evaluated using two widely recognized metrics: the confusion matrix and the ROC curve.

The objective of the current study of approach is to identify the best-performing classifier, one that exhibited both high accuracies according to the confusion matrix and a superior ROC curve. These evaluation techniques played a crucial role in assessing the classifiers' effectiveness and determining the most suitable model for the task at hand. Ultimately, the findings and conclusions drawn from this study contribute to the field of predictive modelling and provide valuable insights for future applications and improvements in the chosen task.

Further, the overall contribution of the paper is discussed as Sect. 2 discuss the past literature review conducted in the field of study, Sect. 3 elaborates on the methodology applied with description of databases and results are formulated in Sect. 4 and finally the conclusion is derived.

2 Literature Review

A literature review's objective is to identify gaps in current knowledge, emphasize the strengths and flaws of existing research, and establish the foundation for future study. Interest in using artificial intelligence (AI) approaches in healthcare, especially in the area of heart disease, has grown during the past several years. Studies have shown that AI systems, particularly deep learning models, are good in accurately detecting and classifying cardiac abnormalities in medical imaging data [4–6]. Moreover, AI has also shown in predicting the prognosis and outcomes of heart disease. By analyzing large datasets that include clinical and demographic information, imaging data, and biomarkers, machine learning models can identify important risk factors and generate predictive models for mortality, heart failure, and other adverse cardiac events. A model created by machine learning effectively identified cardiovascular risk variables showed the potential to enhance risk stratification in heart disease patients.

The use of AI in heart disease therapy optimization and personalized medicine is also gaining ground. AI algorithms can analyze patient-specific data, including genetic profiles, medical histories, and treatment responses, to generate personalized treatment recommendations. This enables clinicians to tailor interventions, medications, and dosages to individual patients, thereby optimizing outcomes and minimizing adverse effects [7–10]. The potential of AI in guiding treatment decisions for heart failure, improving patient outcomes, and reducing hospital readmissions [10–12]. By leveraging AI-driven treatment optimization, clinicians can enhance patient care and contribute to more efficient healthcare resource allocation. While the integration of AI in heart disease management holds great promise, several challenges need to be addressed. Ethical considerations, privacy concerns, and regulatory frameworks are crucial in ensuring the responsible use of AI in healthcare. Additionally, the interpretability and explain ability of AI algorithms remain areas of active research, as the adoption of AI in healthcare requires clinicians to trust and understand the decision-making processes of these models [13–15]. The existing literature indicates the potential of AI in transforming healthcare for heart disease. However, further research is needed to address technical, ethical, and regulatory challenges to ensure the safe and effective integration of AI in clinical practice [16–20].

3   Methodology

The integration of artificial intelligence (AI) in various domains has revolutionized the way we solve complex problems. Due to their widespread adoption in the area of research, AI technique's ability to analyze large datasets, uncover patterns, and generate insights has attracted healthcare practicioners and scientists around the globe. In, correspondence with same the current scope of work focuses on determine the features which can detected by applying the most suitable algorithm for the prognosis of the disease. This study investigates a range of machine learning methods, such as K nearest neighbours (KNN), Adaboost, Xgboost, Gradient Boost, Decision Tree Logistic Regression, and Random Forest Classifiers, to allow practitioners or medical analysts to effectively monitor the prognosis and diagnosis of cardiac disease. Further, we implemented the recommended model and measure its effectiveness and accuracy using a range of performance metrics.

3.1   Dataset Description

The datasets are archived from CDC where it consists of 410,958 rows and 279 columns for the patients suffering from heart disease. The database encompasses of several features such as age, gender, smoking variable, blood pressure, diabetes, alcohol consumption, sleep variables and other features which can be the potential cause of the prognosis of the disease [29]. The datasets covers of 52% female respondents and 48% male, in Age Category variable, which indicates the respondents' age, 79% of them fall into the "Other" category, while the remaining 21% are split between the 60–64 and 65–69 age ranges, 91.4% of them claimed not to have ever experienced myocardial infarction (MI) or coronary heart disease (CHD), while 8.6% have had these diseases which is responsible for various attributes such as BMI represents the respondents' Body Mass Index, which is an important aspect of the study, whereas smoking variable reflects 100 cigarettes have been smoked by the responder over their lives. According to this characteristic, 41% of respondents have smoked at least 100 cigarettes, as opposed to 59% who haven't. According to the alcohol consumption variable, which indicates this if it is true, a respondent is deemed a heavy drinker if they consume more than 14 drinks per week for adult men and more than 7 drinks per week for adult women. Whether respondents have ever been informed that they have suffered a stroke is tracked by the stroke variable. 96% of respondents have not experienced a stroke, but 4% of respondents have gotten this news, according to this variable.

3.2 Pre-processing for Cardiac Data

Data preprocessing is a critical step in determining hidden and knowledgeable patterns in vast scale databases. It performs a specific function in guaranteeing the data's quality, integrity, and usefulness for analysis. Researchers can acquire trustworthy and useful insights from their datasets by using techniques such as data cleansing, integration, transformation, feature selection, and anonymization. This results in more accurate and robust research outcomes [21–24].

3.3 Algorithm Utilized

K-Nearest Neighbors (KNN)
It is a well-liked supervised learning method that is utilized for applications in regression and classification. It makes no assumptions about the distribution of the underlying data because it is non-parametric. KNN is a straightforward technique that locates the k-nearest neighbors to a given data point in the training set using a distance measure. On the basis of the majority class of these neighbors, the forecast is then formed. A hyperparameter that can have a big impact on the algorithm's performance is the choice of k. Overfitting can happen when the k value is too small, whereas underfitting can happen when the k value is too big. KNN is a computationally expensive algorithm, especially when working with large datasets, because it necessitates figuring out how far the test point is from every other training point. Despite this limitation, KNN is widely used in many applications, including image recognition, recommendation systems, and medical diagnosis [25].

AdaBoost
A machine learning technique called AdaBoost (Adaptive Boosting) may be used to a variety of classification and regression problems. AdaBoost is a technique for enhancing the effectiveness of forecasting models. For working with complicated datasets that have high dimensionality, noise, and class imbalance, AdaBoost is especially helpful. The algorithm works by combining several weak classifiers to create a strong classifier that can accurately classify data. AdaBoost has been used in various fields of research, including medicine, finance, and social sciences, for example, in medical research, AdaBoost has been used to predict cancer recurrence and identify genes that are associated with specific diseases. In finance, AdaBoost has been used to predict stock prices and detect fraud. It is used in both binary and multi-class classification problems. AdaBoost is a versatile approach that may also be used with many weak classifiers, such as decision

trees, support vector machines, and logistic regression. In Eq. (1) H(x) in this formula stands for the greatest strong classifier AdaBoost created. It is built as a weighted cumulative of the predictions from many weak classifiers, with the contributions of each weak classifier being weighted by 1, 2, …T. The weighted total is transformed into a binary prediction, +1 or −1, which corresponds to the two class labels in a dichotomy issue, using the sign() function. The summing over all T weakly classifiers in the ensemble is denoted by the summation symbol, (i = 1)T. These weak classifiers, abbreviated as h1(x), h2(x) and hT(x), are frequently uncomplicated models that generate linear forecasts in accordance with simple decision-making procedures. The weights allocated to each weak classifier are represented by the 1, 2, …, T values, which AdaBoost assigns based on the performance of the individual weak classifiers throughout the training phase.

$$H(x) = sign(\sum\nolimits^{T} \alpha_1 \, h_1 \, (x)) \tag{1}$$

XGBoost

XGBoost (Xtreme Gradient Boosting) is the well-known machine learning model for retrieval of information and knowledge from large scale databases. The XGBoost algorithm is based on the boosting method which identifies the output in decision tree where each constantly fix the bugs in the older terminology. The XGBoost has overlaid its performance in varied application domains to retrieve the effective and efficient patterns from large scale databases. However, in the current study of approach the XGBoost algorithm is applied to measure the overall accuracy for the cardiac care patient. The algorithm efficiently determines the overall prediction and confer the healthcare practioners for the future decision making. In addition, the XGBoost algorithm can be discussed in the statistical terminology such as in Eq. 2 The inaugural component, the Loss Term, computes the difference between the actual target values (yi) and the model's predicted values (yi-hat) for each training sample (i). This component assesses how closely the model's predictions match the actual outcomes, serving as a predictability metric. The Normalization Term, the second phase, offers a regularization technique by averaging over K (the quantity of weak learners or trees). Each (fk) word quantifies the complexity of a single tree. It penalizes too complex trees, which is critical for reducing overfitting and providing the model synthesizes well to new data.

$$\text{Objective} = \sum\nolimits_{i=1}^{n} \left[loss \, (yi, y) + \sum\nolimits_{k=1}^{K} \Omega \, (f_k) \right] \tag{2}$$

Gradient Boost

Due to its capacity to increase prediction accuracy in a variety of applications, including image identification, audio recognition, and natural language processing, gradient boosting is a machine-learning technique that has attracted a lot of interest [26–28]. The algorithm works by iteratively building decision trees and adding them to an ensemble to minimize the error of the model. Gradient Boosting has several advantages, such as handling missing data, incorporating categorical variables, and handling outliers. Additionally, the algorithm provides several hyperparameters that allow researchers to customize the algorithm to their specific requirements. Gradient Boosting has been applied in various fields, such as bioinformatics, finance, and marketing. Moreover, several studies have compared Gradient Boosting to other machine learning algorithms and found it to be superior in terms of accuracy and performance. Hence, Gradient Boost can be discussed in Eq. (3) where F(x) represents the predicted output for a given input x, M signifies total number of weak learners or base learners used in the gradient boosting model. γm is calculated as weight assigned to the prediction made by the mth weak learner.hm(x): The prediction made by the mth weak learner for the input x.

$$F(x) = \sum_{n}^{M} \gamma_m h_m(x) \tag{3}$$

Decision Tree

Decision trees are widely applied in several applications domain for making decision and gaining knowledge. In similar, its approach can be applied in several research methodology for analyses and predicting outcomes in various fields, such as finance, medicine, and engineering. We can say that, decision tree is a supervised machine learning approach which builds a tree like structure based on the target class. Hence, the algorithm splits the data into subgroups based on the values of the input variables, then iterates over these subsets until a stopping condition is met. Decision trees provide a number of benefits, including being simple to understand and depict, managing both continuous and categorical data, and offering insights into the most crucial factors. Moreover, decision trees may be used with other machine learning techniques like gradient boosting and random forests to achieve even greater performance. Also, they have been employed in a variety of research investigations, including ones that forecast the risk that a disease would develop, identifying the risk factors for a specific outcome, and detecting anomalies in data. In totality, decision trees tend to be a valuable tool in research methodology which can help researchers to make accurate predictions and gain insights within the data.

Random Forest

In research methodology, Random Forest may be used to assess and predict outcomes in a range of fields, including finance, healthcare, and environmental sciences. Several decision trees are built as part of an ensemble learning technique called Random Forest, and their combined findings are utilized to generate predictions. For each randomly selected subset of the data and attributes, the application builds a decision tree. The average of all the trees' forecasts is used to determine the outcome. Many benefits of Random Forest include its high accuracy, ability to handle both categorical and continuous data, and ability to reveal information about the most crucial factors. Moreover, Random Forest may be used with other machine learning techniques like gradient boosting and deep learning to enhance their performance even more. In several scientific projects, Random Forest has been used to estimate the likelihood of contracting an illness, pinpoint the causes of climate change, and spot data errors. In conclusion, Random Forest is a useful research technique tool that may assist researchers in making precise predictions and learning from their data [27].

4 Results

The correlation coefficients between various variables in a dataset are displayed in Fig. 1. In a correlation matrix diagram, the correlation coefficients are represented as colors or shades, lighter colors denoting weaker or negative correlations and darker colors denoting greater positive correlations. The diagonal of the matrix is usually represented by a histogram or density plot of each variable. Interpreting a correlation matrix plot involves looking for patterns in the colors or shades of the matrix. Variables that are positively correlated will have darker colors or shades, while those that are negatively correlated will have lighter or negative shades. Variables that are not correlated will be represented by neutral colors or white. It is also important to pay attention to the strength of the correlations, as indicated by the color intensity. Positive or negative, stronger correlations are more likely to have practical or predictive relevance, whereas weaker correlations might not be as significant. The size and direction of a linear relationship between two variables are shown in the Correlation matrix, which may take values between 1 and $+1$. The feature of the correlation matrix reveals the correlation between the coefficients. The association between a certain random variable and each of its values is known as correlation. The correlation matrix is efficiently displayed as a heat map while characteristics are being correlated.

An effective technique for finding patterns and links in a dataset, a correlation matrix display can direct additional research and modelling. It's important to remember that correlation does not necessarily imply causation and that other variables might be affecting the correlations that the data shows.

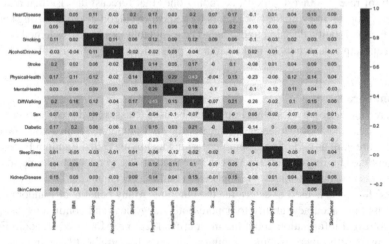

Fig. 1. Correlation Matrix (Heat-map) between attributes

4.1 Model Building

To predict the probability of an individual developing heart disease, we employed six different machine learning classifiers: K-Nearest Neighbors (KNN), AdaBoost, XGBoost,

Random Forest, Gradient Boosting, and Decision Tree. The dataset used had a dependent variable labeled as "Heart Disease", while the other features were considered independent variables. We randomly split the dataset into training and testing subsets, with an 80:20 ratio for the train and test data, respectively. Both classification and regression models were trained on the training data and evaluated using the test data.

For each algorithm, the dataset was divided into two parts, and we applied the necessary data transformation and model fitting codes. During the training phase, the models learned from the feature readings and identified how different variables influenced the outcome. This allowed the models to develop their understanding and prepared them for making predictions.

Following the training phase, we tested the models using the dedicated testing data to assess the accuracy of their predictions. Although we knew the actual outcomes of the test data, the primary goal was to evaluate the models' predictive capabilities. By comparing the models' predicted results with the actual outcomes of the test data, we determined the accuracy of each prediction model.

We repeated these steps individually for each of the six algorithms (KNN, AdaBoost, XGBoost, Random Forest, Gradient Boosting, and Decision Tree). The resulting accuracy scores were recorded for further analysis, providing valuable insights into the performance of each model in accurately predicting the likelihood of developing heart disease. In, Fig. 2 confusion matrix for all models are discussed elaborately.

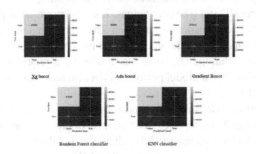

Fig. 2. Confusion Matrix for each model

4.2 Model Validation

To validate the model accuracy, precision and recall have been calculated. We used six different classifiers KNN, AdaBoost, XGBoost, Random Forest, Gradient Boosting, and Decision Tree, AdaBoost with 91.5% is the best-performing model followed by KNN with 90.7%. To further validate we used a confusion matrix and ROC-AUC curve (Table 1).

Table 1. Comparison of accuracy, precision, recall and F1-score between all models

Model	Accuracy	Precision	Recall	F1-score
KNN	90.7	90.7	90.7	90.7
AdaBoost	91.5	91.5	91.5	91.5
XGBoost	89.7	87.7	87.6	87.6
Gradient Boosting	88.6	88	88.4	88.1
Decision Tree	86.6	86.9	85	85.9

The KNN model achieved an accuracy, precision, recall, and F1-score of 90.7%. It shows consistent performance across all metrics, indicating that the model is well-balanced in terms of precision and recall. The AdaBoost model achieved the highest scores across all metrics, with accuracy, precision, recall, and an F1 score of 91.5%. It indicates that the model performed well in terms of correctly classifying instances and avoiding false positives and false negatives. The XGBoost model achieved slightly lower scores than AdaBoost, with an accuracy of 89.7% and slightly lower precision, recall, and F1-score values. It indicates that the XGBoost model has good overall performance but may have a slightly higher number of false positives and false negatives compared to AdaBoost. The Gradient Boosting model achieved an accuracy of 88.6% with relatively balanced precision, recall, and F1-score values. It shows reasonable performance but is slightly lower than the top-performing models. The Decision Tree model achieved an accuracy of 86.6% with comparable precision and F1-score values, but a slightly lower recall value. It indicates that the Decision Tree model has a relatively higher number of false negatives. The AdaBoost model performed the best based on the provided metrics, followed closely by KNN.

The analysis for the model summary is performed with confusion Matrix, F measure and Receiver Operating Characteristics (ROC) curve. In Fig. 3, a comprehensive analysis of ROC curve is conducted for each model such as KNN, AdaBoost, XGBoost, Random Forest, Gradient Boosting, and Decision Tree. The ROC curve tends be the well-known technique to evaluate the performance of each model using the true positive rate and false positive rate, where the true positive rate corresponds the specificity of each value and the overall classification of true data values. Whereas, the false positive rate is predicted with the negative predicted in the positive class, hence it is calculated as (1-specificity). So, in the proposed study of approach we have evaluated ROC for validation of values retrieved. Further, Area Under the curve (AUC) is used as measure to retrieve the best classifier model.

In Fig. 3, overall, ROC curve for each model is classified where X axis contributes to False positive rate and Y as True positive rate. The classifier Adaboost has received the maximum AUC which means it is the best classifier applied with the accuracy of 55% whereas, XGboost and decision tree represent second highest accuracy and then the Random Forest and the gradient Boost and lastly the decision tree has represented the lowest accurate model.

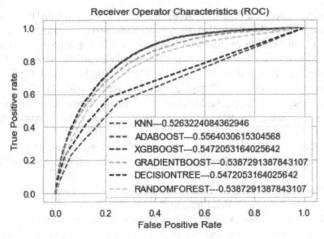

Fig. 3. ROC Curve

5 Conclusion

The Artificial Intelligence has proven to be automatic assistive technology to detect patterns and knowledge from the large-scale databases. The current scope of research focuses on the integrated approach of AI with healthcare databases to discover patterns and information which can benefit healthcare practioners and scientists for the diagnosis and the prognosis of the disease. The aim of the study is to validate the most efficient model which can be applied with heart care databases to potentially detect the features which may or may not responsible for the disease prediction. To validate the above study we have applied models, such as KNN, AdaBoost, XGBoost, Gradient Boost, Decision Tree, and Random Forest. The results retrieved were based on confusion matrix, F-measure and AUC however, the AUC for the models were evaluated to have the performance between 0.54 to 0.56.

References

1. Chen, M., Hao, Y., Zhang, N.: Hospital admission prediction based on healthcare information system data. J. Biomed. Inform. **45**(5), 905–911 (2012)
2. Khera, A.V., et al.: Genome-wide polygenic scores for common diseases identify individuals with risk equivalent to monogenic mutations. Nat. Genet. **50**(9), 1219–1224 (2018)
3. Musunuru, K., et al.: Basic concepts and potential applications of genetics and genomics for cardiovascular and stroke clinicians: a scientific statement from the American Heart Association. Circ. Genom. Precis. Med. **12**(11), e000046 (2019)
4. Bibault, J.E., Giraud, P., Burgun, A., Big Data and Machine Learning in Radiation Oncology Collaboration (B-DaMIC): Big data and machine learning in radiation oncology: state of the art and future prospects. Cancer Lett. **471**, 1–8 (2019)
5. Zhang, Z., Chen, J., Ma, G., Yang, Y., Wu, Y.: Predicting the onset of acute myocardial infarction with a machine learning model using population-based medical databases. Front. Physiol. **10**, 130 (2019)

6. Obermeyer, Z., Powers, B., Vogeli, C., Mullainathan, S.: Dissecting racial bias in an algorithm used to manage the health of populations. Science **366**(6464), 447–453 (2019)

7. Fonseca, C.G., et al.: The cardiac atlas project—an imaging database for computational modeling and statistical atlases of the heart. Bioinformatics **36**(16), 4160–4167 (2020)

8. Maurovich-Horvat, P., Ferencik, M., Voros, S., Merkely, B., Hoffmann, U.: Comprehensive plaque assessment by coronary CT angiography. Nat. Rev. Cardiol. **16**(12), 723–737 (2019)

9. Hazra, A., Mandal, S., Gupta, A., Mukherjee, A.: Heart disease diagnosis and prediction using machine learning and data mining techniques: a review. Adv. Comput. Sci. Technol. **10**, 2137–2159 (2017)

10. Varma, G., Chauhan, R., Singh, D.: Sarve: synthetic data and local differential privacy for private frequency estimation. Cybersecurity **5**, 26 (2022). https://doi.org/10.1186/s42400-022-00129-6

11. Patel, J., Upadhyay, P., Patel, D.: Heart disease prediction using machine learning and data mining technique. J. Comput. Sci. Electron. **7**, 129–137 (2016)

12. Zinat Motlagh, S.F., Chaman, R., Ghafari, S.R., Parisay, Z., Golabi, M.R., Eslami, A.A., et al.: Knowledge, treatment, control, and risk factors for hypertension among adults in Southern Iran. Int. J. Hypertens. **2015** (2015)

13. Noncommunicable diseases Fact Sheet. World Health Organization (2021). https://www.who.int/news-room/fact-sheets/detail/noncommunicable-diseases

14. Mittal, B.V., Singh, A.K.: Hypertension in the developing world: challenges and opportunities. Am. J. Kidney Dis. **55**(3), 590–598 (2010)

15. Gaziano, T.A., Bitton, A., Anand, S., Abrahams-Gessel, S., Murphy, A.: Growing epidemic of coronary heart disease in low-and middle-income countries. Curr. Probl. Cardiol. **35**(2), 72–115 (2010)

16. Buettner, R., Schunter, M.: Efficient machine learning based detection of heart disease. In: 2019 IEEE International Conference on E-Health Networking, Application & Services (HealthCom). IEEE (2019)

17. Roth, G.A., Mensah, G.A., Johnson, C.O., Addolorato, G., Ammirati, E., Baddour, L.M., et al.: Global burden of cardiovascular diseases and risk factors, 1990–2019: update from the GBD 2019 study. J. Am. Coll. Cardiol. **76**(25), 2982–3021 (2020)

18. Hemingway, H., Langenberg, C., Damant, J., Frost, C., Pyörälä, K., Barrett-Connor, E.: Prevalence of angina in women versus men: a systematic review and meta-analysis of international variations across 31 countries. Circulation **117**(12), 1526–1536 (2008)

19. Stanaway, J.D., Afshin, A., Gakidou, E., Lim, S.S., Abate, D., Abate, K.H., et al.: Global, regional, and national comparative risk assessment of 84 behavioural, environmental and occupational, and metabolic risks or clusters of risks for 195 countries and territories, 1990–2017: a systematic analysis for the Global Burden of Disease Study 2017. The Lancet **392**(10159), 1923–1994 (2018)

20. Cardiovascular disease mortality in the developing countries. World Health Statist Quart, vol. 46, pp. 89–150 (1993)

21. Yadav, D.P., Sharma, A., Singh, M., Goyal, A.: Feature extraction based machine learning for human burn diagnosis from burn images. IEEE J. Transl. Eng. Health Med. **7**, 1–7 (2019). Art no. 1800507. https://doi.org/10.1109/JTEHM.2019.2923628

22. Kopp, W.: How western diet and lifestyle drive the pandemic of obesity and civilization diseases. Diabetes Metab. Syndr. Obesity Targets Ther. **12**, 2221 (2019)

23. Lalkhen, H., Mash, R.: Multimorbidity in non-communicable diseases in South African primary healthcare. S. Afr. Med. J. **105**(2), 134 (2015)

24. Hajar, R.: Risk factors for coronary artery disease: historical perspectives. Heart Views Official J. Gulf Heart Assoc. **18**(3), 109–114 (2017). https://doi.org/10.4103/HEARTVIEWS.HEARTVIEWS_106_17

25. Kandaswamy, E., Zuo, L.: Recent advances in treatment of coronary artery disease: role of science and technology. Int. J. Mol. Sci. **19**(2), 424 (2018). https://doi.org/10.3390/ijms19 020424

26. Shahwan-Akl, L.: Cardiovascular disease risk factors among adult Australian-Lebanese in Melbourne. Int. J. Res. Nurs. **1**, 1–7 (2010)

27. Helma, C., Gottmann, E., Kramer, S.: Knowledge discovery and data mining in toxicology. Stat. Methods Med. Res. **9**, 329–358 (2000)

28. Kiyong, N., Heon Gyu, L., Keun Ho, R.: Data mining approach for diagnosing heart dis- ease. Korea Research Institute of Standards and Science, vol. 10, no. 2, pp. 147–154 (2007)

29. Kaggle. https://www.kaggle.com/code/andls555/heart-disease-prediction/notebook

Development of IMU Sensor-Based XGBoost Model for Patients with Elbow Joint Damage

Jae-Jung Kim[1], Ji-Yun Seo[1], Yun-Hong Noh[2], Sang-Joong Jung[1], and Do-Un Jeong[1]([✉])

[1] Dongseo University, 47 Jurye-ro, Sasang-gu, Busan 47011, Korea
92sjjy@naver.com, {sjjung,dujeong}@dongseo.ac.kr
[2] Busan Digital University, 57 Jurye-ro, Sasang-gu, Busan 47011, Korea
yhnoh@bdu.ac.kr

Abstract. Elbow joint damage requires rehabilitation through continuous passive motion (CPM) devices or additional surgery, underscoring the significance of effective and consistent post-surgical rehabilitation. In this paper, we present an IMU sensor-based XGBoost model specifically designed for patients with elbow joint impairments, aiming to overcome the limitations of existing elbow rehabilitation methods. The proposed XGBoost model employs Softmax as its objective function to facilitate multi-class classification, while utilizing an Error function to minimize the error rate. For model training, we compiled a dataset using subjects, adhering to government guidelines for recommended elbow exercise postures and angles. We evaluated the performance of the implemented model by conducting comparisons with the tree-based ensemble model Random Forest, and the commercial model Teachable Machine. The results of this comparison showed that our proposed model achieved an average classification performance of 93.8%, while the Random Forest model exhibited an average of 91.6%, and Teachable Machine demonstrated an average of 94.9%. In conclusion, the performance of our implemented model was found to be comparable to that of existing commercial models. Moving forward, we plan to broaden our research scope by incorporating electromyography (EMG) signals, with the intention of applying our findings to various applications such as lower limb rehabilitation, prosthetic manipulation, and elbow Continuous Passive Motion (CPM) rehabilitation.

Keywords: XGBoost · CPM rehabilitation · IMU sensor

1 Introduction

The number of patients complaining of pain from elbow joint injuries steadily increases yearly [1]. The elbow joint mechanism is constantly exposed to the risk of being damaged because it plays an essential role as an important mediator of the arm mechanism in various sports activities requiring arm movement or manipulating and moving an object. Elbow joint damage can worsen into degenerative arthritis, so preventing joint damage in the early stages is essential. Although it is possible to prevent the disease through appropriate drug treatment and rehabilitation exercises in the early stages, complete recovery

© The Author(s), under exclusive license to Springer Nature Switzerland AG 2024
B. J. Choi et al. (Eds.): IHCI 2023, LNCS 14531, pp. 188–193, 2024.
https://doi.org/10.1007/978-3-031-53827-8_18

is difficult, and many patients are unaware of this and undergo surgery [2]. Treatment of joint injuries after surgery consists of Continuous Passive Motion (CPM) or additional surgery. Joint damage can cause various damage such as fibrosis and edema, so rehabilitation after surgery is essential. In recent years, the excellent performance of artificial intelligence has been demonstrated, and research into rehabilitation technologies such as exercise posture recognition and the usefulness of CPM exercise using artificial intelligence technology continues [3, 4]. In this paper, we developed the XGBoost model for patients with elbow joint damage to enable elbow rehabilitation at home after surgery. The implemented model can help reach a normal joint range of motion. If the user's angle is insufficient, real-time monitoring of the user's motion can determine in what process the angle was insufficient.

2   Development of IMU Sensor-Based XGBoost Model

The usual range of motion of the elbow joint for application to the model implemented in this paper was selected at three angles (35°, 70°, 150°) within the flexion range specified in the disability level judgment criteria published by the Ministry of Health and Welfare [5]. Based on the selected posture, the measurement hardware was attached to the wrist where the greatest range could be reflected and measured the user's acceleration and angular velocity. The measured original signal data may have errors in angle classification due to minute vibrations or drift phenomena. So, it was preprocessed through a complementary filter and used as a learning data set. Figure 1 shows an example of the results of data preprocessing.

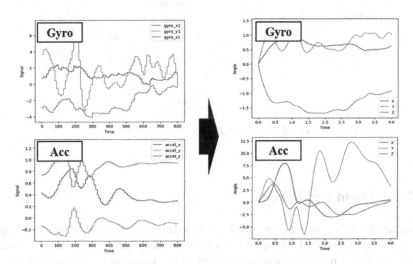

Fig. 1. Example of data preprocessing

In this paper, we implemented an IMU sensor-based XGBoost model for elbow injury patients. XGBoost is primarily used to solve classification and regression problems using tree-based learning methods. The advantage is that you can improve the model by adding new trees to compensate for old tree errors during training. The implemented XGBoost model optimization objective function used Softmax for multi-class classification and Error to reduce the error rate. As a result of learning the implemented model, the rehabilitation posture classification accuracy was confirmed to be 96.8%. Table 1 shows the implemented hyperparameters, and Fig. 2 shows an example of the XGBoost structure.

Table 1. Hyperparameter of XGBoost which use

Hyperparameter	Value
Objective	Softmax
Num_class	2
Eval_metric	Merror
Max_depth	6
eta	0.1

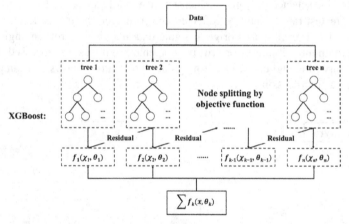

Fig. 2. Example of the XGBoost structure

3 Experiment and Results

To evaluate the performance of the XGBoost-based classification model implemented in this paper, we conducted three experiments using test data that was not used for training. The test data was measured based on the same three angles and consisted of 60 round trips for each angle for 180 measurements. In the first experiment, we applied Precision, Recall, and F1-Score, typically used classification performance evaluation indicators. Precision means the correct amount of data from the data predicted by the correct answer. Recall refers to the amount of data the model recognized as correct compared to correct data, and F1-Score refers to an index that considers both Precision and Recall. As a result of the classification performance evaluation, we confirmed a high accuracy of 0.94% or higher with a Precision of 0.9824, Recall of 0.9491, and F1-Score 0.9655. Table 2 shows the results of the classification performance index.

Table 2. Classification performance metrics results

	Precision	Recall	F1-Score
Propose model	0.9824	0.9491	0.9655

Second, similar to the implemented model, a performance comparison was conducted between a random forest model based on a decision tree and a model implemented using Google's Teachable machine. The random forest model used Grid Search to change hyperparameters at regular intervals to set optimal parameters. Teachable machine is a web-based machine learning development tool developed by Google that can use data such as images, poses, and voices as input [6]. This paper acquired image data of elbow rehabilitation exercise postures through a webcam, converted it to skeleton data, and learned joint information images on a teachable machine. As a result of the performance comparison evaluation, the rehabilitation exercise posture classification performance was confirmed to be 93.8% for the implemented model, 91.6% for the Random forest, and 94.9% for the Teachable machine. Table 3 shows the comparative evaluation results with random forest, and Fig. 3 shows an example of the Teachable machine result screen. The performance evaluation results confirmed that the implemented XGBoost model exhibited performance similar to that of commercial models.

Table 3. Comparison of classification performance between XGBoost, Random Forest, and Teachable machine

Sample	Propose model	RF Model	Teachable machine
A:35	91.6%	90.0%	91.6%
B:70	98.3%	96.6%	95.0%
C:100	91.6%	88.3%	98.3%
Avg	93.8%	91.6%	94.9%

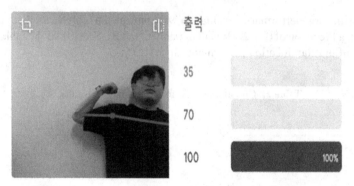

Fig. 3. Example of Teachable machine classification result

4 Conclusion

In this paper, we developed an IMU sensor-based XGBoost model for patients with elbow joint damage. As for the posture for learning, three angles were selected from the judgment of the degree of disability announced by the Ministry of Health and Welfare. Afterward, the data measured through the IMU sensor was processed through a preprocessing process through a complementary filter. The learning results of the implemented model confirmed an excellent accuracy of 95.8%. Two performance evaluation experiments were conducted using 180 test data that were not used for learning. First, high classification performance was confirmed with Precision 0.9824, Recall 0.9491, and F1-Score 0.9655. The second experiment was a comparative evaluation with a Random Forest and Teachable machine. The implemented model had an average classification performance of 93.8%, the Random Forest model had an average classification performance of 91.6%, and the Teachable machine had an average classification performance of 94.9%. In future research, we plan to include electromyography signals so that they can be applied to various fields, such as lower extremity rehabilitation, prosthetic hand manipulation, and elbow CPM rehabilitation. We plan to include electromyography signals so that they can be applied to various fields, such as lower extremity rehabilitation, prosthetic hand manipulation, and elbow CPM rehabilitation.

Acknowledgment. This work was supported by the Technology development Program (RS-2023-00223289) funded by the Ministry of SMEs and Startups (MSS, Korea).

References

1. Kim, S.Y., et al.: Personalized exercise recommender system for rehabilitation using graph neural networks. J. Korean Inst. Commun. Inf. Sci. **47**(4), 644–655 (2022)
2. Cheon, S.J., Moon, K.P., Seo, H.E., Kang, J.H.: Comparison of clinical outcomes between arthroscopic anterior compartment debridement with posterior compartment mini-open debridement and arthroscopic both compartments debridement of the primary elbow osteoarthritis. J. Korean Orthop. Assoc. **57**(1), 44–52 (2022)

3. Kim, J.H., Park, J.H.: The usefulness of AI-based CPM exercise: a study. Ann. Rehabil. Med. **34**(1), 1–10 (2022)
4. Kim, K.G., Chung, W.Y.: A home-based remote rehabilitation system with motion recognition for joint range of motion improvement. J. Korea Inst. Convergence Signal Process. **20**(3), 151–158 (2019)
5. National Knowledge Portal. Disability Assessment Criteria. Ministry of Health and Welfare, 99 (2020)
6. Carney, M., et al.: Teachable machine: approachable web-based tool for exploring machine learning classification. In: Extended Abstracts of the 2020 CHI Conference on Human Factors in Computing Systems, pp. 1–8 (2020)

Application of Daubechies Wavelets in Digital Processing of Biomedical Signals and Images

Hakimjon Zaynidinov[1] , Umidjon Juraev[2](✉) , Sulton Tishlikov[2] ,
and Jahongir Modullayev[1]

[1] Department of Artificial Intelligence, Tashkent University of Information Technologies named
after Muhammad Al Khwarizmi, Tashkent, Uzbekistan
[2] Department of Applied Mathematics and Information Technology, Gulistan State Universty,
Gulistan, Uzbekistan
pingo7520@gmail.com

Abstract. Wavelet analysis of one-dimensional signals has proven effective in
deciphering the electrocardiogram (ECG). Promising results have already been
obtained from their analysis. In particular, it has been shown that anomalous effects
in the ECG are mainly manifested on much larger scales (low frequencies), while
normal structures are characterized by relatively small scales (high frequencies).
Denoising is one of the urgent problems of digital processing of biomedical signals
and tomographic images. Wavelet methods are relatively new and are a method of
denoising using wavelet functions. Wavelets allow for the analysis of various types
of signals and effective noise removal, so it is of particular interest to study their
potential to improve image quality. It is very convenient to use DWT (Discrete
Wavelet Transform) in digital image processing, because it provides deep insight
into the main spatial and frequency features. Wavelets provide excellent time-
frequency localization, meaning they can capture both transient and stationary
features in signals and images. This localization capability is especially valuable
in medical applications where signals may contain abrupt changes or irregular
patterns. In this paper, we will discuss the method of biomedical signal restoration
and denoising in tomographic images using different wavelet functions such as
Haar wavelet, Symlet, Meyer wavelet, Daubechies wavelet.

Keywords: Digital Signal Processing · Low-frequency filter · High-frequency
filter · Scaling (Daubechies) function · Discrete Wavelet Transform · PSNR ·
MSE · ECG signal · tomographic images

1 Introduction

Due to the unique properties of wavelets, they are used in mathematics, functional
analysis, the study of (many) fractal characteristics, singularities and strong local oscillations of functions, solving some differential equations, visual representation, image
and sound compression and digital processing of geometric objects, physics, biology,
medicine, technology and is used to solve many problems in other fields.

B. J. Choi et al. (Eds.): IHCI 2023, LNCS 14531, pp. 194–206, 2024.
https://doi.org/10.1007/978-3-031-53827-8_19

Daubechies Wavelets are a family of orthogonal wavelets with compact carriers calculated by iteration. It was named after Ingrid Daubechies, an American mathematician who first built a family of wavelets. Ingrid Daubechies implemented the idea of multiple-scale analysis. In this case, the signal analyzed at the first stage of transformation is expressed as a sum of low-frequency and high-frequency parts, and the low-frequency part is called approximation coefficients, and the high-frequency part is called detail coefficients.

In this work, we describe a large collection of wavelet transformations discovered by Ingrid Daubechies. Daubechies wavelet transforms are defined in the same way as Haar wavelet transforms - by calculating the mean and difference of the scaled signals and wavelet dot products. The scaling signals and waveforms for Daubechies wavelet transforms have a slightly longer base, meaning they generate means and variances using several additional signal values. However, this small change greatly improves the capabilities of these new transforms and gives us a powerful set of signal processing tools.

The purpose of this work is to explain the following three aspects: First, to construct a mathematical classification and mathematical model of Wavelet Daubechies transform. The second one is about the description of biomedical signal recovery process using Wavelet Daubechies transform. The third involves denoising and improving the quality of tomographic images and verifying its performance based on experimental data. The evaluation involves applying the Wavelet Daubechies (db2) transform to the biomedical signals and images and evaluating the errors by performing a comparative analysis on the tables.

2 Determining Daubechies Discrete Wavelet Transform Coefficients

Due to the continuously varying measurement parameters, the Discrete Wavelet transform becomes more complex in digital processing, and it is important to have a tool that allows for discretization. What we want to bring about is an approximation from continuous scaling to a spectrum or finite set of values by replacing the integral with a sum. Discretization allows to express the signal with coefficients through the theory of elementary functions.

These types of wavelets are called "compact orthogonal wavelets", the name of the Daubechies family is written as dbN, where N is the order (or corresponding filter) and Db is the short name of the wavelet. The db1 wavelet defines the same Haar wavelet [1, 3].

A $D = 2N$ function defined as a Daubechies wavelet transform of class for $N \in \mathbb{N}$ is [4]:

$$\psi(x) := \sqrt{2} \sum_{k=0}^{2N-1} (-1)^k h_{2N-1-k} \varphi(2x - k) \tag{1}$$

Here $h_0, ..., h_{2N-1} \in R$ and these are the filter constant coefficients that meet the following conditions

$$\sum_{k=0}^{N-1} h_{2k} = \frac{1}{\sqrt{2}} = \sum_{k=0}^{N-1} h_{2k+1} \tag{2}$$

as well, for $l = 0, 1, ..., N - 1$

$$\sum_{k=0}^{2N-1+2l} h_k h_{2k-1} = \begin{cases} 1 & if \ l = 0 \\ 0 & if \ l \neq 0 \end{cases} \tag{3}$$

Here: $\varphi = \varphi_N : R \to \mathbb{R}$ Daubechies function given by the recursion equation (Fig. 1):

$$\varphi(x) := \sqrt{2} \sum_{k=0}^{2N-1} h_k \varphi(2x - k) \tag{4}$$

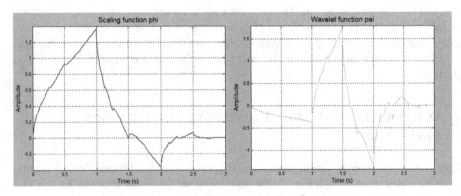

Fig. 1. db2 wavelet's scaling and wavelet functions

The main problem with creating Daubechies wavelets is determining $h_0, ..., h_{2N-1}$ coefficients, which does not deny the scaling function. It is important to note that there are N equations given by the terms of orthogonal. These 2N filter coefficients distinguish a total of N + 2 equations for h_k. db2 Wavelet transformation is actually determined in the same way as the Haar Wavelet transformation. The difference is that it has four h_0, h_1, h_2, h_3 coefficients.

For the function $\phi(t)$ to change wavelet, you need to calculate the coefficients $\{a_i, d_i\}$ These coefficients are found through the following integral:

$$a_k = (f, \phi_k) = \int_R f(x)\overline{\phi_k(x)}dx \tag{5}$$

$$d_k = (f, \psi_k) = \int_R f(x)\overline{\psi_k(x)}dx \tag{6}$$

It should be noted that in (4) and (1) there is a problem calculating a large number of integrals to find the coefficients $\{a_i, \ d_i\}$ To solve this problem, a quick wavelet modification method proposed by Mallat is used [2]. The Mallat algorithm allows you to calculate the wavelet extension coefficients using algebraic options:

$$
\begin{aligned}
a_i &= h_0 f_{2i} + h_1 f_{2i+1} + h_2 f_{2i+2} + h_3 f_{2i+3} \\
d_i &= g_0 f_{2i} + g_1 f_{2i+1} + g_2 f_{2i+2} + g_3 f_{2i+3}
\end{aligned}
\tag{7}
$$

The scaling coefficients of a_i Daubechies, the wavelet coefficients of d_i Daubechies. These (7) equations provide quick algorithms for calculating wavelet coefficients.

In here:

$$
h_0 = \frac{1+\sqrt{3}}{4\sqrt{2}}, \quad h_1 = \frac{3+\sqrt{3}}{4\sqrt{2}}, \quad h_2 = \frac{3-\sqrt{3}}{4\sqrt{2}}, \quad h_3 = \frac{1-\sqrt{3}}{4\sqrt{2}}
$$

it's not hard to take advantage of the fact that it's also a g_0, g_1, g_2, g_3.

$$
g_0 = \frac{1-\sqrt{3}}{4\sqrt{2}}, \quad g_1 = -\frac{3-\sqrt{3}}{4\sqrt{2}}, \quad g_2 = \frac{3+\sqrt{3}}{4\sqrt{2}}, \quad g_3 = -\frac{1+\sqrt{3}}{4\sqrt{2}}
$$

3 Construction of One-Dimensional Wavelet Transforms

An important point in the process of constructing Daubechies wavelets is that they have a limited number of coefficients.

This is due to the scaling criterion. Although there are many Daubechies transformations, they are all very similar. In this section, we focus on the simplest of these, the db2 wavelet transforms.

To calculate the specific level of fast waves based on the orthogonal transformation for a one-dimensional signal, it is necessary to perform the following sequence. Implementation of large-scale signal exchange by means of Discrete Wavelet Daubechies transformation is envisaged. The use of four filter coefficients in this transformation follows from the coefficient classification (3).

In this work, we obtain the following signal x_i, and using these signal values, the Daubechies transform matrices are:

$$
V = \begin{bmatrix}
h_0 & h_1 & h_2 & h_3 & \cdots & 0 & 0 & 0 & 0 \\
g_0 & g_1 & g_2 & g_3 & \cdots & 0 & 0 & 0 & 0 \\
0 & 0 & h_0 & h_1 & \cdots & 0 & 0 & 0 & 0 \\
0 & 0 & g_0 & g_1 & \cdots & 0 & 0 & 0 & 0 \\
\vdots & \vdots & \vdots & \vdots & \ddots & \vdots & \vdots & \vdots & \vdots \\
0 & 0 & 0 & 0 & \cdots & h_0 & h_1 & h_2 & h_3 \\
0 & 0 & 0 & 0 & \cdots & g_0 & g_1 & g_2 & g_3 \\
h_2 & h_3 & 0 & 0 & \cdots & 0 & 0 & h_0 & h_1 \\
g_2 & g_3 & 0 & 0 & \cdots & 0 & 0 & g_0 & g_1
\end{bmatrix}
\tag{8}
$$

The size of the Daubechies transform matrix depends on the number of input signal values, and the implementation of products and operations.

$$x_i = \{7.002,\ 6.0048,\ 5.83,\ 5.300,\ 4.765,\ 5.567,\ 6.630,\ 6.904\}$$

$$V = \begin{bmatrix} h_0 & h_1 & h_2 & h_3 & 0 & 0 & 0 & 0 \\ g_0 & g_1 & g_2 & g_3 & 0 & 0 & 0 & 0 \\ 0 & 0 & h_0 & h_1 & h_2 & h_3 & 0 & 0 \\ 0 & 0 & g_0 & g_1 & g_2 & g_3 & 0 & 0 \\ 0 & 0 & 0 & 0 & h_0 & h_1 & h_2 & h_3 \\ 0 & 0 & 0 & 0 & g_0 & g_1 & g_2 & g_3 \\ h_2 & h_3 & 0 & 0 & 0 & 0 & h_0 & h_1 \\ g_2 & g_3 & 0 & 0 & 0 & 0 & g_0 & g_1 \end{bmatrix} \tag{9}$$

Since the input signal has 8 values, we get our matrix V as an 8×8 matrix. Then we get:

$$V(i) = V \times (x_i) = \begin{bmatrix} 0.48 & 0.84 & 0.22 & -0.13 & 0 & 0 & 0 & 0 \\ -0.13 & -0.22 & 0.84 & -0.48 & 0 & 0 & 0 & 0 \\ 0 & 0 & 0.48 & 0.84 & 0.22 & -0.13 & 0 & 0 \\ 0 & 0 & -0.13 & -0.22 & 0.84 & -0.48 & 0 & 0 \\ 0 & 0 & 0 & 0 & 0.48 & 0.84 & 0.22 & -0.13 \\ 0 & 0 & 0 & 0 & -0.13 & -0.22 & 0.84 & -0.48 \\ 0.22 & -0.13 & 0 & 0 & 0 & 0 & 0.48 & 0.84 \\ 0.84 & -0.48 & 0 & 0 & 0 & 0 & -0.13 & -0.22 \end{bmatrix} \times \begin{bmatrix} 7.002 \\ 6.0048 \\ 5.83 \\ 5.300 \\ 4.765 \\ 5.567 \\ 6.630 \\ 6.904 \end{bmatrix} \tag{10}$$

In the next step, using the matrix (9), we generate the following results from the sequence of initial experimental data.

$$a_0 = h_0 f_0 + h_1 f_1 + h_2 f_2 + h_3 f_3 = 9.027174520098301$$
$$d_0 = g_0 f_0 + g_1 f_1 + g_2 f_2 + g_3 f_3 = 0.06919969424810013$$
$$a_1 = h_0 f_2 + h_1 f_3 + h_2 f_4 + h_3 f_5 = 7.5999662488776$$
$$d_1 = g_1 f_2 + g_1 f_3 + g_2 f_4 + g_3 f_5 = -0.6456464086410003$$
$$a_2 = h_0 f_4 + h_1 f_5 + h_2 f_6 + h_3 f_7 = 7.5536114030466$$
$$d_2 = g_0 f_4 + g_1 f_5 + g_2 f_6 + g_3 f_7 = 0.35278748067139976$$
$$a_3 = h_0 f_6 + h_1 f_7 + h_2 f_0 + h_3 f_1 = 9.774137354732702$$
$$d_3 = g_0 f_6 - g_1 f_7 + g_2 f_0 - g_3 f_1 = 0.5511173190458996$$

Now, we consider the signal reconstruction procedure by using the inverse wavelet transform. To carry out this operation, it seems necessary to use a transposed matrix.

Then the Daubechies inverse transform matrix will be as follows [6–8]:

$$V^T = \begin{bmatrix} 0.48 & -0.13 & 0 & 0 & 0 & 0 & 0.22 & 0.84 \\ 0.84 & -0.22 & 0 & 0 & 0 & 0 & -0.13 & -0.48 \\ 0.22 & 0.84 & 0.48 & -0.13 & 0 & 0 & 0 & 0 \\ -0.13 & -0.48 & 0.84 & -0.22 & 0 & 0 & 0 & 0 \\ 0 & 0 & 0.22 & 0.84 & 0.48 & -0.13 & 0 & 0 \\ 0 & 0 & -0.13 & -0.48 & 0.84 & -0.22 & 0 & 0 \\ 0 & 0 & 0 & 0 & 0.22 & 0.84 & 0.48 & -0.13 \\ 0 & 0 & 0 & 0 & -0.13 & -0.48 & 0.84 & -0.22 \end{bmatrix} \tag{11}$$

We will use Matrix (9) to compute the inverse transform. Using V(i) = [9.02717, 0.06919, 7.59996, −0.64564, 7.55361, 0.352787, 9.7741, 0.5511] as the input vector, we get:

$$f(x) = V^T V(i) = \begin{bmatrix} 0.48 & -0.13 & 0 & 0 & 0 & 0 & 0.22 & 0.84 \\ 0.84 & -0.22 & 0 & 0 & 0 & 0 & -0.13 & -0.48 \\ 0.22 & 0.84 & 0.48 & -0.13 & 0 & 0 & 0 & 0 \\ -0.13 & -0.48 & 0.84 & -0.22 & 0 & 0 & 0 & 0 \\ 0 & 0 & 0.22 & 0.84 & 0.48 & -0.13 & 0 & 0 \\ 0 & 0 & -0.13 & -0.48 & 0.84 & -0.22 & 0 & 0 \\ 0 & 0 & 0 & 0 & 0.22 & 0.84 & 0.48 & -0.13 \\ 0 & 0 & 0 & 0 & -0.13 & -0.48 & 0.84 & -0.22 \end{bmatrix} \times \begin{bmatrix} 9.02717 \\ 0.06919 \\ 7.59996 \\ -0.64564 \\ 7.55361 \\ 0.35278 \\ 9.77411 \\ 0.55113 \end{bmatrix} \tag{12}$$

Applying this matrix to the calculated input spectral sequence, we obtain the following result:

$$x_0 = h_0 a_0 + h_1 d_0 + h_2 a_3 + h_3 d_3 = 6.970065$$
$$x_1 = g_0 a_0 - g_1 d_0 + g_2 a_3 - g_3 d_3 = 6.070065$$
$$x_2 = h_0 a_0 + h_1 d_0 + h_2 a_1 + h_3 d_1 = 5.835327$$
$$x_3 = g_0 a_0 - g_1 d_0 + g_2 a_1 - g_3 d_1 = 5.30059$$
$$x_4 = h_2 d_0 + h_1 a_1 + h_0 d_1 + h_3 a_2 = 4.765852$$
$$x_5 = g_0 d_0 - g_1 a_1 + g_2 d_1 - g_3 a_2 = 5.567959$$
$$x_6 = h_2 d_1 + h_1 a_2 + h_0 d_2 + h_3 a_3 = 6.637434$$
$$x_7 = g_0 d_1 - g_1 a_2 + g_2 d_2 - g_3 a_3 = 6.904803$$

In order to construct a direct transformation matrix in integers, it is necessary to determine the basis of the field in which the matrix transformation and calculation will be performed. As can be seen from the obtained results, the process of restoring the values of incoming signals using db2 is more efficient and accurate (Table 1).

Below is a graphic representation of the recovery process of the ECG signal by db2 (Fig. 2).

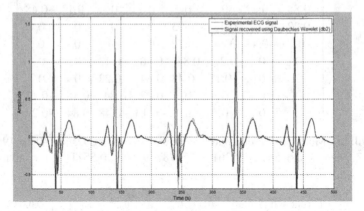

Fig. 2. Graph of the ECG signal recovery process using Daubechies Wavelet (db2)

Graphic representation of the recovery process of the ECG signal on Haar Wavelet (Fig. 3).

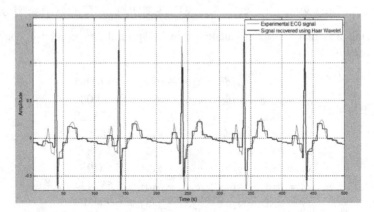

Fig. 3. Graph of the ECG signal recovery process using Haar Wavelet

Graphic representation of the recovery process of the ECG signal on Symlet Wavelet (Fig. 4).

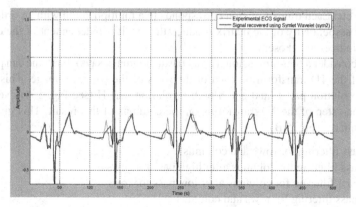

Fig. 4. Graph of the ECG signal recovery process using Symlet Wavelet

Table 1. Evaluation of errors in digital processing of ECG signals using wavelet methods

№	ECG signal	Haar	Δ_1	Symlet	Δ_2	db2	Δ_3
1	7.0026	6.8604	0.1422	7.1702	0.1676	6.9700	0,0326
2	6.0048	6.1705	0.8321	6.2701	0.2653	6.0700	0,0652
3	5.8353	5.7351	0.1002	5.9442	0.1089	5.8353	0,0002
4	5.3005	5.2004	0.1001	5.4501	0.1001	5.3004	0,0001
5	4.7658	4.5652	0.2006	4.9655	0.2006	4.7655	0,0003
6	5.5679	5.3673	0.2006	5.3672	0.2007	5.5677	0,0002
7	6.6374	6.5471	0.0903	6.4371	0.0669	6.6373	0,0001
8	6.9048	6.4032	0.5016	6.7043	0.2005	6.9045	0,0003

In the above table Δ - an absolute mistake.

4 Identification of Wavelet Methods in Denoising in Tomographic Images

An image is a signal that is two-dimensional. The variable is not the time t but the variables are now the x and the y direction. For a gray-scale image, the signal itself gives the value of the "grayness" at position (x, y). One can derive a completely analogous theory for Fourier transform, filters, wavelet basis, etc. in two variables. This leads to a theory of wavelets in two variables which are in general not separable, i.e., $\varphi(x, y)$ cannot be written as a product $\psi_1(x)\psi_2(y)$. A much easier approach is to construct tensor product wavelets which are separable. Since this is the easiest part, we shall start with this approach [4].

Tensor product wavelets.

A wavelet transform of a d-dimensional vector is most easily obtained by transforming the array sequentially on its first index (for all values of its other indices), then on the second etc. Each transformation corresponds to a multiplication with an orthogonal matrix. By associativity of the matrix product, the result is independent of the order in which the indices are chosen.

Let us consider a two-dimensional array (a square image says). First one can perform one step of the 1D transform on each of the rows of the (square) image. This results in a low-resolution part L and and a high-resolution part H (see Fig. 5.A). Next, one performs one step of the 1D transforms on the columns of this result. This gives four different squares (Fig. 5.B):

LL: low pass filtering for rows and columns
LH: low pass filtering for columns after high pass for rows
HL: high pass filtering for columns after low pass for rows
HH: high pass filtering for rows and columns

HH gives diagonal features of the image while HL gives horizontal features and LH gives vertical features.

Thus, if f is the (square) matrix containing the pixels of the image, and if $\tilde{K} = [\tilde{H}\ \tilde{G}]$ is the matrix associated with a single step of the 1D wavelet transform, then

$$f^1 = \boxed{\begin{array}{cc} LL & LH \\ HL & HH \end{array}} = [\tilde{K}^*]f[\tilde{K}^*]^t$$

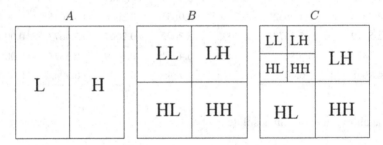

Fig. 5. Separable 2D wavelet transform

There are two possibilities to proceed:

1. the rectangular transform: further decomposition is performed on everything but the HH part (Fig. 6.A).
2. the square transform: further decomposition is performed on the LL part only (Fig. 6.B).

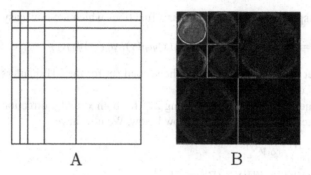

A   B

Fig. 6. Different subdivisions of square

The rectangular transform corresponds to taking a 1-dimensional wavelet transform in x and y independently. Thus with a matrix $\tilde{K}_1$ of smaller size, but of the same form as, $\tilde{K}$, we get for the second step

$$f^2 = \begin{bmatrix} \tilde{K}_1^* & 0 \\ 0 & I \end{bmatrix} f^1 \begin{bmatrix} \tilde{K}_1^* & 0 \\ 0 & I \end{bmatrix}^t$$

and similarly, for all the next steps. Thus if W is the matrix for the complete wavelet transform of a column, then the rectangular transform for the image f is WfW^t.

The rectangular division corresponds to setting

$$W_n = span\{\psi_{nk}(x)\psi_{nl}(y) : k, l \in Z\}$$

while

$$V_n = span\{\varphi_{nk}(x)\varphi_{nl}(y), \varphi_{nk}(x)\psi_{nl}(y), \psi_{nk}(x)\varphi_{nl}(y) : k, l \in Z\}$$

Setting by definition

$$(h \otimes g)(x, y) = h(x)g(y),$$

this gives rise to a wavelet expansion of the form

$$f(x, y) = \sum_{m,l} \sum_{n,k} q_{m,n,k,l}(\psi_{ml} \otimes \psi_{nk})(x, y).$$

Note that the terms in this expansion give a different resolution in x- and y-direction: For the term with $\psi_{ml} \otimes \psi_{nk}$, we get in the x-direction the scaling 2^{-m} while in the y-direction the scaling is 2^{-n} [5–7].

In the square transform we get regions like in Fig. 5B or 5C. At each stage, only the LL quarter is further subdivided. The second step cannot be described by row and column operations on the image f^1. We have to take out the LL part f_{LL}^1 explicitly and we subdivide only this part by an operation of the form $[\tilde{K}_1^*][\tilde{K}_{LL}^1][\tilde{K}_1^*]^t$ etc. This case gives subspaces V_n in the MRA which are now spanned by

$$V_n = span\{\varphi_{nk}(x)\varphi_{nl}(y) : k, l \in Z\} \quad \text{(LL squares)}$$

but the W_n are spanned by mixtures of basis functions which are now easy to describe:

$$W_n = span\{\varphi_{nk}(x)\psi_{nl}(y), \psi_{nk}(x)\varphi_{nl}(y), \psi_{nk}(x)\psi_{nl}(y) : k, l \in Z\}$$

The first set is for the HL quarters, the second set for the LH quarters and the last one for the HH quarters.

Note that there is now only one scaling 2^{-n} for both x- and y-direction.

It is the latter approach we shall follow below. We now have

$$V_{n+1} = V_{n+1}^{(x)} \otimes V_{n+1}^{(y)}$$
$$= (V_n^{(x)} \otimes W_n^{(x)}) \otimes (V_n^{(y)} \otimes W_n^{(y)})$$
$$= (V_n^{(x)} \otimes V_n^{(y)}) \otimes (V_n^{(x)} \otimes W_n^{(y)})) \otimes (W_n^{(x)} \otimes V_n^{(y)}) \otimes (W_n^{(x)} \otimes W_n^{(y)})$$

The projectors are

$$P_n f = \sum_{k,l} v_{nkl}\varphi_{nk}(x)\varphi_{nl}(y)$$

and

$$Q_n f = \sum_{k,l} \left[\omega_{nkl}^{(x)}\psi_{nk}(x)\varphi_{nl}(y) + \omega_{nkl}^{(y)}\varphi_{nk}(x)\psi_{nl}(y) + \omega_{nkl}^{(xy)}\psi_{nk}(x)\psi_{nl}(y) \right]$$

The coefficients are now arranged in matrices ($v_n = (v_{nkl})_{k,l}$ etc.) and they are found by the recursions

$$v_{n-1} = \tilde{H}^* v_n [\tilde{H}^*]^t \quad \text{(LL part)}$$

$$w_{n-1}^{(x)} = \tilde{H}^* v_n [\tilde{G}^*]^t \quad \text{(LH part)}$$

$$w_{n-1}^{(y)} = \tilde{G}^* v_n [\tilde{H}^*]^t \quad \text{(HL part)}$$

$$w_{n-1}^{(xy)} = \tilde{H}^* v_n [\tilde{H}^*]^t \quad \text{(HH part)}$$

Note that each of these matrices is half the number of rows and half the number of columns of v_n: each contains one fourth of the information [7, 8].

As for the 1-dimensional case, the v_{n-1} matrices give the coarser information, while at each level, there are three w-matrices that give the small-scale information. For example, a high value in $w_{n-1}^{(y)}$ indicates horizontal edges, a high value of $w_{n-1}^{(x)}$ indicates vertical edges, and large $w_{n-1}^{(xy)}$ indicates corners and diagonals.

The reconstruction algorithm uses [9, 10]

$$v_{n+1} = Hv_n H^t + Gw_n^{(x)} H^t + Hw_n^{(y)} G^t + Gw_n^{(xy)} G^t$$

Dobeshi filters can be used as real filters. Dobeshi wavelets are wavelets with a compact carrier, which provide good approximation properties of wavelet decomposition.

They do not have an exact expression, but are given by filtering coefficients. Wavelets can be used to extract relevant features from medical signals and images. Different wavelet coefficients can represent different aspects of the data, making it easier to identify certain patterns or anomalies.

Image denoising is one of the most important tasks, especially in digital medical image processing, where the quality is poor due to the noise and artifacts introduced by the original image acquisition systems. In this paper, we present a comparative study of different wavelet denoising methods for CT (computed tomography) images and discuss the results obtained.

Using the above sequences, we will perform table comparisons in order to denoising and increase the compression ratio of tomographic images using wavelet methods. Performance is evaluated using peak signal-to-noise ratio (PSNR) and mean square error (MSE). Table 2 shows the different levels of decomposition by wavelet type using the quadratic image transform method.

Table 2. Comparative analysis of wavelet methods

Wavelets	N decompositions number					
	N = 1		N = 2		N = 3	
	PSNR, dB	MSE	PSNR, dB	MSE	PSNR, dB	MSE
Initial image	35.47	9.34	35.47	9.34	35.47	9.34
haar	45.12	2.01	41.49	4.64	47.58	1.14
sym2	45.97	1.65	43.96	2.63	43.08	3.22
sym3	46.23	1.98	42.99	3.28	43.68	2.81
sym4	45.69	1.77	41.70	4.42	42.16	3.98
dmey	46.43	1.49	43.26	3.09	41.04	5.15
db2	45.97	1.65	47.97	1.04	43.08	3.22
db5	46.97	1.79	44.86	2.14	42.84	3.40
db8	45.21	1.97	44.23	2.47	43.22	3.12

In this table, PSNR-Peak signal-to-noise ratio is an engineering term for the ratio between the maximum possible power of a signal and the power of corrupting noise that affects the fidelity of its representation. As can be seen in the table, the root mean square error of the db2 wavelet in the process of compression and noise removal of the tomographic image is small.

Finally, the Daubechies wavelet filter provides good PSNR and low MSE values. Thus, we conclude that this filter is effective for medical image processing.

5   Conclusion

Previously presented studies have shown that 1D and 2D signal restoration, filtering and image compression methods based on model built using wavelet transform are more suitable than traditional methods.

In the reconstruction of ECG signals using wavelet methods, it was found that Haar wavelets have an absolute error of 0.2709, Symlet wavelets have an absolute error of 0.1638, and Daubechies wavelets db2 has an absolute error of 0.0123. It can be seen that Daubechies db2 wavelets are an effective method for restoring and removing noise from ECG signals.

The analysis of experimental data showed that among the wavelet functions considered in the work, Haar wavelets (haar) and Daubechies (dbN) wavelets are the most suitable for solving the problem of denoising in tomographic images. Haar wavelet level 3 decomposition of tomographic image denoising also performed well. The best denoising performance was achieved in the D4(db2) wavelet with a decomposition depth of N = 2.

The proposed image processing techniques can be used to determine optimal ways of using wavelet functions in solving image quality improvement problems. Also, the results of this work can be used in the design of filters based on wavelet functions.

References

1. Astafieva, N.M.: Wavelet analysis: basic theory and application examples. Phys. Uspekhi **39**, 1085–1108 (1996)
2. Chervyakov, N.I., Sakhnyuk, P.A., Shaposhnikov, A.V., Makokha, A.N.: Neurocomputers in the Residual Classes; Radiotekhnika: Moscow, Russia (2003)
3. Fekri, F., Mersereau, R.M., Shafer, R.W.: Theory of wavelet transform over finite fields. In: Proceedings of the IEEE International Conference on Acoustics, Speech, and Signal Processing ICASSP99, Phoenix, AZ, USA, 15–19 March 1999; Volume 3, pp. 1213–1216 (1999)
4. Zayniddinov, H., Singh, M., Singh, D.: Polynomial Splines for Digital Signal and Systems. LAP LAMBERT Academic Publishing (2016)
5. Xakimjon, Z., Bunyod, A.: Biomedical signals interpolation spline models. In: 2019 International Conference on Information Science and Communications Technologies (ICISCT), pp. 1–3 (2019). https://doi.org/10.1109/ICISCT47635.2019.9011926
6. Singh, D., Singh, M., Hakimjon, Z.: Evaluation methods of spline. In: Singh, D., Singh, M., Hakimjon, Z. (eds.) Signal Processing Applications Using Multidimensional Polynomial Splines. SpringerBriefs in Applied Sciences and Technology, pp. 35–46. Springer, Singapore (2019). https://doi.org/10.1007/978-981-13-2239-6_5
7. Khamdamov, U., Zaynidinov, H.: Parallel Algorithms for Bitmap Image Processing Based on Daubechies Wavelets (2018). https://doi.org/10.1109/ICCSN.2018.8488270
8. Bull, D.R., Zhang, F.: Filter-banks and wavelet compression, pp. 184–186 (2021). https://doi.org/10.1016/C2019-0-00641-3
9. Sweldens, W.: The lifting scheme: a custom-design construction of biorthogonal wavelets. Appl. Comput. Harmon. Anal. **3**, 186–200 (1996)

An Integrated System for Stroke Rehabilitation Exercise Assessment Using KINECT v2 and Machine Learning

Minhajul Islam[1], Mairan Sultana[1], Eshtiak Ahmed[1,2],
Ashraful Islam[1,3]([✉]), A. K. M. Mahbubur Rahman[1,3],
Amin Ahsan Ali[1,3], and M. Ashraful Amin[1,3]

[1] Center for Computational and Data Sciences, Independent University Bangladesh,
Dhaka 1229, Bangladesh
{1730400,1730725,ashraful,akmmrahman,aminali,aminmdashraful}@iub.edu.bd
[2] Faculty of Information Technology and Communication Sciences,
Tampere University, 33100 Tampere, Finland
eshtiak.ahmed@tuni.fi
[3] Department of CSE, Independent University Bangladesh, Dhaka 1229, Bangladesh

Abstract. Stroke-induced physical disabilities necessitate consistent and effective rehabilitation exercises. While a typical regime encompasses 20–60 min daily, ensuring adherence and effectiveness remains a challenge due to lengthy recovery periods, potential demotivation, and the need for professional supervision. This paper presents an innovative home-based rehabilitation system designed to address these challenges by leveraging the capabilities of the KINECT v2 3D camera. Our system, equipped with a graphical user interface (GUI), allows patients to perform, monitor, and record their exercises. By utilizing advanced machine learning algorithms, specifically G3D and disentangled multi-scale aggregation schemes, the system can analyze exercises, generating both primary objective (PO) and control factor (CF) scores out of 100. This scoring assesses the exercise quality, providing actionable feedback for improvement. Our model is trained on the Kinematic Assessment of Movement and Clinical Scores for Remote Monitoring of Physical Rehabilitation (KIMORE) dataset, ensuring robust real-time scoring. Beyond scoring, the system offers pose-correction recommendations, ensuring exercises align with expert guidelines. It can evaluate the efficacy of five distinct exercises, with provision for including more based on individual needs and expert recommendations. Overall, our system offers a streamlined approach to stroke rehabilitation, promising enhanced feasibility, and patient engagement, potentially revolutionizing stroke recovery in the healthcare domain.

Keywords: Stroke · Rehabilitation · Disability · KINECT · Health · Machine learning · Exercise · KIMORE

B. J. Choi et al. (Eds.): IHCI 2023, LNCS 14531, pp. 207–213, 2024.
https://doi.org/10.1007/978-3-031-53827-8_20

1 Introduction

Stroke is a major cause of death globally, resulting from blood vessel bursts or clots impairing brain function, leading to limited motor function and potential paralysis [2]. Rehabilitation exercises are the primary treatment for movement disorders, focusing on tension, stretching, and strengthening [5]. These exercises aim to maintain and improve movement, prevent disability, increase muscle strength, and improve coordination. However, many stroke patients do not engage in prescribed exercises at home effectively, leading to slower recovery. This is due to an inadequate range of motion during treatment sessions, e.g., proper guidelines, expert support, depression, and lack of motivation [3]. Therefore, systems that can address these problematic factors and make rehabilitation exercises more effective might help improve patient comfort and mobility to a great extent.

The use of automatic simulation systems can convey modifications in the field of post-stroke rehabilitation procedures thanks to these systems' capacity to detect body movements and capture motions. By capturing body movements these systems can give us feedback as well. By adapting these automated systems, patients can get proper guidelines for their exercises which can help them to understand how well they are doing and how better they can do by following the guidelines given by the systems. This procedure has the potential to be interactive and interesting enough to encourage patients to motivate their recovery process. This process of rehabilitation is time and cost-convenient because patients do not need to hire an expert physician to guide them. Also, they are doing their exercises at home, and they don't have to travel which is risk-free as well.

We developed a system to detect body movement using the KINECT v2 3D depth sensor, extracting skeletal data for analysis. This skeletal information is incorporated into a machine learning model for assessing the quality of exercises performed by patients. This assessment process is a part of the human movement quality evaluation, a subdomain of human action recognition and analysis [2]. Our system can evaluate five distinct exercises: arm lifting, lateral trunk tilt with extended arms, trunk rotation, pelvic rotations on the transverse plane, and squatting. We employed two methods in our model: multi-scale graph convolution disentanglement and a unified spatial-temporal graph aggregation with a convolution operator named G3D. Using the Kinematic Assessment of Movement and Clinical Scores for Remote Monitoring of Physical Rehabilitation (KIMORE) dataset [1], our model recognizes and analyzes exercises, providing feedback in the form of primary objective scores (PO score) and control factor scores (CF score). The system also features a graphical user interface (GUI) that guides patients through exercises, assesses accuracy, and stores performance data for sharing with physiotherapists. This innovation holds potential value in the rehabilitation healthcare sector for both patients and professionals.

2 Materials and Methods

2.1 Used Device

In our setup, we used the Microsoft KINECT v2, which is a 3D sensor. It records the patient's movements during the exercise. This device comes equipped with two cameras (an RGB camera having 1920 × 1080 pixels resolution and an infrared camera having 512 × 424 pixels resolution) including an infrared emitter. When it comes to assessing a patient's position, posture, and mobility, as well as measuring body composition in 3D, this depth-sensing technology provides valuable insights for the healthcare industry [4]. Such quantitative data can support personalized medical care and cost-effective decision-making, which are increasingly important in clinical practice [1–3]. Moreover, for elderly or medically compromised individuals, KINECT can contribute to maintaining an independent lifestyle [2].

2.2 Machine Learning Model Selection

The KINECT system is utilized to monitor patients' movements during exercises, providing data on 25 skeletal joints. From this data, KINECT constructs a skeletal graph wherein each joint serves as a node and each bone as an edge. Given that exercises are conducted in unique settings and under variable time frames, we developed a model employing skeleton-based action classification algorithms to harness spatial-temporal graphs to emulate human activity dynamics. Our model incorporates two primary techniques to enhance skeleton-based action recognition. The first technique is a disentangled multi-scale aggregation method. This method discerns the significance of nodes in various locations and diminishes superfluous dependencies between neighboring nodes, enhancing the efficiency of long-range modeling. The second technique, termed G3D, is a unified spatial-temporal graph convolution operator. It adeptly models spatial-temporal dependencies in skeleton-based graph sequences and efficiently employs dense cross-space-time edges to expedite information transfer within the spatial-temporal network [4]. By amalgamating these methodologies, we devised a potent feature extractor, MS-G3D, proficient in drawing reliable movement patterns, encompassing spatial-temporal dependence modeling, long-term context aggregation, and other pivotal features. Upon evaluation on the KIMORE dataset, the model demonstrated precision, providing accurate scores and relevant feedback based on the quality of the patients' exercises. Figure 1 illustrates the used framework for extracting features from KINECT v2 where the features from each frame are calculated and then the model predicts scores.

2.3 KIMORE Dataset

For our machine learning model, the KIMORE dataset [1] is used which offers in-depth RGB-3D video and skeleton joint information. KIMORE dataset is valuable for its medical assessments and diverse representation of 78 individuals, both healthy and with motor dysfunction [1]. Physiotherapists identified five exercises for low back pain physiotherapy that are widely used and clinically recognized and these exercises have been included in the KIMORE dataset [1]. Physicians determined the basic aims, known as POs, for each exercise, as well as certain limits known as CFs [1]. The concepts of POs and CFs are used to transform exercise objectives into kinematic properties [1]. POs are described as the upper limb range of motion during lifting, whereas CFs are physical restrictions that must be satisfied [1]. We implemented these 5 exercises in our system and also attempted to compute the POs and CFs to generate scores while keeping the same concepts in mind. The KIMORE dataset offers several chances to deliver sufficient data to train the machine learning model. The usability of the KIMORE dataset encouraged us to decide that it would be the ideal choice for our model, and we utilized it to fine-tune our model.

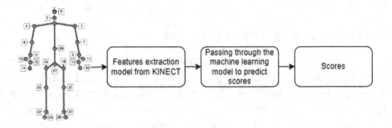

Fig. 1. Framework for extracting features from KINECT v2. Each frame's features are calculated by the feature extraction module, and the machine learning model uses these features to predict scores.

3   Experimental Setup

3.1   Device Setup

To begin the process we first connect the KINECT v2 with the machine that has the GUI Installed in it using a USB 3.0 Cable. The KINECT v2 is placed on a stand of height of 1 m and the patient is required to stand facing the lens of the KINECT v2 at a distance of 2 m.

3.2   Instructions and Guidelines

The exercises performed by the patients are evaluated by our automated system. The system can evaluate 5 major exercises prescribed by physicians or physiotherapists. The GUI allows to interact and operate the whole process of the system. To evaluate a particular exercise, an exercise is selected from a list of 5 exercises from a drop-down list in the GUI. The name and age of the patient is given as an input in the GUI. This information is used to record the patient's information along with a unique ID generated by the system. A demo video of the exercise is played next to the exercise that is to be performed. The patient then takes his/her position in front of the KINECT v2 device. On seeing the graphical skeleton mapped on the patient's body on the monitor screen the patient begins performing the exercise. The exercise is performed for a time period of 20–30 seconds, and a live feed video of the patient's exercise is captured and shown on the GUI. Data from the KINECT v2 device is stored in a skeleton file. After the process ends, a PO and a CF score are generated on a scale of 100 and displayed on the GUI.

4 Results

The institutional review board (IRB) of our home institution has given its approval for this study. We tested our system for the five individual exercises and obtained feedback from two expert physiotherapists to verify the system-provided scores. Based on the evaluation, it is evident that our system can predict a reliable score depending on the quality of the exercise, as well as it can give suggestions for exercise pose corrections. Figure 2 represents a screen capture during one of the exercises named the 'lateral tilt of the trunk with the arms in extension' where the system shows the CF score, PO score, and a suggestion. The participant got 80 as the CF score which indicates that the body control was excellent but the participant got 76 as PO which is a quite appreciating score. However, the participant can improve this score by following the suggestions provided by the system. The system suggests bending the body more towards both sides alternatively. Since the PO score indicates how well the participant follows the demo video and does the exercises, this particular participant here could not bend properly and could not receive a perfect score. The PO and CF scores of the 5 different participants for the same exercise are presented in Table 1.

Table 1. Scores for the 'lateral tilt of the trunk with the arms in extension' exercise

Participants No	Participant's Age	PO Score	CF Score
1	25	77	77
2	26	77	81
3	24	79	80
4	35	79	79
5	25	84	86

Fig. 2. One participant is performing the exercise named the 'lateral tilt of the trunk with the arms in extension'.

5   Conclusion and Future Works

Stroke is a major global cause of paralysis leading to disabilities, and rehabilitation aims to improve patients' health through physical activities. An automated system can streamline this process by providing feedback, analyzing, and assessing activities. In our study, we have created an automatic system utilizing a KINECT v2 and a machine learning model for guiding rehabilitation exercises by providing PO and CF scores. Our system can also make suggestions for patients to make changes, making it participatory and inspiring. Patients can save and share this information with physicians or physiotherapists for feedback. This technology can act as a link between patients and physicians, speeding up the rehabilitation process and keeping patients motivated. However, further research efforts are needed to make the system more robust and scalable. Our future work involves integrating sensors like the inertial measurement unit (IMU) into the current system and conducting an in-situ evaluation of a high-fidelity prototype.

References

1. Capecci, M., et al.: The kimore dataset: kinematic assessment of movement and clinical scores for remote monitoring of physical rehabilitation. IEEE Trans. Neural Syst. Rehabil. Eng. **27**(7), 1436–1448 (2019)
2. Dash, A., Yadav, A., Chauhan, A., Lahiri, U.: Kinect-assisted performance-sensitive upper limb exercise platform for post-stroke survivors. Front. Neurosci. **13**, 228 (2019)
3. Hosseini, Z.S., Peyrovi, H., Gohari, M.: The effect of early passive range of motion exercise on motor function of people with stroke: a randomized controlled trial (2019). https://www.ncbi.nlm.nih.gov/pmc/articles/PMC6428159/

4. Liu, Z., Zhang, H., Chen, Z., Wang, Z., Ouyang, W.: Disentangling and unifying graph convolutions for skeleton-based action recognition. In: Proceedings of the IEEE/CVF Conference on Computer Vision and Pattern Recognition, pp. 143–152 (2020)
5. Megard, C., et al.: Ergotact: including force-based activities into post-stroke rehabilitation. In: Extended Abstracts of the 2019 CHI Conference on Human Factors in Computing Systems, pp. 1–6 (2019)

Gastric Ulcer Detection in Endoscopic Images Using MobileNetV3-Small

T. A. Kuchkorov[✉] [ID], N. Q. Sabitova[ID], and T. D. Ochilov[ID]

Tashkent University of Information Technologies Named After Muhammad Al-Khwarizmi,
Tashkent, Uzbekistan 100084
t.kuchkorov@tuit.uz
http://www.tuit.uz

Abstract. In modern medicine, endoscopy plays a very important role as it allows physicians to detect severe diseases in early development stages. However, diagnosing patients is a challenging duty, as it requires many years of experience from doctors. In this study we proposed a new algorithm based on MobilenetV3-Small architecture to detect lesion areas of the gastrointestinal tract and peptic ulcers by the implementation of deep learning algorithms, including R-FCN, Resnet101, Yolov5, and MobilenetV3-Small, and discuss the possible results of MobilenetV3-Small algorithms in detection of ulcers. The MobilenetV3-Small architecture surpasses other tested algorithms when it comes to the speed of inference. Its efficient utilization of memory resources makes it particularly suitable for deployment on devices with limited resources, thereby improving its operational efficiency. This kind of computer aided systems may potentially help to save time, lower the cost of endoscopic procedures, and lower the risk of such procedures for the patients. To underpin the advantages of MobilenetV3-Small this paper includes a detailed overview and comparison of metrics of other models popular within the two past decades, targeting this field of research.

Keywords: Endoscopic images · endoscopy · gastrointestinal tract · ulcers · object detection · deep learning · CNN · MobilenetV3-Small · neural networks

1 Introduction

The stomach, liver, gallbladder, pancreas, and small intestine constitute the organs responsible for the human digestive system. The digestive system processes food into substances required by the human body. Through this process, humans obtain energy, grow, and maintain cell health. Diseases of the stomach and intestines significantly impact human life. Among the diseases affecting humans globally, gastritis ranks second [2]. Gastritis-related illnesses include erosion, bleeding, ulcers, and inflammation, which disrupt the proper functioning of the digestive system. Around 10% of the global population is affected by ulcers [1] Ulcers are erosions that can occur along the gastrointestinal (GI) tract, with the most common locations being the duodenum, stomach, esophagus, and jejunum. Timely treatment and attention are essential, as these disruptions can contribute to the development of various diseases in the future. According to the Ministry of Health's report, gastritis was the second most common gastrointestinal disease in 2018, affecting a significant number of individuals [6]. Medical professionals typically diagnose GI tract ulcers using endoscopy techniques. The traditional method of observing the GI tract requires much time from an endoscopist to detect and analyze ulcers as they occur in different size and in different stages of erosion. However, it's worth noting that endoscopy, while an effective diagnostic tool for GI tract ulcers, can be a painful procedure and may cause discomfort for patients. Therefore, it is essential to have a skilled and experienced physician who can effectively analyze and detect abnormalities during an endoscopy procedure. In recent years, there has been a significant advancement in image recognition using artificial intelligence (AI) and machine learning. This technology has found increasing application in diagnostic imaging across various medical fields. These areas include skin cancer classification, diagnosis in radiation oncology, diabetic retinopathy, histologic classification of gastric biopsy, and the characterization of colorectal lesions using endocytoscopy [15–19]. Further-more, machine learning algorithms find substantial utility in the medical field owing to their ability to swiftly analyze patient data, surpassing human capabilities [2]. The success of machine learning methodologies largely results from their ability to precisely identify objects in images, making them highly suitable for tasks like object detection, segmentation, and classification. Machine learning branches into multiple categories, one of which is deep learning. Deep learning, based on its core principles, can serve as a valuable tool in assisting the diagnostic procedures of smaller healthcare facilities and clinics that may have limited medical personnel. This approach has proven to be successful [5]. Numerous studies have been conducted in the past regarding the detection of ulcers in the gastrointestinal (GI) tract from endoscopy images. However, the existing techniques for ulcer detection primarily rely on image processing and segmentation methods with a restricted database. Enhancing the accuracy of detection using conventional image processing and segmentation approaches has proven to be challenging. In this study, we introduce a methodology designed for the detection of ulcers in endoscopy images, utilizing MobilenetV3-Small model. Our proposed architecture incorporates prepossessing stages of images in order to improve the effectiveness of ulcer recognition in endoscopic images. To evaluate the potential of our approach for detecting various types of ulcers, we have employed an extensive database that includes ulcers and lesions of different sizes and types.

2 Related Works

Many scholars utilize machine learning and deep learning techniques to create models for the detection of stomach diseases. Over the past three years, as deep learning and convolutional neural networks have become more widely used in the medical field, there has been a surge in research on using artificial intelligence to diagnose and differentiate peptic ulcers. Deep learning techniques built on the foundation of Convolutional Neural Networks (CNN) play a crucial role in overcoming classification challenges [8]. Additionally, it incorporates densely connected layers (dense layers) for training the underlying parameters of convolutional layers, and it employs flattened layers to convert high-dimensional input data into one-dimensional output results [7, 8]. These layers execute operations on data to identify distinctive features. Neurons within each layer of the neural network compute the weighted sum of their inputs and subsequently pass this sum through an activation function, often referred to as the activation function, to determine their average activation weight [8, 9]. To automatically extract image features, CNN suggests various filters aiding in identifying patterns in CNN layers for pattern identification.

In the realm of medical imaging, Convolutional Neural Network (CNN) approaches [23–25] are well-established and have found applications in segmenting images in gastrointestinal (GI) endoscopy [28, 29, 30]. For example, Wu et al. devised a dual neural network that combines the U-Net and ResNet architectures to autonomously segment esophageal lesions [27]. Guo et al. [26] utilized an architecture based on Fully Convolutional Networks (FCN) with atrous kernels to perform the segmentation of polyps. Wang et al. conducted a study to assess how well deep neural networks could identify ulcers in images captured during wireless capsule endoscopy. They tested it on 4,917 ulcer images and 5,007 normal images. The results showed that the neural network could correctly identify ulcers with a sensitivity of 89.7% and had an overall accuracy and specificity of over 90% [21]. Similarly, Alaskarand et al. created a neural network for the diagnosis of gastrointestinal ulcers. They trained the network using 336 images and tested it with 105 images [21].

3   Materials and Methods

3.1   The Dataset

In recent years, a variety of sophisticated methods have been employed in the field of endoscopic image analysis, with special attention to model design and operation. Medical data sets differ from other data sets in that they contain sensitive patient data created by skilled healthcare professionals and are not readily accessible to the general public. To facilitate access to this data for researchers and the research community as a whole, some organizations have created medical data repositories. This research is aimed at the automatic detection of anomalies in the gastrointestinal tract and the identification of lesions. A substantial endoscopic training dataset was created by analyzing patients in real-time at the Republican Specialized Scientific and Practical Medical Center of Cardiology in Uzbekistan. The retrospective analysis involved reviewing endoscopy records of patients who underwent endoscopic examinations at the Republican Specialized Scientific and Practical Medical Center of Cardiology in Uzbekistan's endoscopy center from 2018 to 2022. These records were subsequently reviewed and retrieved for analysis by two expert endoscopists with 20 years of experience in therapeutic endoscopy.

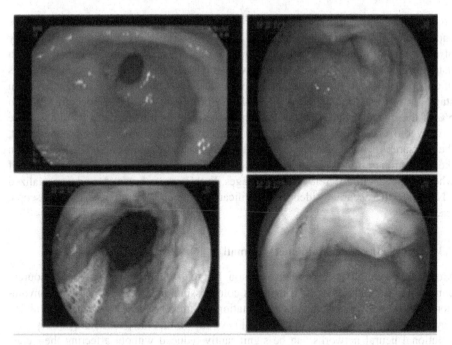

Fig. 1. Dataset samples

Figure 1 shows the samples from dataset, and here we can notice that images in dataset are in different dimensions. The dataset for the research consists of 4000 images, with 2000 used for the training set and 2000 for the testing set. The training set comprises only images of ulcers and ulcerated lesions, while the testing set contains images of other diseases and healthy endoscopic tracts. The dataset consists of images taken at various times, distances, angles, orientations, lighting conditions, and focus levels. These images were acquired during the admission of regular patients at a clinic, who were diagnosed with endoscopic tract diseases. As a result, some images were captured at different moments, under various angles, and lighting conditions within the same object. This implies that the same lesion can occur multiple times in a dataset because of several factors.

3.2 Preprocessing

In this proposed model, the input data is first normalized and the variables are categorized before being trained on a CNN model. Then, the data collection goes through three stages: training, testing, and validation. This process is illustrated in Fig. 2. In this dataset, the dimensions of ulcer images vary between 1600 × 1200 and 3648 × 2736 pixels. To reduce computational costs during training, all images were resized to 480 × 480 pixels. A total of approximately 3000 ulcers were labeled for the training set. Due to the variability in the human gastrointestinal tract and the presence of a large number of ulcers, there exists a disparity between the number of images and the number of ulcers.

To account for this, ulcer size was proportionally allocated to the dimensions of the endoscopic tract image in the 3rd stage. In many cases, we observed that the majority of ulcers (1849 images, 74.08%) occupy less than 15% of the image dimensions. In many instances, ulcer volumes are relatively small, but they can be numerous in terms of quantity. Through subsequent analyses on these images (as shown in the diagram in stage 3), we found that the majority of ulcers (1250 images, 50.08%) occupy less than 5% of the image volume. To mitigate the issue of class imbalance in the dataset, we employed oversampling techniques mentioned above. To enhance the sampling speed, the oversampling method adds zero-valued samples to the actual dataset. Depending on the translation or transformation invariance, CNNs learn a specific feature regardless of where it appears in different types of images. In this research, we created a specialized CNN model that accurately identifies significant lesions and ulcerations in the endoscopic tract.

3.3   Architecture and MobilenetV3-Small

MobilenetV3-Small is a possible solution for real-time applications and resource-constrained portable devices. It involves compressing and accelerating deep convolutional neural networks, reducing computation parameters and values, and minimizing power consumption. Denil and others [7] have shown that the parameters of deep convolutional neural networks can be significantly reduced without affecting their classification accuracy. MobilenetV3-Small introduces several changes. Firstly, depthwise separable convolutions, which reduce computational overhead, and the incorporation of squeeze-and-excitation (SE) blocks, which adaptively scale feature maps on a channel-wise basis to enhance the network's representational power. Furthermore, the integration of Squeeze-and-Excitation (SE) blocks in MobileNetV3-Small plays a crucial role in enhancing the model's capacity to capture essential features. Compared to the VGG-16 model, MobilenetV3-Small is a lightweight network that uses depthwise convolution to reduce computation and parameters. In addition, MobilenetV3-Small achieves classification accuracy of less than 1% on the ImageNet dataset. Lastly, MobileNetV3-Small offers different versions within its family. These versions come with varying hyperparameters, trade-offs, and performance characteristics, designed to cater to specific application requirements.

Structure of layers of MobilenetV3-Small are listed on Table 1 Following the initial preprocessing phase, the dataset proceeds through three primary stages: training, testing, and validation, as outlined below:

Training: During this stage, the preprocessed data is utilized to instruct the MobilenetV3-Small model. The training process entails providing the model with input data, which is then processed through multiple layers within the CNN. Throughout this phase, the model acquires the ability to identify patterns and characteristics in the data, gradually improving its capability to classify gastrointestinal tract anomalies and lesions.

Testing: Subsequent to the training phase, the model is subjected to evaluation using a distinct dataset it has not encountered previously. This testing dataset serves the purpose of assessing the model's overall performance and its capacity to accurately categorize anomalies and lesions in fresh, previously unseen endoscopic images. Metrics like accuracy, precision, recall, and others are typically computed during this testing phase to measure its effectiveness.

Validation: To fine-tune the model and optimize its hyperparameters, a validation step is commonly employed. This process involves adjusting various parameters of the CNN model while evaluating its performance on a validation dataset. This iterative procedure continues until the model attains the desired level of accuracy and robustness.

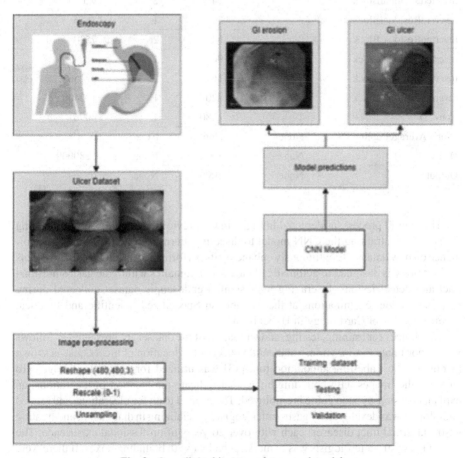

Fig. 2. Overall Architecture of proposed model.

Table 1. Specification for MobileNetV3-Small.

Layer	Filter size	Filters	Stride	Parameters
Input	224	3	2	0
Conv2D	3	32	1	320
InvertedResidualBlock	None	16	1	112
InvertedResidualBlock	3	24	2	920
InvertedResidualBlock	3	32	1	1856
InvertedResidualBlock	3	64	2	7520
InvertedResidualBlock	3	96	1	4864
InvertedResidualBlock	3	160	2	32256
InvertedResidualBlock	3	320	1	102400
Conv2D	1	1280	1	4096
GlobalAveragePooling2D	None	None	None	0
Dense	None	1000	None	1280000
Output	None	1000	None	0

The overall process, as depicted in Fig. 2 in the previous text, entails pre-processing the input data, training the CNN model to discern patterns, testing the model's performance on new data, and continuously enhancing its performance through validation. This methodology is designed to automatically identify anomalies within the gastrointestinal tract and detect lesions, leveraging a substantial endoscopic dataset collected during real-time patient examinations at the Republican Specialized Scientific and Practical Medical Center of Cardiology in Uzbekistan.

All choices for training, testing, and validation of the dataset are presented as follows in the coordinate (xmin, ymin, xmax, and ymax) of the location of ulcers, just as shown in Fig. 1. The LabelImg annotation tool [13] was utilized for annotating images. To annotate the images with bounding boxes that indicate the location of ulcerations, an explanatory step-by-step guide is employed. The ground truth for ulcerations and related pathologies was developed by gastroenterologists specializing in diagnosing and treating gastrointestinal tract diseases, each with over 20 years of professional experience. The task of the gastroenterologists was to mark each ulcer with bounding boxes. If there were discrepancies in their annotations for ulcer identification, a final decision was reached by consensus.

Table 2. Hyperperameters of model.

Learning rate	0.001
Batch size	32
Epochs	500
Optimizer	Adam
Loss function	Categorical crossentropy
Dropout	0.2

The hyperparameters in the Table 2 are values for training a MobileNet V3 model. The learning rate is set to 0.001 to start, and it can be gradually reduced as the model converges. The batch size is set to 32, and the epochs are set to 500. The Adam optimizer is used to optimize the model's parameters. The categorical crossentropy loss function is used to train the model to classify images into different categories. The dropout rate is set to 0.2 to reduce overfitting.

We employed the widely used and lightweight MobilenetV3-Small algorithm for ulcer detection, since we develop real-time detection model. Python programming language, along with TensorFlow, pandas, Numpy, and Tkinter libraries, was used in the creation of the ulcer detection program. The implenetation details of the proposed system model is i9 11900k 64 GB RAM, Nvidia GeForce RTX 3070 TI GPU. 1. Dataset: Gathering data on ulcers or related image information. The dataset should be comprehensive and consist of image data that accurately performs the task of ulcer detection. The creation of the program encompasses the following stages: 2. Data Preparation: Preparing image data for the MobilenetV3-Small architecture, including resizing images, normalizing pixel values, and applying necessary adjustments. 3. Model Training: Training the MobilenetV3-Small model on the ulcer detection dataset, where prepared data is fed into the model, and the model's weights are modified as needed to minimize the difference between detected ulcer markings and ground truth markings. 4. Model Evaluation: Assessing the trained model separately, either through validation or testing sets, using precision, recall, and F1-score metrics to measure accuracy. 5. Model Application: Applying the trained and evaluated model to detect ulcers in new, unseen images or video frames. The created model takes an image or video frame as input and generates results regarding the presence or location of ulcers or probability estimates.

3.4 Performance Metrics

Precision (Accuracy): Precision teaches the model to prioritize high accuracy in its correct responses. It calculates the ratio of correctly identified ulcers to all ulcers identified by the model. In other words, Precision answers the question: "How many of all identified ulcers are genuine?" A high Precision value reduces the number of false positive results.

$$Precision = TP/(TP + FP) \qquad (1)$$

Recall (Sensitivity): Recall assesses the model's ability to correctly identify all positive instances (ulcers) in the dataset. In simple terms, Recall answers the question: "Did the model successfully detect all actual ulcers?" A high Recall value reduces the number of false negatives.

$$Recall = TP/(TP + FN) \tag{2}$$

Dice Coefficient (F1 Score): The Dice coefficient is very similar to Intersection-Over-Union (IoU). They are positively correlated, meaning that if one rates an image's segmentation better than the other, the other will also rate it similarly. Like IoU, they both range from 0 to 1, where 1 indicates the highest similarity between the predicted and ground truth segmentation. True positive results are obtained when the Intersection-Over-Union (IoU) of the detected bounding box and the ground truth bounding box is greater than or equal to 0.5, and it is calculated as follows.

$$F1 = \frac{2 * Precision * Recall}{Precision + Recall} = \frac{2 * TP}{2 * TP + FP + FN} \tag{3}$$

Mean average precision (mAP): a metric commonly used to evaluate object detection and segmentation models. It is calculated by averaging the average precision (AP) scores across all object classes. AP is a measure of the precision-recall performance of a model on a given class.

Precision is the proportion of identified positives that are actual positives, while recall is the proportion of actual positives that are correctly identified as positives. F1 score is a harmonic mean of precision and recall, and is often used as a measure of overall performance.

$$mAP = \frac{1}{N} \sum_{i=1}^{N} AP_i \tag{4}$$

the Eq. (4) presents formula for mAP.

4 Results

Table 3. Inference speed of models on difference devices such as CPU, GPU, Mobile device

Model	CPU (FPS)	GPU (FPS)
FRCNN Inception-v2-ResNet101	1–2	5–10
YOLOv5	10–15	30–40
FRCNN R-FCN	5–7	10–15
FRCNN ResNet-101	7–10	15–20
MobilenetV3-Small	20–30	60–70

As we can see, MobilenetV3-Small is the fastest object detection model on all devices, followed by YOLOv5 and FRCNN R-FCN. FRCNN Inception-v2-ResNet101 and FRCNN ResNet-101 are the slowest object detection models on all devices, but they are also the most accurate (Table 3).

It is important to note that inference speed can vary depending on a number of factors, such as the hardware platform, the size and complexity of the input image, and the number of objects in the image. However, the table above provides the relative speeds of different object detection models on different devices.

Table 4. Results of models on different evaluation metrics

Model	Recall	Precision	F1	mAP
FRCNN R-FCN	0.75	0.77	0.76	74.2
FRCNN ResNet101	0.78	0.80	0.79	78.3
FRCNN Inception-v2-ResNet101	0.79	0.80	0.79	81.4
YOLOv5	0.81	0.83	0.82	84.5
MobilenetV3-Small	0.79	0.77	0.78	78.1

In the Table 4, YOLOv5 achieves the highest recall, precision, F1, and mAP scores, followed by FRCNN Inception-v2-ResNet101, FRCNN ResNet101, FRCNN R-FCN, and MobilenetV3-Small.

The best model for a particular task will depend on a number of factors such as data, the desired accuracy, and the computational resources available. For example, MobilenetV3-Small is a lightweight model that is well-suited for mobile and embedded applications, while YOLOv5 is a more powerful model that can achieve higher accuracy on larger and more complex datasets. As the inference speed is important to our task Mobile Net v3 is more suitable algorithm for real-time ulcer detection model. Since MobilenetV3-Small offers relatively fast inference speed not sacrificing much accuracy.

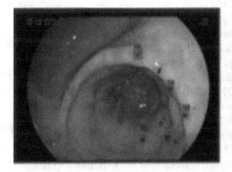 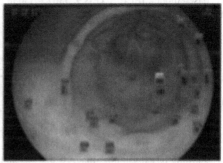

Fig. 3. Detected Lesions and Ulcers by the Model (in percentages).

Figure 3 illustrates the output of our model, and we can see that model can detect ulcer and lesions accurately. Also, models can detect both small and big objects.

5 Conclusion

In conclusion, the MobilenetV3-Small architecture outperforms other tested algorithms in terms of inference speed. Its efficient use of memory resources makes it well-suited for deployment on resource-constrained devices, enhancing its operational efficiency. MobilenetV3-Small competes other CNN-based models, such as FRCNN (FCN, ResNet101) and YoloV5, achieving an above-average accuracy level with a mAP value of 78.1. This model demonstrates its effectiveness not only in detecting large lesions in the gastrointestinal tract but also in identifying smaller lesions and abnormalities, showcasing its versatility and robust performance.

References

1. Kaplan, G.G.: Does breathing polluted air increase the risk of upper gastrointestinal bleeding from peptic ulcer disease? Lancet Planet. Health 1(2), e54–e55 (2017)
2. Isobe, Y., Nashimoto, A., Akazawa, K., Oda, I., Hayashi, K., Miyashiro, I., et al.: Gastric cancer treatment in Japan: 2008 annual report of the JGCA nationwide registry. Gastric Cancer 14(4), 301–316 (2011). https://doi.org/10.1007/s10120-011-0085-6. PMID: 21894577
3. Sandler, M., Howard, A., Zhu, M., Zhmoginov, A., Chen, L.-C.: MobileNetV2: inverted residuals and linear bottlenecks. In: Proceedings of the IEEE Conference on Computer Vision and Pattern Recognition (CVPR) (2018)
4. Howard, A.G., et al.: MobileNets: Efficient Convolutional Neural Networks for Mobile Vision Applications. arXiv, abs/1704.04861 (2017)
5. Denil, M., Shakibi, B., Dinh, L., Ranzato, M.A., De Freitas, N.: Predicting parameters in deep learning. In: Proceedings of the Advances in Neural Information Processing Systems, Lake Tahoe, NV, USA, pp. 2148–2156 (2013)
6. WHO. Gastrointestinal Cancer (2020). https://www.who.int/news-room/fact-sheets/detail/cancer. Accessed 25 Mar 2022
7. Rustam, F., et al.: Wireless capsule endoscopy bleeding images classification using CNN based model. IEEE Access 9, 33675–33688 (2021)
8. Yogapriya, J., Chandran, V., Sumithra, M., Anitha, P., Jenopaul, P., Suresh Gnana Dhas, C.: Gastrointestinal tract disease classification from wireless endoscopy images using pretrained deep-learning model. Comput. Math. Methods Med. 2021 (2021)
9. Zhuang, J., Cai, J., Wang, R., Zhang, J., Zheng, W.-S.: Deep kNN for medical image classification. In: Martel, A.L., et al. (eds.) MICCAI 2020. LNCS, vol. 12261, pp. 127–136. Springer, Cham (2020). https://doi.org/10.1007/978-3-030-59710-8_13
10. Alzubaidi, L., et al.: Towards a better understanding of transfer learning for medical imaging: a case study. Appl. Sci. 10, 4523 (2020)
11. LabelImg Tzutalin. Git code (2015). https://github.com/tzutalin/labelImg
12. Kuchkorov, T.A., Urmanov, Sh.N., Nosirov, Kh.Kh., Kyamakya, K.: Perspectives of deep learning based satellite imagery analysis and efficient training of the U-Net architecture for land-use classification. Developments of Artificial Intelligence Technologies in Computation and Robotics: Proceedings of the 14th International FLINS Conference (FLINS 2020), pp. 1041–1048 (2020)

13. Kuchkorov, T., Ochilov, T., Gaybulloev, E., Sobitova, N., Ruzibaev, O.: Agro-field boundary detection using mask R-CNN from satellite and aerial images. In: International Conference on Information Science and Communications Technologies (ICISCT), pp. 1–3 (2021)

14. Kuchkorov, T.A., Sabitova, N.Q.: Deblurring techniques and fuzzy logic methods in image processing. Descendants of Muhammad al-Khwarizmi Sci.-Pract. Inf. Anal. J. 1(19), 12–16 (2019). ISSN 2181-9211

15. Esteva, A., Kuprel, B., Novoa, R.A., Ko, J., Swetter, S.M., Blau, H.M., et al.: Dermatologist-level classification of skin cancer with deep neural networks. Nature 542, 115–118 (2017)

16. Bibault, J.E., Giraud, P., Burgun, A.: Big data and machine learning in radiation oncology: state of the art and future prospects. Cancer Lett. 382, 110–117 (2016)

17. Gulshan, V., Peng, L., Coram, M., Stumpe, M.C., Wu, D., Narayanaswamy, A., et al.: Development and validation of a deep learning algorithm for detection of diabetic retinopathy in retinal fundus photographs. JAMA 316, 2402–2410 (2016)

18. Misawa, M., Kudo, S.E., Mori, Y., Takeda, K., Maeda, Y., Kataoka, S., et al.: Accuracy of computer-aided diagnosis based on narrow band imaging endocytoscopy for diagnosing colorectal lesions: comparison with experts. Int. J. Comput. Assist. Radiol. Surg. 12, 757–766 (2017)

19. Yoshida, H., Shimazu, T., Kiyuna, T., Marugame, A., Yamashita, Y., Cosatto, E., et al.: Automated histological classification of whole-slide images of gastric biopsy specimens. Gastric Cancer (2017). https://doi.org/10.1007/s10120-017-0731-8

20. Wang, S., Xing, Y., Zhang, L., Gao, H., Zhang, H.: A systematic evaluation and optimization of automatic detection of ulcers in wireless capsule endoscopy on a large dataset using deep convolutional neural networks. Phys. Med. Biol. 64(23), 235014 (2019). https://doi.org/10.1088/1361-6560/ab5086

21. Alaskar, H., Hussain, A., Al-Aseem, N., Liatsis, P., Al-Jumeily, D.: Application of convolutional neural networks for automated ulcer detection in wireless capsule endoscopy images. Sensors 19(6), 1265 (2019). https://doi.org/10.3390/s19061265

22. Kutlimuratov, A., Khamzaev, J., Kuchkorov, T., Anwar, M.S., Choi, A.: Applying enhanced real-time monitoring and counting method for effective traffic management in tashkent. Sensors 23, 5007 (2023). https://doi.org/10.3390/s23115007

23. Borgli, H., Thambawita, V., Smedsrud, P.H., Hicks, S., et al.: HyperKvasir, a comprehensive multi-class image and video dataset for gastrointestinal endoscopy. Sci. Data 7(1), 114 (2020)

24. Enriquez, V.G., Singh, M.: Gender detection using voice through deep learning. In: Kim, J.-H., Singh, M., Khan, J., Tiwary, U.S., Sur, M., Singh, D. (eds.) IHCI 2021. LNCS, vol. 13184, pp. 548–555. Springer, Cham (2022). https://doi.org/10.1007/978-3-030-98404-5_50

25. Chen, L.-C., Zhu, Y., Papandreou, G., Schroff, F., Adam, H.: Encoder-decoder with atrous separable convolution for semantic image segmentation. In: Ferrari, V., Hebert, M., Sminchisescu, C., Weiss, Y. (eds.) ECCV 2018. LNCS, vol. 11211, pp. 833–851. Springer, Cham (2018). https://doi.org/10.1007/978-3-030-01234-2_49

26. Long, J., Shelhamer, E., Darrell, T.: Fully convolutional networks for semantic segmentation. In: 2015 IEEE Conference on Computer Vision and Pattern Recognition (CVPR), pp. 3431–3440 (2015)

27. Guo, Y., Bernal, J., J Matuszewski, B.: Polyp segmentation with fully convolutional deep neural networks extended evaluation study. J. Imaging 6(7), 69 (2020)

28. Wu, Z., Ge, R., Wen, M., Liu, G., et al.: ELNet: automatic classification and segmentation for esophageal lesions using convolutional neural network. Med. Image Anal. 67, 101838 (2021)

29. Sandler, M., et al.: Searching for MobileNetV3. In: ICCV (2019)

Exploring Quantum Machine Learning for Early Disease Detection: Perspectives, Challenges, and Opportunities

Madhusudan Singh[1] (ID), Irish Singh[1] (ID), and Dhananjay Singh[2](✉) (ID)

[1] Oregon Institute of Technology, Klamath Falls, OR 97601, USA
madhusudan.singh@oit.edu
[2] ReSENSE Lab, School of Professional Studies, Sanit Louis University, St. Louis, MO 63103, USA
dan.usn@ieee.org

Abstract. This study examined the transformative influence of Quantum Machine Learning (QML) in redefining medical applications. The paper provides a thorough analysis of the convergence of quantum computing and healthcare, thoroughly examining the potential benefits, challenges, and prospects. The paper extensively examines the difficulties faced when implementing QML in healthcare applications, specifically focusing on concerns related to noise, circuit design, and the conversion of conventional data into formats suitable for quantum systems. At the same time, it highlights the wide range of opportunities that QML provides, especially in tackling important medical concerns like cancer, neurological problems, and cardiovascular diseases, where prompt intervention is crucial. The study explicitly explores the potential of QML to fundamentally transform approaches to early disease diagnosis. This paper delves into many prospective, challenges in growth, and the numerous possibilities it offers for updating healthcare methodologies. The results emphasize the significant impact that QML can have on improving the accuracy and effectiveness of medical applications, opening possibilities for creative solutions in diagnosing and treating diseases.

Keywords: Quantum Machine Learning · Neural Networks · Support Vector Machine · Disease Detection

1 Introduction

Early disease diagnosis is crucial in modern healthcare as it plays a vital role in improving patient outcomes and effectively managing complex illnesses [1]. The presence of diseases like cancer, neurological disorders, and cardiovascular ailments highlights the urgent requirement to uncover subtle biomarkers that indicate the beginning of a disease. This will allow for earlier intervention to reduce the advancement of the disease and improve the quality of life for patients. Nevertheless, traditional analytical methods frequently struggle to identify these enigmatic molecular patterns among a multitude of genetic, proteomic, metabolic, and physiological factors [2]. Patient data is transformed into quantum states, analyzed by quantum models, and used to predict and guide clinical decisions. Figure 1 has shown the QML medical applications in early disease detection.

© The Author(s), under exclusive license to Springer Nature Switzerland AG 2024
B. J. Choi et al. (Eds.): IHCI 2023, LNCS 14531, pp. 226–242, 2024.
https://doi.org/10.1007/978-3-031-53827-8_22

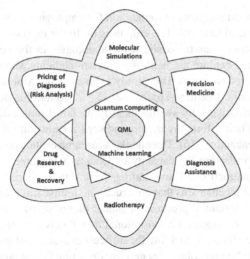

Fig. 1. QML Medical Applications for early disease detection.

Quantum computing has evolved as a breakthrough technology paradigm to address this urgent need. It offers a compelling solution due to its unmatched ability to quickly analyze complex, high-dimensional datasets. Quantum machine learning (QML) is a powerful tool that utilizes the distinctive properties of quantum physics, such as superposition and entanglement, to analyze large datasets and reveal hidden connections that can indicate early pathophysiological changes in Fig. 1. The potential of QML resides in its capacity to identify important diagnostic cues, similar to identifying difficult-to-find items, which enables early risk screening and precise medical diagnostics [3].

This study explores the growing interdisciplinary topic of Quantum Machine Learning (QML) as it relates to the early detection of diseases. Section 2 delves into the examination of QML and its associated medical technologies and applications. Section 3 offers a succinct summary of the QML hardware, software, and crucial technological elements used in medical applications. Section 4 explores essential QML algorithms and their diagnostic applications, including quantum neural networks, quantum Boltzmann machines, and quantum support vector machines. Section 5 delineates the difficulties linked to QML in the medical domain. Section 6 examines the viewpoints, difficulties, and possibilities in the early identification of diseases in the medical domain utilizing QML. Finally, we have concluded in Sect. 7.

2   Related Work

A paradigm changes in medical filed has been brought about by the combination of quantum computing and machine learning, opening the door for novel methods of precision medicine and early disease detection. Drawing from foundational works and research advances, this article provides a brief study of the emerging QML applications in medical fields.

Schuld's research establishes the foundation for employing quantum algorithms to unravel complex medical datasets. The emphasis on linear regression offers a foundational comprehension of quantum-assisted methodologies in the healthcare field [4]. Biamonte and his team explore the field of precision medicine, showcasing the potential of quantum-powered analytics. Their study focuses on tailoring treatment strategies using Quantum Machine Learning approaches [5]. Zhao and his colleagues expedite the process of drug discovery by swiftly identifying novel molecules using Quantum Machine Learning. Their efforts lay the groundwork for quantum-assisted progress in the creation of pharmaceuticals [6]. Tkatchenko's research centers on elucidating molecular structures, demonstrating the capacity of Quantum Machine Learning to decode complex biological interactions. These findings have significant ramifications for comprehending molecular mechanisms that are crucial in disease situations [7]. Lamata and his team have made substantial progress in cancer diagnosis by enhancing both accuracy and efficiency using quantum convolutional neural networks and quantum support vector machines. They focus on tackling the difficulties associated with analyzing medical imaging data [8]. Yu and colleagues employ Quantum Graph Neural Networks and Quantum Generative Adversarial Networks to provide early predictions of Alzheimer's illness. Their study centers around utilizing a wide range of patient data to make more accurate forecasts [9]. Liang and Li employ Quantum Recurrent Neural Networks and Quantum Boltzmann Machines to advance the field of diabetes prognosis. Their effort seeks to improve the precision and dependability of forecasting the advancement and ramifications of diabetes [10].

The works of Li and Akbar jointly improve the precision of diagnosing cardiovascular illness by utilizing Quantum Feedforward Neural Networks and Quantum Probabilistic Models. Their research enhances the comprehension of cardiovascular problems through the utilization of quantum-assisted approaches [10, 11].

We have investigated and highlights the significant impact that Quantum Machine Learning can have in several healthcare fields. It provides novel methods and tackles the obstacles associated with early disease identification. Finally, it provides a comprehensive viewpoint on how to use quantum-powered machine learning to revolutionize precision medicine, expedite medication discovery, and interpret complex medical data with crucial steps in promoting proactive healthcare practices and enhancing patient outcomes.

3   Quantum Machine Learning (QML) in Medical Applications

Quantum computing introduces radical departures from classical binary logic, intrinsically leveraging quantum mechanical properties to represent and manipulate information [9]. By encoding data within quantum bits (qubits) that can exhibit the phenomena of superposition and entanglement, quantum systems can perform massively parallel computations to analyze exponentially large datasets represented by a finite number of qubits. Harnessing this volumetric capacity promises to benefit data-intensive techniques like machine learning for discovering valuable correlations and patterns within vast, high-dimensional biomedical data [10] as represents in Fig. 2.

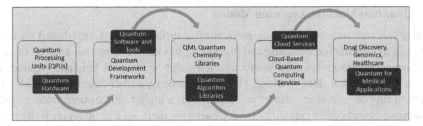

Fig. 2. Quantum Machine Learning for Medical Applications

3.1 Quantum Software Platforms and Tools

Practical realization of quantum-enhanced machine learning necessitates robust software infrastructure to bridge classical and quantum computing resources [11]. Hybrid classical-quantum algorithms must integrate modular quantum subroutines within classical workflows for data loading, preprocessing, storage, and visualization. High-level open software stacks developed by providers like IBM, Rigetti and Google now integrate QML modules with classical Python libraries like NumPy, SciPy and scikit-learn for streamlined application development. These include Qiskit, Forest and TensorFlow Quantum respectively, with tools like QuTiP and PennyLane offering additional frameworks agnostic across quantum hardware platforms [12]. Languages like Silq also facilitate writing algorithms directly in quantum pseudo-code. Concurrently, low-level quantum compilation toolchains like ScaffCC, XACC and t|ket> compile algorithmic instructions into optimized circuits tailored to backend quantum processors. Together these software advances provide an emerging ecosystem to readily implement QML innovations. Figure 3 has shown that software stack and it's feature for Quantum Computing.

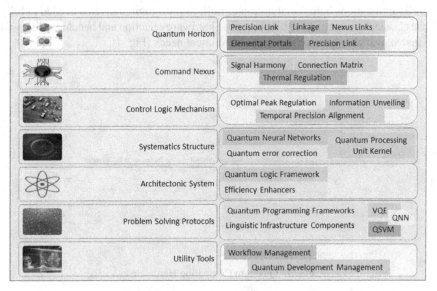

Fig. 3. Quantum computing software stack

3.2   Quantum Machine Learning Models

Diverse quantum machine learning models offer different affordances based on their approach to information encoding and processing workflows. Quantum analogs of artificial neural networks leverage interconnected qubit layers with trainable parameters similarly to classical deep learning [13]. Alternatively, kernel-based techniques like quantum support vector machines map inputs into a higher dimensional feature space to improve classification capability [14]. Generative methods like quantum Boltzmann machines use stochastic qubit interactions to model multidimensional distributions [15]. Meanwhile emerging quantum reservoir computing approaches exploit complex recurrent qubit dynamics as a computational resource without needing trainable parameters [16]. Each model balances tradeoffs regarding extensibility, noise resilience and algorithmic complexity. Selecting optimal QML architectures matched to clinical application requirements and hardware constraints will enable exploiting quantum advantages. Figure 4 represents the process of hybrid approach of Classical Machine Learning (CML) and Quantum Machine Learning (QML).

3.3   Quantum Computing Hardware

Quantum computing hardware underpins model abstractions by physically manifesting quantum phenomena for information processing [17]. Multiple technological platforms demonstrate qubit implementations, with photonics, superconducting circuits, trapped ions and other designs exhibiting rapid iterative advances [18]. Key metrics of coherence times, gate fidelities, qubit connectivity and count continue rising across prototypes by IBM, Google, Rigetti and IonQ - although substantial challenges remain reaching fault-tolerance required for scalable, error-corrected operation. Integration with cloud API access also enables exploring quantum capabilities without specialized infrastructure. Near-term systems with ~100 qubits anticipated by 2025 promise crossover viability for QML applications before maturation of million-plus qubit platforms needed to fully eclipse classical computing [19]. Appropriate hardware selection and benchmarking will help match available quantum resources to clinical needs [Fig. 4].

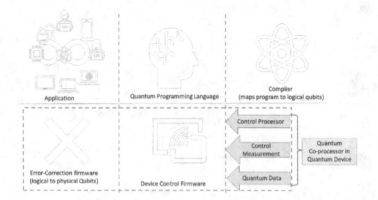

Fig. 4. Quantum Computing Hardware

4   QML Algorithms

Realizing quantum advantages for biomedical learning tasks involves tailoring algorithms to leverage quantum properties within problem-specific contexts. We survey prominent QML models which show strong promise for advancing disease detection capabilities given amenable clinical data types as represents in Fig. 5.

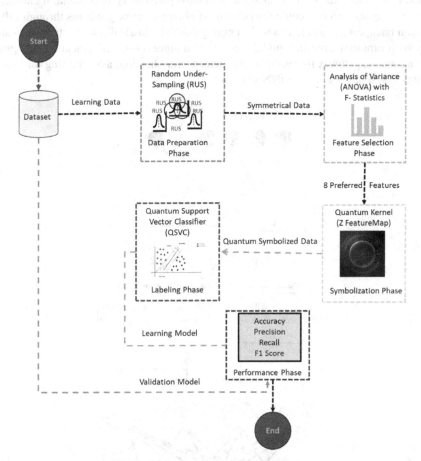

Fig. 5. Quantum Support Vector Machines

4.1   Quantum Neural Networks (QNN)

Quantum neural networks (QNN) analogously apply qubit-mediated information flows through interconnected node layers to mirror deep learning for pattern recognition, nonlinear regression and other machine learning tasks [20]. QNN models implement familiar network primitive operations like perceptrons, convolution and pooling using reversible quantum gates and entanglement mechanisms to enable richer representations

[21]. This facilitates learning within raw input datasets and high-dimensional feature spaces favorable for clinical biosignal analysis applications (Fig. 3).

Various QNN architectures offer differentiated strengths - recurrent QNN can capture temporal dynamics in physiological timeseries [22], convolutional QNN enable integrated feature extraction from medical images [23], while feedforward QNN allow robust deep analysis of omics data [24]. QNN-based disease classifiers have demonstrated high accuracy for cancer diagnosis using gene expression patterns. Hybrid quantum-classical training schemes also overcome restrictions on obtaining error gradients through fully quantum backpropagation to enable leveraging existing classical optimization software [25]. With rapidly increasing model capacity and algorithmic refinements, QNN provides a versatile pathway for multiple early detection applications as supporting hardware matures Fig. 6 has drawn the QNN structure.

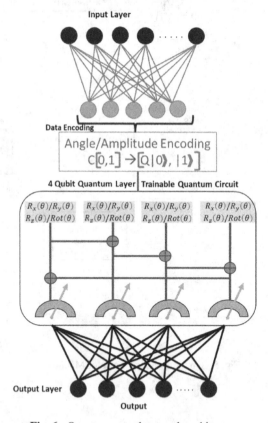

Fig. 6. Quantum neural network architecture

4.2   Quantum Boltzmann Machines

Quantum Boltzmann machines (QBM) offer a generative approach to model multidimensional distributions using qubit-mediated stochastic neural networks [26]. QBM leverage sampling and quantum interference effects to capture complex correlations within training data for density modeling and information retrieval applications, avoiding limitations of classical Boltzmann machines [27]. QBM have shown capability for EEG bio signal modeling by learning compressed representations of spatiotemporal brain dynamics indicative of pathological neurophysiology for early neurological disorder diagnosis [28]. Quantum annealing processors like by D-Wave systems have demonstrated QBM implementations at industrially relevant scales using over 5000 superconducting qubits. While coherence and connectivity constraints persist, ongoing rapid chip iteration promises economically viable quantum advantage regimes for QBM-based medical data modeling using quantum annealing technology within coming years [29]. Figure 7 represents the hidden and visible nodes characteristics of quantum Boltzmann machines with machine learning.

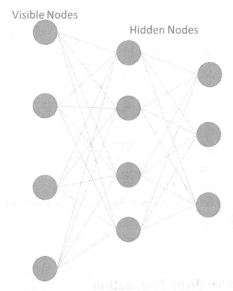

Fig. 7. Machine-learning-with-quantum-Boltzmann-machines

4.3 Quantum Support Vector Machines

Quantum support vector machines (QSVM) apply established kernel methods for statistical pattern recognition by efficiently mapping classical input data into enhanced quantum feature spaces as presents in Fig. 8 [30]. Quantum kernel functions leveraging parameterizable quantum circuits can project clinical datasets into exponentially large Hilbert spaces relative to classical SVMs to better resolve diagnostic boundaries even under noise. QSVM classifiers have shown accurate pathology detection from metabolomic and genomic biomarkers for cancer and Alzheimer's disease versus classical counterparts [31]. Photonic implementations have also enabled QSVM proof-of-principle demonstrations. As algorithms and hardware mature to support more flexible data encoding and model optimization, QSVM are poised to deliver quantum advantages for boosted early detection accuracy from subtle molecular cues [Fig. 8].

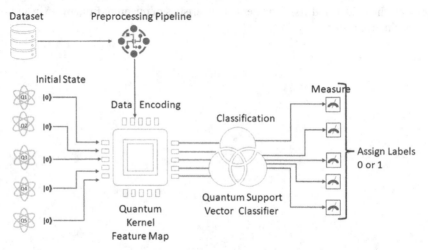

Fig. 8. Established Process between Quantum Support Vector Machine Classifier and Quantum Kernel Feature Map [30]

5 Challenges for Medical Translation

The potential of QML to revolutionize healthcare is evident in its application to specific medical datasets, genomic information, and intricate biochemical interactions. However, realizing this potential necessitates addressing critical challenges, such as developing noise-resistant algorithms, optimizing models to align with hardware limitations, and seamlessly integrating QML into existing clinical workflows.

5.1 Noise Tolerance and Mitigation

Delicate quantum states requisite for advantage easily degrades from environmental noise and errors which currently limit most hardware platforms. Developing noise-resilient algorithms and circuits is critical for maintaining model fidelity and prediction reliability necessitated in medical settings [32]. Hybrid quantum-classical co-processing can also help correct errors by using redundant qubit encoding and majority voting schemes. Advances in quantum error correction codes and fault-tolerant logic operations will additionally enable scalable, reliable QML implementations as qubit counts grow [33].

5.2 Problem-Aware Model Optimization

Adapting QML algorithms like neural networks or support vector machines to best leverage available quantum resources remains non-trivial, requiring cross-disciplinary perspective [34]. Factors like qubit connectivity constraints, gate primitive libraries and coherence decay profiles vary widely across candidate backend processors. Co-designing circuit layouts, feature encoding and learning optimization tailored for target clinical applications and hardware combinations can better exploit quantum volume limits than naive adaptation of classical algorithms. Problem-specific benchmarking will identify when quantum advantages manifest over classical equivalents given practical technological constraints [35].

5.3 Workflow Integration Complexities

Integrating QML solutions like predictive diagnostic assistants into clinician workflows requires considering complex sociotechnical factors beyond just model accuracy improvements. User trust in quantum-derived insights, transparent handling of algorithmic limitations, and interpretable explanation of machine decisions will enable clinical adoption [36]. Additional clinical validation studies measuring real-world performance across diverse patient populations are also critical to demonstrate meaningful improvements over incumbent diagnostic modalities before replacement. Ultimately realizing QML advances for patient benefit requires deliberately bridging disciplinary boundaries between medical, quantum science and ethics/policy communities [37].

6 Final Discussion

Quantum Machine Learning (QML) has emerged as a powerful tool in the field of medical applications, with the ability to introduce innovative approaches that could fundamentally change the healthcare industry. We have investigated QML in the medical applications space in the Table 1.

Table 1. Quantum Machine Learning (QML) Perspective, Challenges and Opportunities in Early Disease Detection in Healthcare

Medical Applications	QML Methodology of Medical fields	Perspective of Medical Data Study	QML Challenges of Medical fields	QML Opportunities of Medical fields
Healthcare	Quantum-assisted linear regression model	Specific medical datasets: patient demographics, medical history, and test results	Identify the target variable you want to predict	To improve the efficiency and accuracy in predicting specific health outcomes
Precision Medicine	Quantum support vector machines, quantum neural networks model	Patient health, including genomic, clinical, lifestyle, and other relevant information	Define metrics for evaluating medicine analytics	To guide decisions on medication, dosage, and intervention strategies
Drug Discovery	Quantum chemistry algorithms suitable for simulating molecular structures and properties	Predicting molecular properties relevant to drug discovery: binding affinity, toxicity, and solubility	To predict molecular properties, assess drug-likeness, and prioritize compounds	To the development of novel medications, addressing unmet medical needs
Alzheimer's Prediction	Leveraging quantum graph neural networks (QGNN) and quantum generative adversarial networks (QGAN)	Including electronic health records, cognitive assessments, genetic information, and imaging data	Define metrics for evaluating the accuracy, sensitivity, and specificity	To predict earlier Alzheimer's disease onset using patient data
Cancer Detection	Leveraging quantum convolutional neural networks (QCNN) and quantum support vector machines (QSVM)	Analyze medical imaging data: MRI, X-ray film, CT scans, and pathology slides	To improve the precision, sensitivity, and speed of early cancer diagnosis	To improve the accuracy and efficiency of detecting various cancer types from medical imaging data

(continued)

Table 1. (*continued*)

Medical Applications	QML Methodology of Medical fields	Perspective of Medical Data Study	QML Challenges of Medical fields	QML Opportunities of Medical fields
Diabetes Prognosis	Quantum recurrent neural networks (QRNN) and quantum Boltzmann machines (QBM)	Analyze patients' diverse datasets of time-series: glucose levels, insulin usage, lifestyle factors, and medical history	To enhance the accuracy and reliability of diabetes prognosis	To analyze temporal patient data and improve the precision of predicting diabetes-related outcomes
Cardiovascular Disease Diagnosis	Leveraging quantum feedforward neural networks (QFNN) and quantum probabilistic models (QPM)	Analyze diverse datasets: cholesterol levels, blood pressure, and cardiovascular imaging data - MRI, CT scans	Define metrics for evaluating the accuracy, sensitivity, and specificity	To improve patient outcomes, optimize treatment plans, and advance the understanding of cardiovascular conditions
Parkinson's Disease Prediction	Quantum support vector machines, quantum neural networks model	Analyze diverse datasets encompassing patient symptoms, Genetic information, neuroimaging data, and treatment responses	To improve prediction accuracy and treatment personalization	Enhance the analysis of neuroimaging data and more accurate insights into the structural and functional changes

- *Healthcare:* Applying a Quantum-assisted linear regression model to specific medical datasets helps solve the problem of determining the target variable for predicting health outcomes. The potential is in improving the effectiveness and precision of forecasting particular health outcomes, providing a route to more accurate and tailored healthcare interventions.

  Objective: *Determining the dependent variable for forecasting health results.*

Opportunity: *Enhancing efficacy and precision in forecasting distinct health out-comes, hence facilitating tailored healthcare interventions.*

- **Precision Medicine:** By employing Quantum support vector machines and quantum neural networks, it is possible to evaluate patient health data, which includes genomic, clinical, and lifestyle information. This approach offers a chance to establish metrics for assessing the effectiveness of medicine analytics. This not only informs decisions regarding medication and dose, but also establishes a basis for more customized and efficient intervention tactics in the field of precision medicine.

  Objective: *Establishing criteria for assessing pharmaceutical analytics.*
  Providing guidance for making informed choices on medication, dosage, and intervention approaches, so helping to the development of individualized healthcare solutions.

- **Drug Discovery:** The use of quantum chemistry algorithms to simulate chemical structures poses a problem in accurately predicting features that are important for drug discovery. The potential for advancement is significant, resulting in the creation of innovative drugs that target unfulfilled medical requirements through the prediction of molecular characteristics, evaluation of drug similarity, and prioritization of compounds with enhanced precision.

  Objective: Forecasting molecular characteristics pertinent to pharmaceutical development.
  Opportunity: Developing innovative drugs by predicting molecular features, measuring drug-likeness, and ranking molecules.

- **Alzheimer's Prediction:** Utilizing quantum graph neural networks and quantum generative adversarial networks to examine a range of patient data, such as electronic health records and imaging, presents the difficulty of establishing precise accuracy measures. The potential lies in accurately forecasting the early occurrence of Alzheimer's disease, a crucial stride towards proactive patient care and administration.

  Objective: *Establishing criteria for measuring accuracy, sensitivity, and specificity.*
  Opportunity: *Utilizing patient data to forecast the start of Alzheimer's disease at an early stage, facilitating proactive healthcare for patients.*

- **Cancer Detection:** Utilizing quantum convolutional neural networks and quantum support vector machines to analyze medical imaging data presents the problem of enhancing precision, sensitivity, and speed. This opportunity is extensive, with the goal of improving the precision and effectiveness of identifying different types of cancer from intricate medical imaging data. This might potentially result in earlier diagnosis and better patient outcomes.

Objective: *Enhancing the accuracy, responsiveness, and efficiency of early cancer detection.*

Opportunity: *Improving the precision and effectiveness of cancer detection across different forms of cancer through the analysis of medical imaging data.*

- **Diabetes Prognosis:** The use of Quantum recurrent neural networks and quantum Boltzmann machines to assess various patient time-series data presents a difficulty in improving the accuracy and dependability of diabetes prognosis. The possibility is in utilizing temporal patient data to conduct in-depth analysis, resulting in more accurate predictions, and ultimately enhancing the treatment strategies for diabetes-related outcomes.

  Objective: *Improving the precision and dependability of diabetes prognosis.*

  Opportunity: *Utilizing temporal patient data to enhance accuracy in predicting diabetes-related outcomes.*

- **Diagnosis of Cardiovascular Disease:** Applying quantum feedforward neural networks and quantum probabilistic models to various cardiovascular datasets poses difficulties in determining accurate metrics. The promise lies in enhancing patient outcomes, optimizing treatment strategies, and advancing our comprehension of cardiovascular problems, so promoting a comprehensive approach to the diagnosis and therapy of cardiovascular illness.

  Objective: *Defining metrics for accuracy, sensitivity, and specificity.*

  Opportunity: *Enhancing patient results, enhancing therapeutic strategies, and advancing knowledge of cardiovascular disorders.*

- **Predicting Parkinson's Disease:** Utilizing Quantum support vector machines and quantum neural networks to assess various datasets associated with Parkinson's disease presents a difficulty in enhancing prediction precision and tailoring treatment. The objective is to improve the processing of neuroimaging data, yielding more precise observations regarding the structural and functional alterations linked to Parkinson's disease.

  Objective: *Improving the precision of predictions and tailoring treatments to individual needs.*

  Objective: *Enhancing the examination of neuroimaging data to achieve more precise understanding of the anatomical and physiological alterations linked to Parkinson's disease.*

To summarize, the use of Quantum Machine Learning in medical applications presents numerous opportunities to tackle obstacles and enhance precision, productivity, and customization in diverse healthcare fields. Amidst these problems, the potential of quantum techniques in medicine is highly promising, presenting novel solutions that could revolutionize healthcare procedures.

7 Conclusions

Quantum machine learning holds genuine disruptive potential to profoundly expand early disease detection capabilities if key challenges around robustness, design and adoption can be overcome. While current hardware restrictions preclude immediate widespread implementation, projections based on prototype scaling anticipate industrially relevant quantum advantage demonstrations for biomedical applications within 5–10 years (IBM, 2020). This intermediate milestone will validate pathfinder QML solutions sufficiently to warrant clinical translation focus. Subsequent maturation toward high qubit-count, fault-tolerant processors later in the decade will then cement quantum capabilities within diagnostic healthcare. Capitalizing on this window of opportunity will require marshaling intensified, dedicated efforts across stakeholders from quantum engineers to medical scientists to policy experts. The resultant interdisciplinary insights generated from codesigning specialized QML techniques with clinical validation feedback loops promises to catalyze innovation not otherwise possible using conventional approaches alone. Concerted nurturing of this emerging quantum medicine ecosystem will drive progress toward next-generation early screening tools, ushering cutting-edge QML applications into patient impact by 2030.

References

1. Amin, M.H., Andriyash, E., Rolfe, J., Kulchytskyy, B., Melko, R.: Quantum Boltzmann machine. Phys. Rev. X **8** (2018)
2. Benedetti, M., Garcia-Pintos, D., Perdomo, O., Leyton-Ortega, V., Nam, Y., Perdomo-Ortiz, A.: A generative modeling approach for benchmarking and training shallow quantum circuits. NPJ Quantum Inf. **5** (2019)
3. Bergholm, V., et al.: PennyLane: automatic differentiation of hybrid quantum-classical computations (2021)
4. Schuld, M., Bocharov, A., Svore, K.M.: Quantum machine learning for healthcare and medicine. Nat. Rev. Phys. **4**(9), 535–556 (2021)
5. Biamonte, J., Wittek, P., Pancotti, N., Rebentrost, P., Wiebe, N., Lloyd, S.: Quantum machine learning. Nature **549**, 195–202 (2017)
6. Zhao, J., Guo, X., Bucyk, T., Dougherty, E.R.: Quantum machine learning for drug discovery: state-of-the-art and future prospects. Wiley Interdiscip. Rev.: Comput. Mol. Sci. **12**(3), e1620 (2022)
7. Tkatchenko, A., Müller, K.R.: Quantum machine learning in chemistry and materials. Nat. Comput. Sci. **2**(8), 581–587 (2022)
8. Roy, S., Roy, D., Bhattacharjee, C.: Quantum machine learning models for drug discovery. Drug Discov. Today **26**(8), 1868–1884 (2021)
9. Lamata, L., Alvarez-Rodriguez, U., Martín-Guerrero, J.D., Sanz, M., Solano, E.: Quantum convolutional neural networks for early cancer diagnosis. Quantum Sci. Technol. **6**(3), 034003 (2021)
10. Yu, S., Xin, T., Li, H., Li, J., Zhang, Y., Chen, G.: Quantum generative adversarial network for Alzheimer's disease assessment. Inf. Sci. **504**, 232–248 (2019)
11. Zheng, M., Feng, Z., Chen, C., Li, J., Zhou, D.: Quantum graph neural network: application to prediction of Alzheimer's disease. Inf. Sci. **592**, 349–361 (2022)

12. Liang, X., Li, Y.: Quantum recurrent neural networks for diabetes detection and prediction. In: 2019 IEEE International Conference on Bioinformatics and Biomedicine (BIBM), pp. 550–555. IEEE (2019)

13. Li, R., Perez-Salinas, A., Alcalde, J.I., Ying, E., Peng, X.: Quantum machine learning for cardiovascular diseases: a survey. IEEE Rev. Biomed. Eng. (2022)

14. Akbar, S., Ammad-Ud-Din, M., Khan, F.A., Javed, A., Ishfaq, M., Baig, M.M.A.: Quantum machine learning models for the diagnosis of heart disease. IEEE Access 9, 108637–108652 (2021)

15. Bruzewicz, C.D., Chiaverini, J., McConnell, R., Sage, J.M.: Trapped-ion quantum computing: progress and challenges. Appl. Phys. Rev. 6 (2019)

16. Cao, Y., et al.: Quantum chemistry in the age of quantum computing. Chem. Rev. 121, 10856–10915 (2021)

17. Ferrante di Ruffano, L., Hyde, C.J., McCaffery, K.J., Bossuyt, P.M., Deeks, J.J.: Assessing the value of diagnostic tests: a framework for designing and evaluating trials. BMJ: Br. Med. J. (Online) 362 (2018)

18. Fujii, K., Nakajima, K.: Harnessing disordered-ensemble quantum dynamics for machine learning. Phys. Rev. Appl. 8, 024030 (2017)

19. Govindarajan, S., Lloyd, S., Lubinski, P., Maslov, D., Roetteler, M.: Quantum machine learning using data encoding algorithms. NPJ Quantum Inf. 5 (2019)

20. Havlíček, V., et al.: Supervised learning with quantum-enhanced feature spaces. Nature 567, 209–212 (2019)

21. IBM: IBM predicts quantum computing to reach a tipping point at "about 1000 qubits" moving from scientific discovery to engineering advantage|Quantum Computing Report [WWW Document] (2020). https://quantumcomputingreport.com/ibm-predicts-quantum-computing-to-reach-a-tipping-point-at-about-1000-qubits-moving-from-scientific-discov ery-to-engineering-advantage/. Accessed 2 Oct 2023

22. https://aliceliu2004.medium.com/quantum-support-vector-machines-a-new-era-of-ai-126 2dd4b2c7e

23. Kennedy, B., et al.: Factors associated with lack of awareness and treatment of heart failure in primary care: cross-sectional analyses of a large primary care cohort. Open Heart 3, e000476 (2016)

24. Kieferová, M., Wiebe, N.: Tomography and generative training with quantum Boltzmann machines. Phys. Rev. A - At. Mol. Opt. Phys. 96, 1–13 (2017)

25. Ladd, T.D., Jelezko, F., Laflamme, R., Nakamura, Y., Monroe, C., O'Brien, J.L.: Quantum computers. Nature 464, 45–53 (2010)

26. LaRose, R.: Overview and comparison of gate level quantum software platforms. Quantum 3, 130 (2019)

27. Li, R., Perez-Salinas, A., Cruz-Irisson, M., Torres-Herrera, E.J.: A quantum support vector machine algorithm based on qubit encoding and single-qubit rotations. Quantum Inf. Process. 14 (2015)

28. Mohseni, M., et al.: Commercialize early quantum technologies. Nat. Phys. 13, 414 (2017)

29. Oh, D., Kim, J., Kim, J., Chung, Y.: Mixed quantum-classical convolutional neural networks. In: 2019 IEEE 31st International Conference on Tools with Artificial Intelligence (ICTAI) (2019)

30. https://medium.com/@roysuman088/advancing-machine-learning-with-quantum-boltzmann-machines-a-new-paradigm-dcf5d3aa7e74

31. Schuld, M., Sinayskiy, I., Petruccione, F.: Prediction by linear regression on a quantum computer. Phys. Rev. A 89 (2014)

32. Schuld, M., Sinayskiy, I., Petruccione, F.: An introduction to quantum machine learning. Contemp. Phys. 56, 172–185 (2015)

33. Stokes, A., Izaac, J., Killoran, N., Carleo, G.: Quantum natural gradient. Quantum **4**, 269 (2020)
34. https://www.techscience.com/iasc/v36n1/50016/html
35. Wan, K.H., Dahlsten, O., Kristjánsson, H., Gardner, R., Kim, M.S.: Quantum generalisation of feed forward neural networks. NPJ Quantum Inf. **3** (2017)
36. Wittek, P.: Quantum machine learning: what quantum computing means to data mining (2014)
37. Yoo, S., Bang, J., Lee, C.: A review of quantum machine learning: Schrödinger meets machine learning. Rep. Prog. Phys. **84** (2021)

Human-Robot Interaction
and Intelligent Interfaces

Subject-Independent Brain-Computer Interfaces: A Comparative Study of Attention Mechanism-Driven Deep Learning Models

Aigerim Keutayeva and Berdakh Abibullaev[✉]

Nazarbayev University, Astana, Kazakhstan
{aigerim.keutayeva,berdakh.abibullaev}@nu.edu.kz

Abstract. This research examines the employment of attention mechanism driven deep learning models for building subject-independent Brain-Computer Interfaces (BCIs). The research evaluated three different attention models using the Leave-One-Subject-Out cross-validation method. The results showed that the Hybrid Temporal CNN and ViT model performed well on the BCI competition IV 2a dataset, achieving the highest average accuracy and outperforming other models for 5 out of 9 subjects. However, this model did not perform the best on the BCI competition IV 2b dataset when compared to other methods. One of the challenges faced was the limited size of the data, especially for transformer models that require large amounts of data, which affected the performance variability between datasets. This study highlights a beneficial approach to designing BCIs, combining attention mechanisms with deep learning to extract important inter-subject features from EEG data while filtering out irrelevant signals.

Keywords: Attention Mechanism · Brain-Computer Interface (BCI) · Deep Learning (DL) · Vision Transformers (VT) · Motor Imagery (MI)

1 Introduction

Brain-computer interfaces (BCIs) have emerged as a promising system for allowing users to operate objects with their mental activities. Electroencephalogram (EEG) recordings are one of the most frequently utilized techniques for acquiring brain signals [2,3]. Although EEG is a non-invasive data acquisition technique, the data collected is noisy, and most true brain signals are either masked by background activity or attenuated and smoothed by head layers, encompassing the skull and tissues [3]. The low signal-to-noise ratio poses challenges, compounded by significant heterogeneity in signal configurations between different

This work was supported by Nazarbayev University under the Faculty Development Competitive Research Grant Program (FDCRGP), Grant No. 021220FD2051.

B. J. Choi et al. (Eds.): IHCI 2023, LNCS 14531, pp. 245–254, 2024.
https://doi.org/10.1007/978-3-031-53827-8_23

participants. As a result, most BCIs currently in use employ subject-specific (subject-dependent) training that requires users to allocate substantial time for system calibration and training, facilitating tailored future usage [17]. This process is particularly inconvenient and time-consuming for individuals with disabilities. Consequently, new research aims to advance towards the creation of systems that are calibration-free or subject-independent [6,7,17,21].

This study explores subject-independent classification of motor imagery (MI) tasks from EEG through the attention mechanism-based transformer models. The transformer framework, proposed in the seminal paper by Vaswani et al. in 2017 [20], has made progress in the domain of natural language processing. This architecture served as the foundation for models such as BERT, GPT, and numerous others, extending the frontiers of machine learning endeavors in various domains, not limited to just NLP [13,15,21].

Transformers are distinguished by their unique and innovative parts, primarily the self-attention mechanism. This mechanism enables the model to assign significance to varying segments of the input data in relation to each other. Additionally, the positional encoding furnishes the model with a chronological framework, ensuring its capacity to account for data positions within a sequence. The inclusion of multi-head attention in transformers ensures that the model can concentrate on diverse segments of the input accordingly [22]. Key advantages of the transformer architectures encompass parallel processing, scalability, and adaptability. Transformers handle all data inputs concurrently, resulting in expedited training durations. Their scalability is evident in larger models' ability to discern complex patterns in extensive datasets, thus achieving benchmark performance across various challenges. Although originally conceptualized for NLP objectives, the versatility of transformers is demonstrated by their efficacy in other fields, including computer vision and BCIs [4,10]. The success of the original transformer architecture facilitated the development of specialized variants tailored to diverse tasks. Paramount among these is Vision Transformers [5,13].

A review of existing classification models in the domain of BCIs indicates that transformer models outperform conventional deep learning models in EEG classification [20,21]. In particular, research conducted by Sun et al. (2021) highlighted the effectiveness of transformer-based models for the classification of EEG tasks, achieving state-of-the-art performance on a benchmark EEG dataset using self-attention and cross-attention techniques to learn subject-specific and subject-independent features [17]. In another study, Tao et al. [19] implemented a gated Transformer for multi-class classification on the EEG data. These and related studies provide compelling motivation for using attention mechanisms in BCI system development.

This study investigates subject-independent BCIs by utilizing deep learning models founded on attention mechanisms, namely ViT, Hybrid Spatial CNN and ViT, and Hybrid Temporal CNN and ViT. Our approach involves using attention mechanisms and deep learning models to determine the most informative features universal across subjects, concurrently filtering out irrelevant data and noise. The findings show that the Hybrid Temporal CNN combined with ViT model excels beyond other models when evaluated on the BCI competition IV 2a dataset, highlighting its robustness against inter-subject variations. However, in the case of the BCI competition IV 2b dataset, it did not show any improvement in the average of best accuracies.

2   Materials and Methods

2.1   Datasets

This research used open-access EEG datasets, namely BCI IV 2a and BCI IV 2b, to study motor imagery tasks performed by diverse participants. The primary objective of the investigation was to decode motor imagery data related to tasks centered on the visualized movement of the left and right hands. The continuous EEG data was divided into distinct 4-second intervals that aligned with the mental imagery onset for both left-hand and right-hand imagery tasks. [1]. All datasets underwent preprocessing to reduce the output classes to only two categories: left-hand and right-hand. Furthermore, a segmentation and reconstruction (S&R) method [16] was applied in the time domain, along with Z-score normalization. The details on each dataset are provided in Table 1.

Table 1. Details on EEG datasets are denoted with "L", "R", "F", and "T" which correspond to the motor imagery tasks of the left-hand, right-hand, feet, and tongue, respectively.

Dataset	Subjects	EEG Channels	SR (Hz)	Trials/Session	Sessions	MI class
BCI IV 2a [18]	9	22	250	288	2	L/R/F/T
BCI IV 2b [12]	9	3 (C3, Cz, C4)	250	120	5	L/R

2.2   Proposed Transformer Model Architecture

Transformers utilize the advantages of parallel processing, allowing for the simultaneous processing of entire data sequences, which notably accelerates the training stages, particularly when executed on modern hardware designed for parallel computations. Central to the Transformer design is the attention mechanism, with self-attention being of particular significance. This unique mechanism adeptly allocates varied weights to elements within a sequence, contingent upon their relevance to the specific task at hand. Consequently, this allows the model to place selective emphasis on particular segments of the input, thereby enhancing the model's overall understanding and representation of data sequences.

In this study, the following models were investigated: ViT, Hybrid Spatial CNN and ViT, and Hybrid Temporal CNN and ViT.

– **Vision Transformer (ViT):** ViT represents a neural network architecture predominantly customized for the purpose of image classification [8]. Instead of resorting to conventional convolutional layers, ViT makes use of the transformer encoder, conceived initially for Natural Language Processing (NLP) tasks [20].

In deploying ViT to motor imagery data, it is imperative to properly configure the data. A viable approach is to view the data Z_s as a sequence of "tokens" or "patches" P.

$$Z_{patch} = W_p Z_s + b_p \tag{1}$$

In this context, W_p represents the weight for the patch embedding, while b_p denotes its corresponding bias. Subsequently, these embeddings undergo processing via the transformer architecture, which employs self-attention, feedforward networks, and other integral components.

In its foundational form, self-attention functions as follows. Three main matrices are discerned throughout the training phase, namely the query W^q, key W^k, and value W^v matrices, as cited in [20]. For each element of an embedded input series presented to the self-attention layer, these matrices facilitate the creation of three separate vectors for each input, designated as the query Q, key K, and value V vectors.

$$Q_i = \vec{x}_i \cdot W^q \tag{2}$$

$$K_i = \vec{x}_i \cdot W^k \tag{3}$$

$$V_i = \vec{x}_i \cdot W^v \tag{4}$$

The "Scaled Dot-Product Attention" is derived using the equation presented in the Eq. 5.

$$\text{Attn}(Q, K, V) = Softmax\left(\frac{QK^T}{\sqrt{d_k}}\right) V \tag{5}$$

Through the incorporation of "Scaled Dot-Product Attention" layers, which constitute the multi-head attention, the model is enabled to focus concurrently on diverse representational fragments from various positions. The "Multi-Head Attention" is illustrated in formula 6:

$$\begin{aligned} \text{MH}(Q, K, V) &= [head_1, ..., head_h], \\ head_i &= \text{Attention}(Q_i, K_i, V_i) \end{aligned} \tag{6}$$

- **Spatial Attention:** Spatial Attention is a mechanism that determines and assigns weights to the contributions deriving from distinct spatial positions of the EEG electrodes. This becomes particularly advantageous when the precise electrode placement holds significance for the given task.

Given an EEG input matrix X of size $T \times E$ where T is the number of time points and E is the number of electrodes, the spatial convolution operation can be defined as:

$$Y_{spatial} = f(X * W_{spatial} + b_{spatial}) \tag{7}$$

Here, $*$ denotes the convolution operation, $W_{spatial}$ is the weight matrix associated with spatial convolution, $b_{spatial}$ is the bias term, and f is an activation function such as ReLU.

The spatial attention weights can be calculated using:

$$\alpha_{spatial} = \text{softmax}(W_{a_s} Y_{spatial} + b_{a_s}) \tag{8}$$

Here, W_{a_s} is the attention weight matrix for spatial attention, and b_{a_s} is its corresponding bias. The softmax function ensures the attention weights sum to 1.

The weighted spatial feature matrix is then:

$$Z_{spatial} = \alpha_{spatial} \odot Y_{spatial} \tag{9}$$

Here, $\odot$ represents element-wise multiplication.

- **Temporal Attention:** Temporal Attention serves as a technique enabling the model to concentrate on specific time intervals within the EEG signal. This mechanism proves valuable when the temporal progression of EEG features plays a pivotal role in the task at hand.

  Given an EEG input matrix X of size $E \times T$ where E is the number of electrodes and T is the number of time points, the temporal convolution operation can be defined as:

$$Y_{temporal} = f(X * W_{temporal} + b_{temporal}) \tag{10}$$

  Here, $W_{temporal}$ is the weight matrix associated with temporal convolution, $b_{temporal}$ is the bias term, and f is an activation function like ReLU.

  The weighted temporal feature matrix is then:

$$Z_{temporal} = \alpha_{temporal} \odot Y_{temporal} \tag{11}$$

These attention strategies can be integrated into multiple deep learning architectures, such as Transformers and Convolutional Neural Networks (CNNs), to enhance the model's efficacy in processing EEG information. In our study, we use the combination of CNNs and ViTs as illustrated in Fig. 1.

2.3   Model Selection

We used the Leave-One-Subject-Out (LOSO) approach for model evaluation and selection. The LOSO is a cross-validation technique widely adopted within the realm of BCI research, particularly when working with limited datasets or examining the subject-independent properties of the models. In the LOSO approach, data from all participants except one are allocated for training and validation purposes. Specifically, of this data, 80% is designated for training and the remaining 20% for the fine-tuning of model hyperparameters through validation. The excluded subject's data is then utilized exclusively for model evaluation.

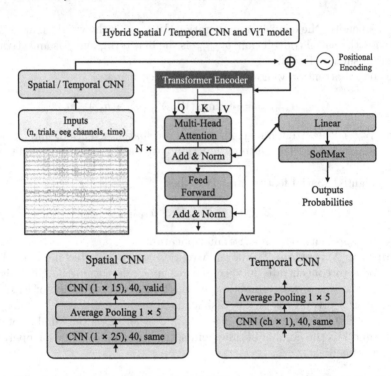

Fig. 1. The detailed architecture of original ViT model from [8], Hybrid Spatial CNN and ViT model, and Hybrid Temporal CNN and ViT model for EEG classification.

In the process of optimizing our models, several key hyperparameters were meticulously selected based on both empirical evidence and existing literature. Firstly, the model training was limited to a maximum of 150 epochs and an early stopping criterion was integrated to prevent overfitting. For the batch size, a mini-batch size of 72 was selected for the Transformer models after a comprehensive exploration of various batch sizes. In terms of optimization strategies, we utilized the mini-batch gradient descent method in conjunction with the Adam optimizer. The parameters for the learning rate and decay were established at 0.001 and 0.0001, respectively, in line with the suggestions presented by [11]. Moreover, cross-entropy was chosen as the loss function, consistent with the guidelines provided by [9].

3 Results and Discussion

This section evaluates the efficacy of diverse deep learning techniques applied to BCI datasets, specifically BCI competition IV 2a and 2b. Prior research efforts have investigated these datasets using various methods; however, it is crucial to acknowledge the disparities in experimental setups and testing procedures observed across these studies.

The performance evaluation included the following methods: Deep CNN [1], EEGNet [14], ViT [8], Hybrid Spatial CNN and ViT, and Hybrid Temporal CNN and ViT. The effectiveness of each method for individual subjects was assessed by employing the Leave-One-Subject-Out (LOSO) methodology. To ensure comparability with prior studies, modifications were applied to all the methods. Tables 2 and 3 present the performance outcomes of these diverse methods on the Motor Imagery (MI) datasets, highlighting the top LOSO results.

After examining the differences between the parameter sizes (Params) in Tables 2 and 3, it is pertinent to note that the discrepancy arises from the varying input dimensions. As per Table 1, the BCI IV 2a dataset uses 22 EEG channels, while the BCI IV 2b dataset utilizes only 3 EEG channels. Given that the architecture of the models stays consistent across datasets, the difference in the number of input channels significantly affects the number of parameters, especially in the initial layers of the models. This leads to the variations observed in the Params columns for the same models across the two tables.

The results show that models founded on attention mechanisms, including the ViT and its spatial, and temporal derivatives, demonstrated varying performance contingent upon the specific dataset.

Table 2. Comparison with representative methods on BCI IV 2a dataset.

Methods	S1	S2	S3	S4	S5	S6	S7	S8	S9	Average	Params
CNN[64, 32, 16, 8]_K(H = 3, W = 8)	0.72	0.59	0.78	0.65	0.57	0.61	0.58	0.86	0.78	0.68	1.01M
CNN[64, 32, 16, 8]_K(H = 3, W = 24)	0.74	0.58	0.81	0.65	0.56	0.63	0.64	0.77	0.76	0.68	1.15M
CNN[8, 16, 32, 64]_K(H = 3, W = 8)	0.70	0.51	0.80	0.63	0.52	0.68	0.55	0.77	0.75	0.66	7.64M
CNN[8, 16, 32, 64]_K(H = 3, W = 24)	0.66	0.52	0.74	0.58	0.53	0.69	0.56	0.74	0.72	0.64	7.76M
EEGNet	0.80	0.53	0.86	0.55	0.58	0.63	0.58	0.87	0.67	0.67	19.95k
ViT	0.53	0.56	0.55	0.56	0.55	0.57	0.55	0.55	0.54	0.55	2.73M
Hybrid Spatial CNN and ViT	0.59	0.59	0.58	0.56	0.55	0.59	0.60	0.54	0.61	0.58	989.43k
Hybrid Temporal CNN and ViT	0.73	0.59	**0.92**	**0.70**	**0.59**	**0.73**	0.56	**0.95**	0.77	**0.73**	769.83k

For the BCI IV 2a dataset, it was observed that the base Vision Transformer (ViT) underperformed than most Deep CNN models and EEGNet, registering an average accuracy of 0.55. However, its integration with alternative modeling techniques led to marked enhancements in performance. The Hybrid Temporal CNN and ViT model achieved significantly higher accuracy of 0.73, thereby surpassing the other evaluated models. In the context of the BCI IV 2b dataset, a similar trend was noted with the standalone ViT model yielding a comparatively lower average accuracy of 0.53. However, the confluence of ViT with other modeling methods again resulted in remarkable improvements. Despite these improvements, Deep CNN models held a dominant position, securing the highest average accuracy at 0.74. From a computational perspective, all ViT-based models showcased a notably reduced parameter count relative to Deep CNN models, offering advantages for model deployment and computational efficiency. Specifically, in the BCI IV 2a dataset, the Hybrid Temporal CNN and ViT model

Table 3. Comparison with representative methods on BCI IV 2b dataset.

Methods	S1	S2	S3	S4	S5	S6	S7	S8	S9	Average	Params
CNN[64, 32, 16, 8]_K(H = 3, W = 8)	0.70	0.57	0.57	0.92	0.73	0.78	0.71	0.79	0.79	0.73	195.96k
CNN[64, 32, 16, 8]_K(H = 3, W = 24)	0.71	0.57	0.55	0.92	0.76	0.82	0.74	0.77	0.78	0.74	328.06K
CNN[8, 16, 32, 64]_K(H = 3, W = 8)	0.72	0.55	0.54	0.91	0.71	0.78	0.66	0.78	0.78	0.72	1.10M
CNN[8, 16, 32, 64]_K(H = 3, W = 24)	0.67	0.56	0.54	0.91	0.72	0.79	0.73	0.77	0.79	0.72	1.23M
EEGNet	0.67	0.60	0.57	0.84	0.72	0.79	0.72	0.77	0.80	0.72	19.35k
ViT	0.54	0.53	0.52	0.52	0.55	0.52	0.55	0.51	0.53	0.53	1.42M
Hybrid Spatial CNN and ViT	0.67	0.58	0.54	0.84	0.62	0.65	0.65	0.73	0.71	0.67	405.75k
Hybrid Temporal CNN and ViT	0.68	0.60	0.54	0.85	0.75	0.72	0.68	0.78	0.79	0.71	768.61k

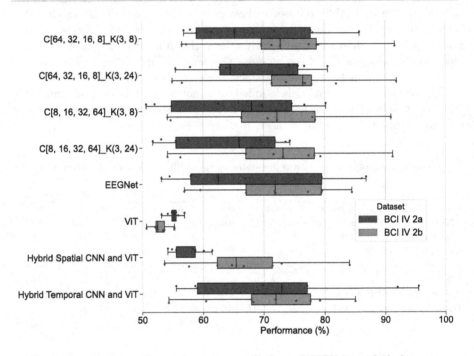

Fig. 2. Comparison with representative methods on BCI IV 2a and 2b datasets.

not only exhibited superior performance but also stood out by having the least number of parameters among all models evaluated (Fig. 2).

4   Conclusion

In this research, we evaluated various models, including ViT, Hybrid Spatial CNN and ViT, and Hybrid Temporal CNN and ViT, to determine their accuracy in Brain-Computer Interfaces. We analyzed the advantages and disadvantages of each model, using the LOSO approach to ensure the validity of our findings. Additionally, we considered two different Motor Imagery EEG datasets to gauge the performance of these attention-centric models under different conditions.

Our observations indicated that the BCI IV 2a dataset, encompassing 22 EEG channels, yielded better results with the Hybrid Temporal CNN and ViT model in comparison to the BCI IV 2b dataset with a mere 3 EEG channels. This highlights the challenges associated with utilizing data-intensive models such as transformers when the spatial information in the dataset is limited. Broadly, our findings suggest that integrating attention mechanisms with deep learning offers a promising avenue for the development of adaptable BCIs.

References

1. Abibullaev, B., Dolzhikova, I., Zollanvari, A.: A brute-force CNN model selection for accurate classification of sensorimotor rhythms in BCIS. IEEE Access **8**, 101014–101023 (2020). https://doi.org/10.1109/ACCESS.2020.2997681
2. Abibullaev, B., Zollanvari, A.: A systematic deep learning model selection for p300-based brain-computer interfaces. IEEE Trans. Syst. Man Cybern. Syst. **52**(5), 2744–2756 (2021)
3. Ball, T., Kern, M., Mutschler, I., Aertsen, A., Schulze-Bonhage, A.: Signal quality of simultaneously recorded invasive and non-invasive EEG. Neuroimage **46**(3), 708–716 (2009)
4. Dai, Y., et al.: MultiChannelSleepNet: a transformer-based model for automatic sleep stage classification with PSG. IEEE J. Biomed. Health Inform. 1–12 (2023). https://doi.org/10.1109/JBHI.2023.3284160
5. Devlin, J., Chang, M.W., Lee, K., Toutanova, K.: BERT: pre-training of deep bidirectional transformers for language understanding (2019). arXiv:1810.04805 [cs]
6. Dolzhikova, I., Abibullaev, B., Sameni, R., Zollanvari, A.: An ensemble CNN for subject-independent classification of motor imagery-based EEG. In: 2021 43rd Annual International Conference of the IEEE Engineering in Medicine & Biology Society (EMBC), pp. 319–324. IEEE (2021)
7. Dolzhikova, I., Abibullaev, B., Sameni, R., Zollanvari, A.: Subject-independent classification of motor imagery tasks in EEG using multisubject ensemble CNN. IEEE Access **10**, 81355–81363 (2022)
8. Dosovitskiy, A., et al.: An image is worth 16 × 16 words: transformers for image recognition at scale (2021)
9. Goodfellow, I., Bengio, Y., Courville, A.: Deep Learning. MIT Press, Cambridge (2016)
10. Khan, S., Naseer, M., Hayat, M., Zamir, S.W., Khan, F.S., Shah, M.: Transformers in vision: a survey. ACM Comput. Surv. (CSUR) **54**(10s), 1–41 (2022)
11. Kingma, D.P., Ba, J.: Adam: a method for stochastic optimization. CoRR abs/1412.6980 (2014)
12. Leeb, R., Lee, F., Keinrath, C., Scherer, R., Bischof, H., Pfurtscheller, G.: Brain-computer communication: motivation, aim, and impact of exploring a virtual apartment. IEEE Trans. Neural Syst. Rehabil. Eng. **15**(4), 473–482 (2007). https://doi.org/10.1109/TNSRE.2007.906956
13. Lu, J., Batra, D., Parikh, D., Lee, S.: ViLBERT: pretraining task-agnostic visiolinguistic representations for vision-and-language tasks. In: Advances in Neural Information Processing Systems, vol. 32 (2019)

14. Peng, R., et al.: TIE-EEGNet: temporal information enhanced EEGNet for seizure subtype classification. IEEE Trans. Neural Syst. Rehabil. Eng. **30**, 2567–2576 (2022). https://doi.org/10.1109/TNSRE.2022.3204540

15. Raffel, C., et al.: Exploring the limits of transfer learning with a unified text-to-text transformer (2020). arXiv:1910.10683 [cs, stat]

16. Song, Y., Zheng, Q., Liu, B., Gao, X.: EEG conformer: convolutional transformer for EEG decoding and visualization. IEEE Trans. Neural Syst. Rehabil. Eng. **31**, 710–719 (2023). https://doi.org/10.1109/TNSRE.2022.3230250

17. Sun, J., Xie, J., Zhou, H.: EEG classification with transformer-based models. In: 2021 IEEE 3rd Global Conference on Life Sciences and Technologies (LifeTech), pp. 92–93 (2021). https://doi.org/10.1109/LifeTech52111.2021.9391844

18. Tangermann, M., et al.: Review of the BCI competition IV. Front. Neurosci. **6** (2012). https://doi.org/10.3389/fnins.2012.00055

19. Tao, Y., et al.: Gated transformer for decoding human brain EEG signals. In: 2021 43rd Annual International Conference of the IEEE Engineering in Medicine & Biology Society (EMBC), pp. 125–130 (2021). https://doi.org/10.1109/EMBC46164.2021.9630210

20. Vaswani, A., et al.: Attention is all you need. In: Advances in Neural Information Processing Systems, vol. 30 (2017)

21. Xie, J., et al.: A transformer-based approach combining deep learning network and spatial-temporal information for raw EEG classification. IEEE Trans. Neural Syst. Rehabil. Eng. **30**, 2126–2136 (2022). https://doi.org/10.1109/TNSRE.2022.3194600

22. Zhang, A., Lipton, Z.C., Li, M., Smola, A.J.: Dive into deep learning (2023). arXiv:2106.11342 [cs]

PPHR: A Personalized AI System for Proactive Robots

Bailey Wimer, M. I. R. Shuvo, Sophia Matar, and Jong-Hoon Kim[✉]

Advanced Telerobotics Research Lab, Computer Science, Kent State University,
Kent, OH, USA
jkim72@kent.edu
http://www.atr.cs.kent.edu

Abstract. This paper proposes the Personalized Proactive Home
Robotic (PPHR) system: an intelligent robotics system that enables
proactive behavior in the context of a home environment. With the implementation of the PPHR system, robots will be able to predict actions
that users would want by gathering data about both the users and the
environment. Then, once confident predictions are made, the robot can
perform the actions either actively or proactively depending on the situation. The system also uses personalized learning models to adapt the
experience to each of its users, and federated learning to improve data
privacy and train models faster. To prove the viability of the system,
we have designed and implemented the learning capabilities using data
gathered internally. Additionally, we have shown the transfer learning
capabilities of the system, allowing users to actively add new actions to
the robot at any time. With promising results, the system will serve as
a large step towards improved human-robot interaction.

Keywords: Home Robotics · Personalized Machine Learning ·
Behavior Prediction

1 Introduction

The integration of robots into our daily lives has significantly increased, emphasizing the necessity for more personalized and proactive human-robot interactions. One path for improving these interactions comes in the form of artificial
intelligence. With new artificial intelligence models and concepts being discovered frequently, there is significant potential for improvements to be made within
the field of robotics.

Personalized Machine Learning (PML) is utilized in robotics, HRI, and several medical fields [1–4]. Rudovic et al. [2] introduced a customizable deep learning approach for robots to assess the emotions of children with autism. Authors
used multimodal data to adapt their strategies for each child and get expert
results. Researchers collected data from individuals with anxiety problems using
smartphone sensors in the study [3]. They then utilized deep learning algorithms

---

This research was a partially supported by the ODHE RAPIDS-5 Grant.

B. J. Choi et al. (Eds.): IHCI 2023, LNCS 14531, pp. 255–267, 2024.
https://doi.org/10.1007/978-3-031-53827-8_24

to examine this data and create individualized models that accurately predicted anxiety symptoms through ecological momentary assessments (EMA). Chiu et al. [4] developed a machine learning-based approach to test for dysglycemia (abnormal blood glucose levels) using ECG data.

The study [5] proposed a customized machine learning approach for tracking professional volleyball players' injuries. The study collected data on players' training, injury histories, movement, and performance in practice and competition. Lopez-Martinez et al. [6] created a tailored pain detection approach combining fNIRS-recorded brain activity, wavelet transforms, Bayesian hierarchical models, and Dirichlet process priors. Ren et al. [7] created a tailored machine learning strategy to predict more negative impacts in teenagers. The approach uses GPS, accelerometer, microphone, and smartphone screen usage data to construct a unique model for each participant.

Federated Learning (FL) has been studied in robotics, HRI, and autonomous systems. A novel cognitive architecture based on FL was introduced in [8] for multi-agent learning from demonstration (LfD), involving human participation in robot self-learning loops for collaborative learning. Later work demonstrates adaptive improvement of large-scale multi-agent LfD using short- and long-term human behavior analysis and cognitive robot learning architecture [9]. In a multi-robot system, trajectory forecasting (spatiotemporal predictions) has been done using FL versions, including the standard FL approach where a cloud server aggregates local models and the serverless version. The authors found that the aforementioned techniques perform similarly for trajectory predicting, and provided the first federated learning dataset based on multi-robot behaviors [10].

In social situations, robots can predict human behavior and recognize the situation. Human-robot interaction reveals many human behaviors, such as gestures, gazes, vocal responses, postures, and expressions, that might predict future behavior [11]. Z. Shen et al. [12] developed a system to determine human personality traits by extracting nonverbal characteristics during social encounters using robots' perceptual functions. The robot asked each participant personality questionnaires in addition to documenting their usual tasks. Liu et al. [13] developed a method to learn reactive and proactive behavior through social interaction.

By combining these technologies, our team has designed the Personalized Proactive Home Robotics system (PPHR), is an innovative machine learning system designed to enable proactive behaviors in robots. The PPHR system uses personalized deep machine learning and federated learning to make smart decisions about proactive robot behaviors. In this paper, we will first discuss the proposed architecture for the entire system. We will then briefly describe the methodology powering the learning component. This is followed by an examination of implemented parts of the system and their performance. We will conclude with a discussion of future works for the PPHR system.

2   System Architecture

PPHR is broken into three major components, which can be seen in Fig. 1. The sensing component is responsible for identifying and processing real-world data

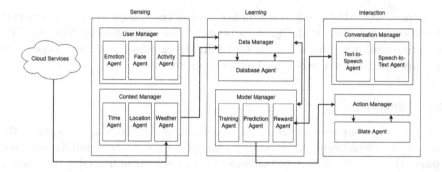

Fig. 1. The PPHR system is broken into three major components: a sensing component, a learning component, and an interacting component. The arrows in the diagram illustrate the flow of data between the separate systems.

from users and the environment around the robot. The learning component then takes that data and uses it to train prediction models. Once the models make a prediction, the interaction component is responsible for acting on that prediction and receiving feedback from the user. The following sections will describe each component in further detail.

2.1 Sensing

The sensing component of the system analyzes the environment and people around the robot to produce quality data for use in the other sections of the system. This is accomplished by utilizing advanced machine learning and artificial intelligence technology. These technologies are managed by two separate subcomponents: the User Manager and the Context Manager.

The User Manager examines the people in the vicinity of the robot. The Activity Agent, within the User Manager, tracks what each user's activities are. It also determines which person is interacting with the robot if multiple people are present. Once the user has been selected, the Face Agent and the Emotion Agent are used to identify who the user is and the current emotions they are displaying respectively.

The Context Manager consists of three agents which provide detailed information regarding the environment in which the robot exists. The first agent is the Time Agent, which gathers important information which relates to the current time in which actions are being performed. This is accomplished by the internal clock of the robot as well as brightness sensors to gather data on local lighting. The Location Agent uses mapping techniques to construct a virtual layout of the household. Then, through localization, this agent reports the current room in which the robot is at the time actions take place. Finally, the Weather Agent leverages cloud services to determine and report the area's weather conditions, including precipitation and temperature.

Importantly, this is only one simple design of the sensing system. Every part of the overall system is designed to be extensible with new or different data collection methods. This will allow for future development to involve as complex or simple of data as is necessary for the situation.

2.2 Learning

The learning component is the most important of the three components of the system. It is responsible for everything related to the artificial intelligence that powers the system. Within the component, sub-components enable the storage of data and training of models.

The Data Manager manages the fetching and storing of data through the Database Agent, as well as any necessary preprocessing for data. The Model Manager's job is much more complex than the other managers', and it happens in three stages. Before the system can make any actions or predictions, an initial model must be trained. This is done through the Training Agent, which takes data previously gathered from the sensing component and uses it to train the model. Once a model has been trained, then the Prediction Agent takes over to begin feeding the model live data to create predictions for actions that should occur. Once a prediction is made, both the prediction and the confidence in that prediction are sent to the Action Manager for processing. Finally, once the robot has acted on a prediction, the Reward Agent determines the accuracy of the prediction, as explained in Sect. 3.1, Reinforcement Learning. The outcome of this processing is used to further inform the system during future training.

2.3 Interacting

The interacting component is the least well-defined of the three components due to the variability of implementation across different robotic systems. Within the interacting component, the conversations are managed by the Conversation Manager. It utilizes a Text-to-Speech Agent and a Speech-to-Text Agent to do so. In the context of this system, conversations are the primary source of data related to what actions a user would want the robot to perform.

The second form of interaction in the system is performing actions. When a prediction is made, the information relating to it is sent to the Action Manager. From there, the State Agent uses the confidence of the prediction to determine whether the robot should be in a passive, active, or proactive state. If the confidence is low, the robot will enter the passive state, where it will not act on the prediction and instead wait for the user to make a request. When the confidence is very high, the robot will proactively complete the action without interacting with the user at all. If the confidence is between passive and proactive, then the robot will be in the active state and approach the user first to ask if an action should be performed.

3   Methodology

3.1   Reinforcement Learning

To facilitate continued learning from the predictions that the system makes, we are leveraging reinforcement learning [14]. The reward for each prediction will be derived from the actions that the user takes after the prediction is acted on. For example, if the robot brews a pot of coffee, and then the user gets a glass of water, there will be a negative reward associated with that action. However, if the user were to instead get a cup of coffee, then a positive reward would be given. Additionally, if more information is required to determine the value of the actions taken, the robot could prompt the user to rate the action on a scale from one to ten.

3.2   Personalization of Models

Due to the wide array of preferences held by individuals who could potentially use the system, using one model for an entire household is unreliable. By using facial recognition technology, the system is able to create a unique model for each user by training on data specific to that user. Not only does this increase the accuracy of predictions for each user, but it also allows new users to enter the system without jeopardizing other users' models. Similar concepts have seen success in applications across other fields such as the medical field [2,3,6].

3.3   Federated Learning

When introducing a new user into the system, there is a large overhead of gathering new data and training a model from scratch each time. To remedy this, the system uses federated learning - a technique that involves aggregating learning from multiple models together, then transferring that learning to a model for a new user [9]. The structure of the federated learning system is shown in Fig. 2. Once local models are trained, they are sent to the Central Server and aggregated to form several global models. Each global model represents one group of users, determined by the User Classification system explained below.

Within the central server, the accuracy of all of the client models is normalized as shown in (1), where a_k is the accuracy of model k and M is the number of client models.

$$\alpha_k = \frac{a_k - a_{min}}{a_{max} - a_{min}} \tag{1}$$

$$\alpha^* = \sum_{k=1}^{M} \alpha_k \tag{2}$$

Algorithm 1. Aggregation of client models

Require: $\Theta^1...\Theta^M, L > 0, N_1...N_L > 0$
Ensure: $\overline{\Theta} \leftarrow$ Aggregated Model
  for $k = 1$ to M **do**
    $a_k \leftarrow$ accuracy of model k
  end for
  $A_k \leftarrow (a_k - a_{min})/(a_{max} - a_{min})$
  $A^* \leftarrow$ the sum of A
  for $l = 2$ to L **do**
    for $i = 1$ to N_L **do**
      for $j = 1$ to N_{L-1} **do**
        $\Theta_{temp} \leftarrow 0$
        for $k = 1$ to M **do**
          $\Theta_{temp} \leftarrow \Theta_{temp} + \Theta_{ij}^{lk} * A^k$
        end for
        $\overline{\Theta}_{ij}^l \leftarrow (\Theta_{temp}/A^*)$
      end for
    end for
  end for

For each node, the average of the weights of the node are taken, weighted by the normalized accuracy of each model. The weights of model k for layer l are Θ_{ij}^{lk}, where i is a node in the lth layer and j is a node in the $l^{th} - 1$ layer.

$$\overline{\Theta}_{ij}^l = \frac{1}{\alpha^*} \sum_{k=1}^M \Theta_{ij}^{lk} \cdot \alpha_k \tag{3}$$

User Classification. When a new user is added to a robot in the system, they are first presented with a set of questions related to their background and preferences. Based on the answers to these questions, the user is classified into one of several groups of users through K-Means clustering on the central server. Once their group has been determined, their federated learning will occur with the models related to that classification. A similar system of classification was used in [2].

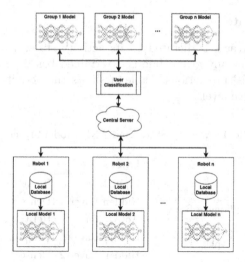

Fig. 2. The federated aspect of the system involves any number of local models aggregating together to form the basis of future learning across global models.

3.4 Extensibility of Models

The PPHR system also utilizes transfer learning techniques to create an extensible environment for users. This allows for a model to be trained on n classes, then expanded to classify $n+1$ (or more) classes. It does so without losing information from the previous model. This method allows for an expansion of the system to predict new actions that a user may want while maintaining a high accuracy and fast training speed. Algorithm 2 shows an implementation of this method. In it, Θ is the weights of the original model, C_{old} is the current number of classes, C_{new} is the new number of classes that the model will be expanded to, and N_i is the number of nodes in layer i.

Algorithm 2. Transfer Learning for Expanding Models

Require: $\Theta, L > 0, C_{new} > C_{old} > 0$
Ensure: $\tilde{\Theta} \leftarrow$ Expanded Model
    for $k = 1$ to $L - 1$ **do**
        $\tilde{\Theta}^k \leftarrow \Theta^k$
    end for
    $\tilde{\Theta} \leftarrow$ New dense layer of size C_{new}
    for $i = 1$ to N_{L-1} **do**
        for $j = 1$ to C_{old} **do**
            $\tilde{\Theta}_{ij}^L \leftarrow \Theta_{ij}^L$
        end for
        for $j = C_{old}$ to C_{new} **do**
            $\tilde{\Theta}_{ij}^L \leftarrow$ random number in range $(-1, 1)$
        end for
    end for

3.5 Activity State

The way the robot acts is dictated by its activity state. The current activity state of the robot at any given time is determined based on the confidence of personalized models' predictions. Table 1a shows the activity states and their associated confidence levels.

Table 1. Activity States and ANN Model Structure

Layer	Type	Size
Input Layer	Input	6 Nodes
Hidden Layer 1	Dense	24 Nodes
Activation	ReLU	N/A
Regularization	Dropout	20% Rate
Hidden Layer 2	Dense	128 Nodes
Activation	ReLU	N/A
Output Layer	Dense	n Nodes
Activation	Softmax	N/A

State	Confidence Range
Passive	< 85%
Active	85% - 95%
Proactive	> 95%

(a) Activity States

(b) ANN Model Structure

4 Implementation

For the purposes of this study, we will specifically examine the personalized models and transfer learning aspects of the system. In doing so, we will build a solid foundation and proof of the future capabilities of the system. For our implementation, we assume that any features that are robot-specific (i.e. conversations and actions) or related to data collection (e.g. facial recognition and location analysis) are functional and reliable.

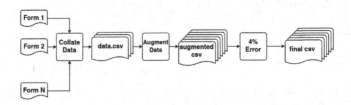

Fig. 3. Data requires multiple preprocessing steps before use in the system

4.1 Model Structure

The machine learning model for the PPHR system is an artificial neural network with multiple layers, shown in Table 1b. This model has been implemented in

Python, using Keras [15]. For our test case, the data contains 6 inputs (see Sect. 4.2, Data Collection), reflected as the "Input Layer" on Table 1b. The data passes through two hidden layers with "ReLU" activation functions. Between the two hidden layers, a dropout layer exists to help regulate the model to avoid overfitting. Finally, the last hidden layer has n nodes, representing the number of classes in our model. For this implementation, that number varies between 3 and 6. This layer uses the softmax activation function to return the model's confidence in each prediction.

4.2 Data Collection

We utilized a survey consisting of 32 unique scenarios to train our model, with each scenario defined by variables such as season, day of the week, time of day, weather, and emotion. Participants were presented with six possible actions to choose from for each scenario: Bring me water, Bring me coffee, Prepare an outfit, Tell me the weather, Clean the table, and Open the window.

To enhance our data collection, we broadened our general categories and segmented them into more specific topics, allowing us to collect more data and improve the accuracy of our models. We multiplied each variable by its desired level of specificity, thereby increasing the amount of data collected. For instance, we expanded the "season" variable by breaking the seasons into "Winter" and "Summer", then further divided them into months, with December, January, and February for Winter and June, July, and August for Summer. This increased the specificity of the season variable by a factor of three compared to our initial scenario. We followed a similar approach for the other variables, resulting in a significant growth of our data collection by over 31 times. Additionally, after augmenting the data collected, a 4% error rate was introduced to more accurately model a real-world scenario. The data preprocessing is shown in Fig. 3.

5 Results

This section shows the proposed system performance on the above-stated dataset. Figure 4 depicts the accuracy and loss curve of the proposed model for both the training and testing stages. Figure 4a shows the training and validation accuracy of the proposed model for 6 classes with Batch Size (BS) value 8 for a specific user. It is observed from the figure that the model achieved the best training and validation accuracy of 94.28% and 93.71% at epochs 96 and 71, respectively. As mentioned before, we have used a 4% error rate in generating the data, so the validation accuracy is justified with this value (93.71%).

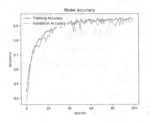 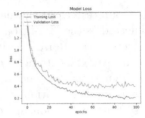 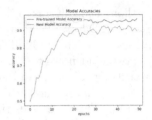

(a) Training and valida-
tion accuracy of the pro-
posed model with a Batch
Size of 8.

(b) Training and valida-
tion loss of the proposed
model with a Batch Size
of 8.

(c) The accuracy curve of
a pre-trained model us-
ing transfer learning compared to a model trained
from scratch.

Fig. 4. Performance of the proposed model

Similarly, Fig. 4b demonstrates the training and validation loss of the proposed model for the same configuration. It also observed that as the epoch
increases, the loss becomes smaller, which indicates that the proposed model
improved over time and made a more accurate prediction, as it is already shown
in Fig. 4a. And Fig. 4c shows the validation accuracy of the proposed system for
two different configurations for a specific user where in one case (New Model),
the model is trained with the 6 classes like the previous and the other case
(Pre-trained Model) uses the concept of the transfer learning where the model is
first trained with 5 classes and used the weights of this model to train the next
model with 6 classes. As it uses the pre-trained weight, it got more accurate
compared to the New Model and got higher accuracy more rapidly than the
New Model. For example, the New model got the highest validation accuracy
of 92.28% at epoch 39, whereas the best-achieved accuracy for the Pre-trained
model is 96.32% at epoch 50. In general, it can be seen from the beginning that
the accuracy of the Pre-trained model is much higher than the new model that
justified the uses of the transfer learning concept.

Figure 5 shows the same concept of transfer learning where the model is first
trained with 3 classes and then used its weight to train with 4 classes and do
this thing on an incremental basis up to 6 classes. As each of the models for
4 to 6 classes use the weight of the pre-trained model with one less number of
classes, it trains faster than the previous class and gives better accuracy than the
previous one. For example, the model with the 6 classes gives higher accuracy
compared to all the models. But as each time the number of classes increases,
it is not guaranteed that at all iterations, the model with the highest number of
classes will show better results because as the number of classes increases, there
would be more chance for wrong prediction. But in the general case, the model
with the highest number of classes showed the best result.

The performance comparison of a model generated by incorporating all the
data from A to K and all the personalized models based on validation metrics is
shown in Table 2. Each model shown in the table operates independently from

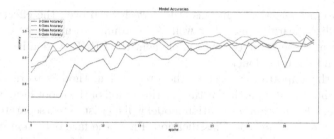

Fig. 5. Validation accuracy of the proposed model with Transfer Learning Concept with incremental class number.

the other models, showing the improved performance of personalized models over a generalized model. It can be seen from the table that a general model attempting to make predictions for multiple people can be inaccurate due to a wide variety of preferences. On a rainy morning, one person may prefer coffee while another prefers water. To solve this issue, we turn to personalized learning models. The individual models A through K, show a much higher accuracy and confidence when compared to the combined model. This is consistent with studies on personalized machine learning in other contexts [2,6].

Table 2. Combined Model vs. Personalized Models

Model	Accuracy	Loss	Confidence
Combined A-K	0.7691	0.8006	0.6956
Model A	0.9381	0.2943	0.8814
Model B	0.9617	0.1742	0.9475
Model C	0.9481	0.3713	0.8830
Model D	0.9573	0.3120	0.8963
Model E	0.9249	0.3519	0.8669
Model F	0.9194	0.3418	0.8791
Model G	0.9231	0.4723	0.8376
Model H	0.9198	0.4871	0.8541
Model I	0.9569	0.2279	0.9297
Model J	0.9100	0.3402	0.8888
Model K	0.9043	0.3258	0.8922

6   Discussion

Within robotics, there is a rising need for improved human-robot interactions through active and proactive behaviors. The PPHR system presented is a promising approach to achieving this goal. By enabling more personalized robot behaviors, the system has the potential to improve the user's experience and meet their needs more effectively. In this study, we have proven the effectiveness of the personalized learning model as well as the ability to introduce new actions to the system. However, in order to implement the PPHR system in its entirety, more work needs to be done.

One of the important aspect of the proposed method is to detecting the emotion of the user at a particular time as the action of the robot largely depends on that data. The emotion recognition model will consist of two separate sections: Face Detection and Emotion Classification. First, the face detection model will be employed to detect the face portion from the robot vision and then output of the model will be fed into the emotion classification model to recognize the final emotion. We are planning to use YOLOv5 model to detect the face and then use any deep learning model like RepVGG [16] for the emotion classification tasks.

Another aspect of the system proposed in this paper is federated learning. There is a large precedent for the use of federated learning systems within the medical field. We intend to implement a version of the proposed federated learning system using the data we have collected internally as well as data that we collect in our future studies. By doing so, we will develop a centralized intelligence for the system, resulting in faster learning and integration into the PPHR system in the future. We have begun work on an implementation of PPHR for SoftBank Robotics' "Pepper" robot [17]. This robot is designed primarily for human-robot interaction and will be suitable as a platform for this system. Once fully implemented, we will begin studying the application of the system in the real world through conversations and interactions with different groups of test subjects.

7   Conclusion

The extent to which robots are now a part of our daily lives emphasizes the need for more proactive and individualized human-robot interactions. This research proposed a robotic system, PPHR, that will be able to gather information about users and their surroundings in order to predict and execute the actions that users would like, either actively or proactively, depending on the situation by developing personalized learning models and also using federated learning concept in order to generalize the model. This study has so far developed the system's capacity for learning and applied the transfer learning principle to let the user add new actions whenever they choose and obtain good results, which is a crucial component of the proposed system.

References

1. Rudovic, O.O.: Personalized machine learning for human-centered machine intelligence. In: Proceedings of the 1st International on Multimodal Sentiment Analysis in Real-Life Media Challenge and Workshop (2020)
2. Rudovic, O., et al.: Personalized machine learning for robot perception of affect and engagement in autism therapy. Sci. Robot. **3** (2018)

3. Jacobson, N.C., Bhattacharya, S.: Digital biomarkers of anxiety disorder symptom changes: personalized deep learning models using smartphone sensors accurately predict anxiety symptoms from ecological momentary assessments. Behav Res Ther. **149**, 104013 (2022). Epub 11 Dec 2021. PMID: 35030442; PMCID: PMC8858490. https://doi.org/10.1016/j.brat.2021.104013

4. Chiu, I.M., Cheng, C.Y., Chang, P.K., Li, C.J., Cheng, F.J., Lin, C.R.: Utilization of personalized machine-learning to screen for dysglycemia from ambulatory ECG, toward noninvasive blood glucose monitoring. Biosensors (Basel) **13**(1), 23 (2022). PMID: 36671857; PMCID: PMC9855414. https://doi.org/10.3390/bios13010023

5. de Leeuw, A.-W., van der Zwaard, S., van Baar, R., Knobbe, A.: Personalized machine learning approach to injury monitoring in elite volleyball players. Eur. J. Sport Sci. **22**, 1–14 (2021). https://doi.org/10.1080/17461391.2021.1887369

6. Lopez-Martinez, D., Peng, K., Lee, A., Borsook, D., Picard, R.: Pain detection with fNIRS-measured brain signals: a personalized machine learning approach using the wavelet transform and Bayesian hierarchical modeling with Dirichlet process priors. In: 2019 8th International Conference on Affective Computing and Intelligent Interaction Workshops and Demos (ACIIW), Cambridge, UK, pp. 304–309 (2019). https://doi.org/10.1109/ACIIW.2019.8925076

7. Ren, B., et al.: Predicting states of elevated negative affect in adolescents from smartphone sensors: a novel personalized machine learning approach. Psychol. Med. 1–9 (2022). https://doi.org/10.1017/S0033291722002161

8. Papadopoulos, G.T., et al.: Towards open and expandable cognitive AI architectures for large-scale multi-agent human-robot collaborative learning. arxiv:2012.08174 (2020)

9. Papadopoulos, G.T., et al.: User profile-driven large-scale multi-agent learning from demonstration in federated human-robot collaborative environments. arxiv:2103.16434 (2021)

10. Majcherczyk, N., et al.: Flow-FL: data-driven federated learning for spatiotemporal predictions in multi-robot systems. arxiv:2010.08595 (2020)

11. Breazeal, C.: Social interactions in HRI: the robot view. IEEE Trans. Syst. Man Cybern. Part C (Appl. Rev.) **34**(2), 181–186 (2004). https://doi.org/10.1109/TSMCC.2004.826268

12. Shen, Z., Elibol, A., Chong, N.Y.: Nonverbal behavior cue for recognizing human personality traits in human-robot social interaction. In: 2019 IEEE 4th International Conference on Advanced Robotics and Mechatronics (ICARM), Toyonaka, Japan, pp. 402–407 (2019). https://doi.org/10.1109/ICARM.2019.8834279

13. Liu, P., Glas, D., Kanda, T., Ishiguro, H.: Learning proactive behavior for interactive social robots. Auton. Robot. **42** (2018). https://doi.org/10.1007/s10514-017-9671-8

14. Kaelbling, L.P., Littman, M.L., Moore, A.W.: Reinforcement learning: a survey. J. Artif. Intell. Res. **4**, 237–285 (1996)

15. Gulli, A., Pal, S.: Deep Learning with Keras. Packt Publishing Ltd. (2017)

16. Ding, X., Zhang, X., Ma, N., Han, J., Ding, G., Sun, J.: RepVGG: making VGG-style convnets great again. In: Proceedings of the IEEE/CVF Conference on Computer Vision and Pattern Recognition, pp. 13733–13742 (2021)

17. Pandey, A.K., Gelin, R.: A mass-produced sociable humanoid robot: pepper: the first machine of its kind. IEEE Robo. Autom. Mag. **25**(3), 40–48 (2018)

Classifying Service Robots in Commercial Places Based on Communication: Design Elements by Level of Communication

Karam Park[✉] [ID] and Eui-Chul Jung[ID]

Department of Design, Seoul National University, Seoul 08826, South Korea
{karam,jech}@snu.ac.kr

Abstract. Service robots have been traditionally classified by technology and labor type. However, service robots communicate with human employees and consumers to collaborate and deliver service on the frontline. The purpose of this research was to propose a framework for classifying service robots used in commercial places based on the communication that is essential for collaboration and service delivery. The framework classifies service robots into six types, with the x-axis representing the level of communication and the y-axis representing the labor method that service robots use to provide services. Through the framework, the study explained 1) how the service robot and the employee communicate when they collaborate and 2) how the service robot and consumer communicate by communication levels. In addition, it identified the design elements needed for each section of the framework. This will help robot developers and designers consider design elements at different levels of communication when designing service robots in commercial places to enhance the experience of collaboration and service delivery. It is expected that communication will be considered important when developing service robots using the classification of service robots based on the framework proposed in this paper.

Keywords: Service robots · Human-Robot Interaction Design · Human-Robot Communication

1 Introduction

Various companies are using service robots to provide their service to customers in commercial places. Serving robots deliver food and guide robots take orders, replacing the service labor of human employees in restaurants. Employers are hiring service robots because they are cheaper than personnel expenses and tireless. Accordingly, human employees fear that robots will take their jobs. Most classification approaches to service robots focus on 'technology' and 'labor' because robots require a lot of technical skills to produce. Historically, they have been developed to replace labor. Karabegovic [1] categorized service robots into five criteria according to their 1) degree of autonomy, 2) environment, 3) movement technology, 4) role in the community, and 5) type of service. 1) Degree of autonomy and 3) movement technology classify service robots based on

© The Author(s), under exclusive license to Springer Nature Switzerland AG 2024
B. J. Choi et al. (Eds.): IHCI 2023, LNCS 14531, pp. 268–278, 2024.
https://doi.org/10.1007/978-3-031-53827-8_25

their technology. 2) Environment, 4) role in the community, and 5) type of service categories classify robots based on what kind of service labor they provide, where, and to whom, which can be seen as a classification of robots based on labor. In general, people categorize service robots by naming them according to the type of labor they perform, such as 'serving robot,' 'guide robot,' 'education robot,' and the like, because robots tend to be developed for specific functions and purposes.

Currently, service robots provide services on the frontline in commercial places with human employees. The restaurant industry utilizes two main types of service robots: those that serve and those that take orders. They serve food, take orders, and ring up payments on behalf of human workers; thus, they are 'replacing' human labor. Nevertheless, consumers can have a negative reaction to service robots and the brand that uses them when these robots completely replace human employees [2]. As a result, even though service robots are partnering with humans and taking over some human tasks, replacing human workers entirely is still challenging. Humans and robots are good at different tasks [3]. When service robots substitute human employees, consumers may develop a negative image of the brand [2, 4]. Therefore, service robots should 'collaborate' with humans in the service industry rather than substitute humans [5, 6]. As service robots collaborate with service employees and provide services to consumers, communication between robots and humans becomes essential. Hence, it is necessary to further classify service robots based not only on the technology and labor classification but also on 'communication,' which is a crucial factor underlying communication between service robots and employees and effective delivery of services to customers.

This study proposed a framework to classify service robots used in commercial places by type of service and level of communication. The framework was described using the example of commercial places. A restaurant that is actively using serving and guide robots was used to explain collaboration between the service robot and the employees. This study aimed to 1) present the types of collaboration that service robots have with human co-workers on the service frontline and 2) explain how service robots communicate with consumers to provide services with the proposed framework. Additionally, the study discusses design elements that must be considered when developing a service robot, depending on the level of communication. Through the proposed framework, it hopes to increase the likelihood of understanding the importance of communication with service robots, in addition to technology and labor, and thus develop them with a communication focus.

2 Related Works

2.1 Collaboration with Human Employees and Service Robot

Previous research has shown that consumers expect robots and human employees to collaborate during service experiences. Human employees who may be working with service robots are also worried about losing their jobs, although they are positive about working with robots when the roles of robots and humans are clearly distinguished [7]. Service robots are good at physical tasks because they do not get tired, while human employees are better at dealing with difficult emotions [3]. For this reason, service robots need to collaborate with humans and create new service values based on their abilities instead of replacing humans [8]. Verbal and non-verbal communication is essential for service delivery, even among human employees. The exchange of information between service robots and human workers during service delivery should take place through both verbal and non-verbal means, including displays, gestures, and behaviors.

2.2 The Role of Communication

Service robots in commercial places communicate with customers through verbal and non-verbal means to provide service. Eye contact between humans and robots, as one element of non-verbal communication, influences the user's natural attitude towards the robot and its purpose [9]. Differences in communication and personality between humans and robots affect how the user performs in the game as well [10]. Although these studies were not conducted with service robots in commercial places, they show that communication is critical for successful service delivery, which is the purpose of service robots.

Service robot also affects the image of the brand [11]. The more consistent the voice assistant's answers, which is part of the function of the service robot, the easier it is for users to recognize the brand persona and image [12]. It is assumed that brands communicate only when they want, but they communicate through everything they usually say and behave in reality [13]. Therefore, service robots can be used as a brand communication strategy to demonstrate brand differentiation. Likewise, frontline service employees should receive a manual on the actual brand communication strategy to deliver services according to this differentiation.

2.3 Communication Levels for Service Robots

Powell [14] described five levels of interpersonal communication, from Cliché at level 1 to Intimacy at level 5, to reflect the extent to which individuals communicate their private information and feelings. Table 1 extends Powell's framework to the relationship between service robots and users. Level 1 Cliche is a casual greeting, whereas Level 2 Reporting refers to interactions based on the basic functionality of the service robot. Level 3 Opinion concerns interactions with the user to provide a more personalized service. Levels 1–3 are common service robots that provide an objective service rather than exchange emotions with the robot. Levels 4 and 5, which involve sharing personal private information and emotions with robots, are common in companion robots used in a personal space. Thus, this paper used Level 1, Level 2, and Level 3 to build a framework for service robots in commercial places.

Table 1. Five Service Robot Levels and User's Communication

Level of Communication	Description of Service robot & user's Communication	Type of Robot
Level 1 Cliche	Greeting everyone (users and non-users) before they started using the service	Service robot in commercial places
Level 2 Reporting	The user interacts with the service robot to receive a service The service robot provides to the user with the inputted service	
Level 3 Opinion	The user provides different information to robots to collaborate (employees) and receive services (consumers) that are more relevant to them The service robots provide the labor the user needs	
Level 4 Emotional	The robot gets a lot of personal information about the user, so it can have appropriate conversations and exchange emotions based on the user's situation	Companion robot
Level 5 Transparency	The user becomes attached to the robot and considers it as a friend and companion. The user tells the robot everything he cannot tell others	

3 Communication Classification of Service Robots

The most common communication levels of service robots, Level 1, Level 2, and Level 3, as shown in Table 1, are set as the X-axis to build a classification according to the service robots' communication framework. The Y-axis represents physical labor-information provision as the service robots' labor method. Service robots that perform physical labor use non-verbal communication, and those that provide information primarily use verbal communication to interact and communicate with users. Since service robots communicate with users differently depending on the type of service they provide, the Y-axis is set to labor types. Figure 1 is a framework for classifying service robots, which is divided into six sections based on the communication level of service robots and the type of labor.

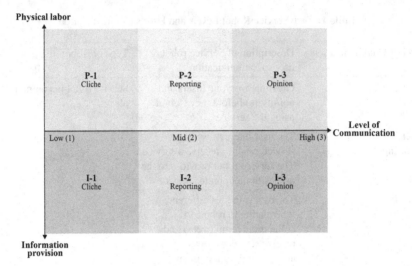

Fig. 1. A Framework for Classifying Service Robots based on Communication

3.1 Categorizing Communication for Collaboration Between Service Robots and Human Employees

Table 2. Communication Classification of Service Robots – Collaboration Perspective

Physical labor- Level 1 Cliche	P-2 Reporting	P-3 Opinion
Clearly separate service robot and human employee tasks.	Service robots and employees perform the same service separately.	Service robots and employees work together on the same service labor.
Service robots and human employees work independently, so collaboration and communication levels are low.	Both service robots and employees collaborate by sharing with each other when they have physically completed providing a service.	Service robots and employees collaborate and communicate. Human employees perform tasks that are difficult for robots. Service robots perform tasks that are difficult for humans to perform.
Information provision- Level 1 Cliche	**I-2 Reporting**	**I-3 Opinion**
Service robots' and human employees' work is clearly separated by space.	Service robots and employees perform the same service separately.	Service robots and employees work together on the same service labor.
Service robots and humans may have separate jobs and workplaces, so communication frequency is low.	Service robots and employees collaborate by giving each other information about the services they provide non-physically.	Service robots recap information that employees do not remember, and employees collaborate with them by delivering new information.

Table 2 categorizes how service robots and frontline service employees collaborate by communication level in the framework. In communication level 1, the service robot and human employees have clearly delineated tasks, so they communicate only briefly. The employees can also manipulate the interface of the service robot to allow them to perform service tasks. In particular, workplaces can differ, with service robots greeting customers outdoors and humans providing the actual service indoors in the I-1 (Information provision-Level 1) section. In communication level 2, the service robot and the human employee each deliver the same service. Hence, the communication between the service robot and the human employee is about objective content. They communicate with each other to share what they have done to avoid providing the same service twice. In communication level 3, robots and humans exchange opinions because they complement each other and provide the same service together. For example, if a human employee needs to service heavy stuff, the robot can help or do something that is better suited for this task. In the I-3 section, service robots provide consumers with vast amounts of information that humans find hard to remember, and employees notify robots about new information from the company, complementing each other. In other words, rather than replacing each other's work, synergy occurs through communication when service robots and employees distribute work clearly and boost each other's capabilities. Through their collaboration, consumers will be able to receive more effective and satisfactory services.

3.2 Categorizing Communication Between Service Robots and Consumers in the Commercial Places

Table 3. Communication Classification of Service Robots – Consumer Perspective

P-1 & I-1 Cliche	P-2 Reporting	P-3 Opinion
Consumers are interested in brands and service robots. Service robots provide physical and non-physical services to consumers in one direction.	Consumers recognize that service robots can provide services but only engage in simple communication with greetings and reporting about service provision.	Consumers recognize that the service robot can provide personalized service and request additional desired physical services.
	I-2 Reporting	**I-3 Opinion**
	Consumers recognize that the service robot can provide services, so they ask the robot questions and try to have simple conversations. The communication is mainly about basic service provision.	Consumers recognize that service robots can provide personalized service and are willing to share their preferences and needs with robots in order to receive personalized services.

Table 3 describes the communication between the service robot and the consumer at different levels of the framework. Figure 2 shows examples of serving robots by level. The P-1 section describes the physical labor that the service robot should perform while receiving consumers. One example is CLOi ServeBot from LG [15], which smiles and says hello through its display while its primary task is to serve. The P-2 section is where the service robot is primarily responsible for providing basic services for the brand and holding a brief conversation. For example, BellaBot, a serving robot from Pudu Robotics [16], performs serving services and reports to the customer on whether the service was completed. Service robots from the P-3 section provide customized services and branded ancillary services and provide physical labor for additional services desired by consumers. In this section, communication and services from P-2 serve as a base; however, this time, the user can invoke the robot to provide additional physical services. Alternatively, the service robot predicts which services are likely to be needed first, asks for them, and provides them.

Service robots in the I-1 section welcome consumers by promoting the brand. A typical example is using Pepper in various industries to entertain customers and promote brands [17]. In the I-2 section, service robots provide only representative and basic service information. Typically, the guide robots at airports and subway stations provide information on directions and places through voice and non-voice communication. As another example, Pepper can provide banking service information to customers at HSBC [18]. Service robots from the I-3 section identify consumer needs, deliver customized services, and determine optimal services to customers. They aim to properly identify consumers' needs through communication and supply appropriate services that consumers have not recognized.

P-1 level	P-2 level	I-1 level	I-2 level
CLOi ServeBot (LG)	**Bellabot** (Pudu Robotics)	**Pepper** (Softbank)	**Pepper in HSBC** (Softbank)

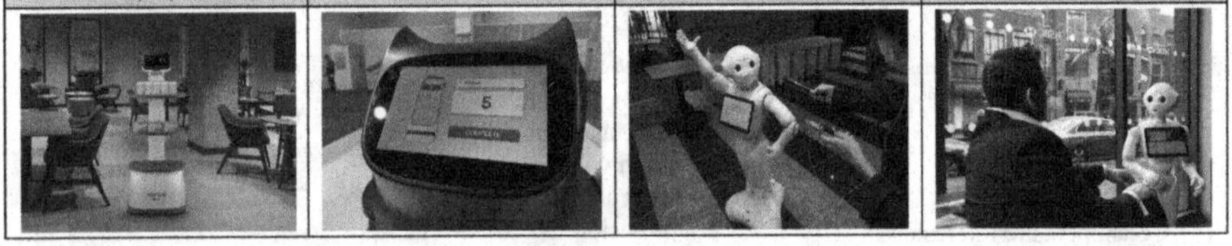

Fig. 2. Example of Robots

To sum up, Level 1 physical labor and information delivery service robots make consumers aware of the brand. In Level 2, service robots are used to communicate and deliver existing services to consumers. Level 3 is where the service robot interacts with the consumer to introduce and personalize various services that the consumer may not be aware of. Ultimately, the level of communication between the service robot and the customer determines the personalization and effectiveness of the service.

4 Design Elements for Communication of Service Robot

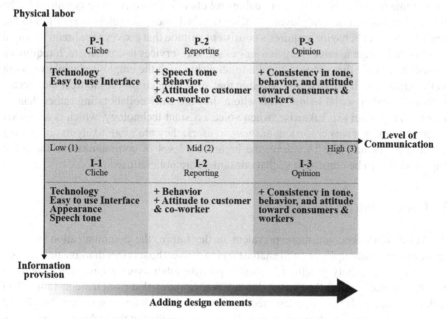

Fig. 3. Design Elements for Communication of Service Robot

While technology improves service performance, design improves the collaborative experience of co-workers working together and proper service delivery. Designing service robots based on design elements that are needed for communication will increase the positive experience of collaboration and service delivery for human employees and consumers. In HRI, six factors affect interactions: appearance, voice, height, non-verbal behavior, approach distance, and emotion [19]. Figure 3 depicts the design elements needed for service robots to communicate with human employees and consumers at each stage of communication to increase the effectiveness of service delivery and collaboration based on six factors that influence interaction. The required design elements are gradually added from Level 1 to Level 3. In Level 1, accurate service delivery is paramount, so the service robot should be equipped with the technology to do this well. Its interface should be easy to use for employees and consumers. In particular, it is important to drive differentiated interaction because it can show the brand to the consumer and convey information in the case of I-1.

For this reason, the appearance, which can vary depending on the role of the robot [20], and the voice and its tone, which are the most basic communication elements that can convey information to people [21], should be modified to effectively convey brand and service information. Through tone of voice, Level 2 service robots could express to the customers the smoothness of the collaborative relationship with human employees when reporting about finish their work with human employees and this can be affect brand image. [3]. Especially since P-2 is more collaborative than P-1, speech tones can be added as a necessary communication element for collaboration. In addition, tone of voice [22] and behavior [23] are important, as they can demonstrate a brand's unique service image and personality that enables robots to effectively serve consumers, like human frontline service employees. Unlike Levels 1 and 2, Level 3, in addition to the tone of voice and behavior, requires a consistent attitude that conveys the brand's unique services and image to effectively deliver customized services to consumers. It requires a consistent attitude that shows a smooth relationship with the employees while working on the same task. It also requires a design that allows the employees to treat the service robot as "another social being" with whom they are truly collaborating rather than as a tool to help them [3]. Likewise, when voice assistant technology, which is also used in service robots, gives consistent answers to users, they are more likely to understand the brand persona [12]. Therefore, the higher the level of communication, the more important it is to be consistent so that consumers are not confused.

5   Conclusion

If service robots become more prevalent in the future, the communication between service robots and employees will enable them to deliver the services that consumers want quickly and accurately. It will also provide positive labor experiences, such as helping employees reach their full potential through collaboration that complements rather than replaces humans. This study presented the framework for defining service robots based on "communication" with employees and customers instead of the technology and labor types of service robots since communication is a crucial component of this process. With the framework presented, service robot developers and designers can try to think about how the service robot they want to develop will communicate with its users, collaborators, and consumers.

Generally, user experience designers have focused on designing a serving robot that can do a human's labor well and the usability of the robot for the human operating with it. However, if designers recognize the importance of communication between the robot and human employees through the proposed frame, robot experience designers can focus on the experience of collaborating and working with the robot rather than the experience of operating the robot. UX designers can design experiences by recognizing the communication levels of the service robots they are designing and understanding design elements that are essential for each communication level. A robot that is designed with a full understanding of robot-human worker communication will increase human employees' positive experiences of collaboration. Moreover, if designers were to design a Level 3 robot with the framework and specific design elements of each level in mind, they would be able to unite the various design elements to provide a consistent experience.

While there are no Level 3 service robots today, incorporating robotic experiences that consider design elements and experience consistency would facilitate the development of Level 3 robots. This will allow designers to create differentiated service robots for each brand and use service robots as a brand communication factor. In addition, many human employees find collaborating with service robots difficult. Collaboration training methods or content can be created before collaborating with a service robot [2]. The training approach will vary depending on how users need to communicate and collaborate with the robot. For example, in Level 2, they could continue to do their work but be trained to communicate with the service robot about task distribution and collaboration. In Level 3, employees would learn to cooperate with service robots to deliver services together and to inform them of difficult tasks. With this training, more human employees will understand how service robots communicate and recognize them as specific social beings that work with them.

Since different brand images employ different brand communication strategies, a future study could utilize the framework provided in this study to show how the communication involving service robots varies depending on the brand image. The limitation of the proposed framework is that it introduced communication methods of service robots in commercial places but did not propose a classification of communication methods within the company or private places. Future research should categorize the communication methods of other types of service robots, such as education robots.

Acknowledgments. This work was supported by the Ministry of Education of the Republic of Korea and the National Research Foundation of Korea (NRF-2023S1A5A2A03084950).

Reference s

1. Karabegović, I.: Classification of service robots. In: Karabegović, I. (ed.) Service Robots: Advances in Research and Applications, pp. 1–11. Nova Science, New York (2021)
2. MaLeay, F., Osburg, V.S., Yoganathan, V., Patterson, A.: Replaced by a robot: service implications in the age of the machine. J. Ser. Res. **24**(1), 104–121 (2021)
3. Wirtz, J., et al.: Brand new world: service robots in the frontline. J. Ser. Man. **29**(5), 907–931 (2018)
4. Kim, Y.S.: The effect of customer orientation of service robots on perceived trust and expectation of hospitality experience. Int. J. Tou. Hos. Res. **36**(3), 157–173 (2022)
5. Fong, T., et al.: A preliminary study of peer-to-peer human-robot interaction. In: 2006 IEEE International Conference on System, Man, and Cybernetics, Taipei, Taiwan, 8–11 October 2006 (2006)
6. Marble, J., Bruemmer, D., Few, D.A., Dudenhoeffer, D.D.: Evaluation of supervisory vs. peer-peer interaction with human robot teams. In: 37th Hawaii International Conference on System Sciences, Big Island, HI, USA, 26 February 2004 (2004)
7. Palunch, S., Tuzovic, S., Holz, H.F., Kies, A., Jorling, M.: "My colleague is a robot" - exploring frontline employees' willingness to work with collaborative service robots. J. Ser. Man. **33**(2), 363–388 (2022)
8. Kaupp, T., Makarenko, A., Durrant-Whyte, H.: Human robot communication for collaborative decision making a probabilistic approach. Rob. Auto. Sys. **58**(5), 444–456 (2010)
9. Huang, C., Mutlu, B.: Robot behavior toolkit: generating effective social behavior for robots. In: 7th ACM/IEEE International Conference on Human-Robot Interaction (HRI), Boston, MA, USA, 30 July 2012 (2012)
10. Andriella, A., et al.: Do i have a personality? endowing care robots with context-dependent personality traits. Int. J. Soc. Rob. **13**(8), 2081–2102 (2021)
11. Choi, S., Liu, S., Choi, C.: Robot–brand fit the influence of brand personality on consumer reactions to service robot adoption. Mar. Let. **33**(1), 129–142 (2022)

12. Kim, N.Y., Lee, C., Hong, H., Yun, J.: A study on personas of artificial intelligence speakers. J. Com. Des. **75**, 242–253 (2021)
13. Dowling, G.R.: Measuring corporate image: a review of alternative approaches. J. Bus. Res. **17**(1), 27–34 (1988)
14. Powell, J.: Why Am I Afraid To Tell You Who I Am? Argus Communications, Chicago (1969)
15. LG: CLOi ServeBot (Shelf Type) (2023). https://www.lg.com/global/business/cloi/lg-cloi-ser vebot-shelf. Accessed 24 Oct 2023
16. PUDU: BettaBot (2023). https://www.pudurobotics.com/products/bellabot. Accessed 24 Oct 2023
17. New York Post: Robots ready to greet Japanese coronavirus patients in hotels. https://nypost.com/2020/05/01/robots-on-hand-to-greet-japanese-coronavirus-patients-in-hotels/ (2020). Accessed 24 Oct 2023
18. Robotics Business Review: Pepper Heads to Beverly Hills for Latest Assignment (2019). https://www.roboticsbusinessreview.com/retail-hospitality/pepper-heads-to-beverly-hills-for-latest-assignment/. Accessed 24 Oct 2023
19. Park, D., Pan, Y.: The proposal of the personality and behavioral design of the Robot Persona according to the role and use context of the Robot. J. Dig. Cont. Soc. **22**(11), 1843–1853 (2021)
20. Lee, E., Lee, J.: Expected face design, personalities and emotional images according to the role of social robot. J. Kor. Des. For. **25**(4), 37–48 (2020)
21. Buczynski, J., Nass, C., Brave, S.: Wired for speech: how voice activates and advances the human-computer relationship. Libr. J. **130**(15), 88 (2005)
22. So, H.S., Kim, M.S., Oh, K.M.: People's perceptions of a personal service robot's personality and a personal service robot's personality design guide suggestions. In: The 17th IEEE International Symposium on Robot and Human Interactive Communication, Munich, Germany, 15 August 2008 (2008)
23. Keller, K.L.: Brand synthesis: the multidimensionality of brand knowledge. J. Con. Res. **4**, 595–600 (2003)

Cascade Mentoring Experience to Engage High School Learners in AI and Robotics Through Project-Based Learning

Sophia Matar[1], Bailey Wimer[1], M. I. R. Shuvo[1], Saifuddin Mahmud[2], Jong-Hoon Kim[1(✉)], Elena Novak[3], and Lisa Borgerding[3]

[1] ATR Lab, Computer Science, Kent State University, Kent, OH, USA
jkim72@kent.edu
[2] Computer Science and Information Systems, Bradley University, Peoria, IL, USA
[3] School of Teaching, Leadership, and Curriculum Studies, Kent State University, Kent, OH, USA

Abstract. This paper introduces a novel AI and Robotics education model featuring cascade mentoring and project-based learning. It empowers junior members to become active mentors, fostering reciprocal mentoring experiences and enhancing their understanding of computer science, AI, and Robotics. We establish a collaborative pedagogy, connecting K-12 students with a university research lab using standards-based curricula. Our research demonstrates the positive impact of cascade mentoring on both mentor students and mentees in computer science and physical science education, improving their attitudes toward these subjects. Moreover, we explore the potential of educational robotics to address societal concerns such as green energy and sustainability.

Keywords: ML · Computational Thinking · Cascade Mentoring · Project-Based Learning

1 Introduction

Artificial Intelligence (AI) and Robotics have emerged as game-changing technologies in our rapidly changing world, reshaping many industries. Their impact affects nearly every aspect of our lives. As we journey deeper into the 21st century, the importance of having a workforce proficient in Science, Technology, Engineering, and Mathematics (STEM) has become increasingly evident. High schools are crucial in preparing the next generation for this technologically advanced future.

A pressing concern arises when we examine the current state of high school education concerning AI and Robotics. Despite the growing significance of these fields, there needs to be more engagement among high school students. Many students need help to connect with these subjects, which results in missed opportunities for skill development and readiness for future careers. Within this context, we focus on exploring the potential of project-based learning (PBL) to bridge this gap and elevate student engagement.

B. J. Choi et al. (Eds.): IHCI 2023, LNCS 14531, pp. 279–294, 2024.
https://doi.org/10.1007/978-3-031-53827-8_26

First, we aim to investigate the impact of project-based learning (PBL) on high school learners' engagement in AI and Robotics. PBL has gained recognition as a pedagogical approach encouraging active, hands-on learning, making it a promising method to capture students' interest in these complex domains. Additionally, we will explore how PBL can create a mutually beneficial learning experience for mentors and mentees, thus highlighting the potential for mentor growth and development within the mentorship structure. Second, we seek to examine the effectiveness of cascade mentoring within PBL settings, investigating how mentorship structures can further enhance student engagement and learning outcomes. In doing so, we will emphasize the dual impact of mentorship, shedding light on how it supports students' academic and professional growth and fosters mentor learning experiences, ultimately contributing to a more holistic understanding of the PBL approach in AI and Robotics education.

This research holds substantial significance for the field of STEM education. By explaining the potential of PBL and cascade mentoring in AI and Robotics education, we contribute valuable insights to educators and researchers alike. Our findings have the potential to kick-start curriculum development and teaching practices in high schools, ultimately strengthening the pipeline of future talent in these crucial fields.

To delve deeper into these research objectives, this paper is organized as follows: Sect. 2 comprehensively reviews the existing literature on AI and Robotics education, project-based learning, and mentoring. Section 3 details the pedagogical approach used for this research. Our Implementation is discussed in Sect. 4. In Sect. 5, we present our findings and discuss their implications. Finally, Sect. 6 offers conclusions and recommendations for future research and educational practices.

1.1   Background

Cascade mentoring is a structured approach to mentorship that involves a hierarchical distribution of mentors and mentees. In education, cascade mentoring typically involves experienced individuals mentoring less experienced learners, creating a mentorship cascade. This approach has gained prominence in recent years as an effective means of fostering skill development, knowledge transfer, and personal growth within educational settings, promoting collaborative learning.

Pedagogical approaches encompass educators' methods, strategies, and philosophies to facilitate learning. In this research, our pedagogical approach combines elements of project-based learning with cascade mentoring. This approach aims to leverage the benefits of both methodologies to enhance high school learners' engagement in AI and Robotics. By integrating hands-on projects and mentorship, we sought to create an immersive and supportive learning environment that promotes active participation and knowledge acquisition.

Project-based learning (PBL) is a pedagogical approach that emphasizes active, hands-on learning experiences for students. In PBL, students engage in open-ended projects that require them to apply their knowledge and skills to solve complex problems. This method encourages critical thinking, creativity, and the development of practical skills. PBL is well-suited for STEM education as it allows students to explore concepts in a tangible and meaningful way. It promotes student autonomy, encourages inquiry, and fosters a deeper understanding of the subject matter. In the context of AI and Robotics education, PBL provides a promising route for enhancing student engagement and learning outcomes.

To further enhance our educational project's immersive and innovative learning environment, we integrated cutting-edge technology, particularly the mBot robot. The mBot robot, developed by Makeblock, is a versatile and user-friendly educational robot designed to introduce students to robotics and programming. By incorporating the mBot into our project, we were able to offer students the opportunity to work with a physical robot, fostering a tangible connection to their learning experience. Through extensions, we extended the capabilities of the mBot, allowing students to explore AI and machine learning concepts in practical ways.

2   Related Work

The ongoing underrepresentation of historically marginalized groups, especially women, in computer science (CS) is a significant concern in STEM education. Despite efforts to improve CS accessibility, barriers like limited early exposure, negative stereotypes, and a lack of relatable role models persist. Novak's research [13] aims to tackle these challenges through cascading peer-mentoring, connecting high school students with a University CS research lab using educational robotics. This approach exposes students to CS concepts, engages them in robotics problem-solving, and aligns with the trend of hands-on, project-based STEM learning. By integrating cascade mentoring and project-based learning within AI, Robotics, and educational robotics, our research offers a comprehensive model to address diversity and engagement issues in CS education.

In a related study by Jeon [7], middle school students' AI literacy was examined through a project-based summer camp. The study showed significant improvements in student's conceptual understanding of AI and their ability to apply AI concepts in collaborative problem-solving. Our proposed research builds upon these insights by introducing cascade mentoring to enhance further engagement and learning outcomes in AI and Robotics education, demonstrating how mentorship can enrich the educational experience and skill development in these fields.

Chistyakov et al.'s article [3] explores project-based learning (PjBL) in science and STEAM education, emphasizing its capability to enhance student learning outcomes, critical thinking, and problem-solving skills. Our approach builds upon these insights by integrating cascade mentoring into the educational framework, providing a structured mentorship hierarchy that enhances student learning outcomes and empowers students to take on teaching and mentoring roles. This expansion of PjBL by incorporating mentorship dynamics offers a comprehensive and engaging learning environment. Furthermore, our research explores the integration of educational robotics to address real-world challenges, providing students with practical applications for their AI and Robotics knowledge. This unique combination of PjBL, cascade mentoring, and educational robotics holds the potential to enhance STEM education significantly.

In the realm of peer mentoring and its impact on student learning, Aderibigbe et al.'s prior research [1] explored the effectiveness of peer mentoring in fostering understanding and emotional stability among students. Our proposed research extends the understanding of peer mentoring by introducing a novel approach that integrates cascade mentoring and PBL in the context of AI and Robotics education. This approach not only explores the benefits of mentorship but also enhances student engagement and learning outcomes in complex STEM domains, providing a holistic framework that addresses the evolving needs of students in the 21st century.

In the context of increasing participation in CS education, a prior study focused on providing relatable role models to spark youth's interest in CS. Clarke-Midura et al. [4] introduced a mentoring model where high school students were trained as near-peer mentors for middle schoolers attending summer programming camps. Our proposed approach goes beyond this by introducing a cascade mentoring and PBL model that leverages mentor relatability and promotes reciprocal mentoring experiences, active participation, and hands-on learning in AI and Robotics. This approach provides a more holistic and immersive educational experience, ultimately enhancing engagement and learning outcomes in these complex domains.

In another related work, Lee's paper [9] presents an AI Methods in Data Science (DS) curriculum and professional development (PD) program for high school teachers. Our approach takes it further by introducing cascade mentoring and PBL into high school AI and Robotics education. This innovative combination enhances engagement and offers a sustainable mentorship and skill development model, contributing to a more holistic AI and Robotics education ecosystem.

In entrepreneurship education, active and experiential learning combined with student peer mentoring has shown promising outcomes in enhancing student engagement and academic success. Kari Håvåg Voldsund et al. [16] highlight the integration of student peer mentoring into an entrepreneurship course. Our proposed approach extends this concept into AI and Robotics education by integrating PBL with cascade mentoring, creating a more comprehensive and engaging learning environment.

In the context of AI education for K-12 students, prior research has made notable strides in developing tools and curricula to impart knowledge about artificial intelligence (AI) and its societal implications. Brummelen [15] focused on assessing an AI curriculum and conversational agent interface for MIT App Inventor, emphasizing eight AI competencies. Our study introduces a novel AI and Robotics education model, combining cascade mentoring and PBL, which addresses AI knowledge and actively engages students in hands-on AI and Robotics projects using the mBot robot. By integrating mentoring within the pedagogical framework, our approach enhances AI understanding and fosters teaching and mentoring skills among students, bridging the gap between theory and practice in AI education.

Sun et al.'s paper [14] explores the effectiveness of a near-peer mentoring model for engaging youth in computer science. Our proposed approach takes the concept of mentorship to a new level by introducing a cascade mentoring system within a project-based learning framework, connecting K-12 students with a university research lab. This approach leverages near-peer mentoring and extends mentorship across multiple levels, allowing junior members to mentor younger students. Our model encourages reciprocal mentoring, providing opportunities for participants to teach and mentor others in computer science, AI, and Robotics, offering a more comprehensive and immersive educational experience that addresses the engagement gap in AI and Robotics education.

In computer science education, current curricula have primarily focused on formal computing aspects for IT professionals and researchers, potentially overlooking the development of innovative thinking. Project-based learning (PBL) is recognized for cultivating students' initiative and experiential learning. This aligns with FIU's Discovery Lab objectives, where undergraduates are mentored and exposed to cutting-edge technology [8]. Our approach expands on these concepts, involving undergraduates mentoring high school students and creating a cascade mentorship structure. This model enhances PBL by integrating mentorship and reciprocal learning. Our research demonstrates the impact of this approach on student engagement, learning outcomes, and attitudes in computer science and science education, improving existing methodologies.

Zadok's [17] paper explores the implementation of project-based learning (PBL) in junior high school robotics education, emphasizing the transformation of teacher-student relationships and PBL usage. However, our project goes further by introducing a cascading mentorship model, connecting K-12 students with a University CS research lab through standards-based K-12 curricula. This approach promotes PBL and empowers students to take on mentorship roles, improving the overall learning experience and addressing the underrepresentation of marginalized groups in computer science and robotics education.

Coufal's [5] research focuses on using educational robots in STEM education and the influence of project-based teaching on student competencies. While this aligns with the abstract's emphasis on educational robotics and student learning, our project extends the concept by incorporating cascade mentoring. By introducing cascading mentorship into the educational framework, we provide a structured mentorship hierarchy that enhances student learning outcomes and empowers students to become mentors. This mentorship dynamic not only improves student competencies but also fosters a supportive learning environment, aligning with the project's aim to address diversity and engagement challenges in computer science and robotics education, ultimately enhancing students' experiences and outcomes in these fields.

The research by Nathan R. Dolenc et al. [6] offers insights into mentoring dynamics within a high school robotics team. This study is closely related to our paper, as it explores mentor-student interactions in the context of robotics competitions. Both papers share a common interest in mentorship and its impact on student development. Our work extends these ideas by introducing a cascading peer-mentorship model in AI and Robotics education, connecting K-12 students with a University CS research lab through project-based learning. Dolenc et al.'s research is a valuable reference for understanding mentorship dynamics and their relevance to our proposed cascade mentoring model.

In educational robotics and computational thinking, Cervera's study [2] highlights the positive impact of mentoring by older students on younger students' motivation and skills in computer programming with Bee-bot robots. Our research builds on Cervera's work by introducing a cascade mentoring model connecting K-12 students with a University CS research lab for project-based learning using the mbot robots. Unlike Cervera's study, our approach incorporates a diverse mentoring structure involving peers and university researchers. This new model enhances mentorship dynamics and offers a comprehensive framework to improve student engagement and learning outcomes in AI and Robotics education, contributing to the field's development.

3   Pedagogical Approach

A pedagogical approach encompasses education's philosophy, methods, and strategies to facilitate effective learning and instruction. It includes the principles and theories that guide how teachers design and deliver their lessons, interact with students, and enable a conducive learning environment. Pedagogical approaches span a broad spectrum, from traditional teacher-centered methods to more student-centered and refined approaches emphasizing active engagement and critical thinking.

3.1   Project-Based Learning

Project-Based Learning (PBL) is a distinctive instructional strategy within pedagogy. PBL is an immersive, hands-on approach where students actively delve into problems or questions by engaging in projects over an extended period. In the context of PBL, students collaborate and apply their knowledge to tackle complex challenges. This approach strongly emphasizes subject matter knowledge while caring for developing critical thinking, problem-solving skills, creativity, and collaboration - all essential skills for students to succeed in the modern world. PBL encourages students to take ownership of their learning, frequently resulting in deeper comprehension and long-lasting knowledge retention.

The project under discussion in this context used a Project-Based Learning (PBL) approach. This project immersed students in a hands-on, real-world challenge: creating a functioning trash can robot.

Initially, students were given an identified problem related to trash management, encouraging people to throw waste in the proper container and prevent recyclable waste from ending in landfills, marking the project's first phase. This phase corresponds with PBL's emphasis on addressing real-world issues. Subsequently, the students started the project's design process within their formed groups. Working together to create a list of functionalities they wanted for their robot.

The design and construction of the Trash Can Robots represented the project's central focus, providing a hands-on component that integrated various subjects such as science, engineering, and technology. This phase also enabled students to exercise their creativity and innovation, which are fundamental aspects of the PBL approach. Furthermore, the project allowed students to delve deeply into the subject, fostering a profound understanding of robotics and environmental issues.

As students collaborated within their groups, they also gained valuable teamwork and communication skills, aligning with PBL's emphasis on collaborative learning. This project exemplified Project-Based Learning by immersing students in a real-world challenge, promoting active learning, integrating multiple subjects, encouraging problem-solving, and fostering collaboration while emphasizing the practical application of knowledge and skills.

4   Implementation

The project's central idea revolved around developing an intelligent trash can robot with AI functionalities and interactive robotic capabilities to promote environmental education. The mentees work on creating a unique design and list of functionalities for their robots. Every project has the base communication implemented, depicted in Fig. 2. Each group is responsible for developing their designs using the mBot as the basis for their robot and using mBlock and the different AI and machine learning extensions to program the robot.

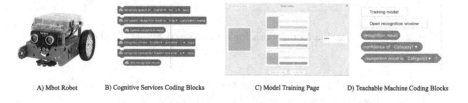

A) Mbot Robot   B) Cognitive Services Coding Blocks   C) Model Training Page   D) Teachable Machine Coding Blocks

Fig. 1. Key Components of mBot Robot kits and Machine Learning concepts

This project used the mBot robotics kit [12], a beginner-friendly programmable robot made by Makeblock [10], an entry-level educational robot designed for beginners, making it a suitable tool for students to learn about robotics and programming interactively. Alongside the mBot, the mBlock application is a block programming platform [11]. It provides an interface allowing beginners to learn programming and robotics concepts without the need for complex code and concepts. In this project, mentees used the mBlock extensions in their projects. Cognitive Services is the first extension used by the mentors. The extension integrates advanced AI capabilities into mBlock projects. It enables users to use AI services for tasks such as image recognition, voice recognition, and natural language processing. The second extension was the teachable machine extension. The extension integrates machine learning into the mBlock projects. Allows for custom object recognition, gesture interpretation, and voice command understanding. Some key components were depicted in Fig. 1.

4.1   Student Project Theme and System Architecture

In their Smart Trashcan mBot project, students designed their robot to follow lines and interact with users and sensors to create an engaging experience. However, the incorporation of machine learning is the main event of this project, enabling the robot to recognize and sort trash into recyclable and landfill categories. Throughout this project, students worked on diverse science challenges, ultimately cultivating a profound appreciation for the intersection of technology, sustainability, and human engagement. Figure 2 describes the primary connections between the mBot, AI components on the student's laptop, and the user:

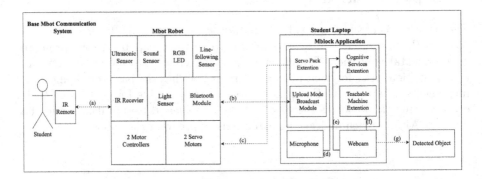

Fig. 2. Communication Framework for mBot Based Smart Trashcan Robots

(a) The infrared (IR) receiver enables the mBot to receive signals from an IR remote, allowing users to control the robot remotely.
(b) The mBot features a built-in Bluetooth module that facilitates wireless communication with external devices like laptops. This feature enabled students to deploy code to the mBot robot from their laptops.
(c) Mblock offers a servo pack extension kit with additional coding blocks for controlling servo motors. Students utilized this extension to hold the two servo motors added to the mBot for their project.
(d) With the cognitive services module, students could utilize the laptop's microphone as an additional component for their customized mBot project. This addition allowed students to use voice commands to control their robot.
(e) Using the cognitive services module, students could incorporate what the laptop's webcam captures as an additional component for their customized mBot project.

(f) This addition enabled students to leverage the teachable machine extension to train and build their photo database. This allowed the system to recognize objects within the database and execute customized functions developed by the students.

(g) This connection links the various objects that must be detected and the entire system.

4.2 Mentor Training and Mentorship Networking

The faculty mentors organized a workshop to discuss the foundations of mentoring and the expectations for the upcoming semester. Figure 2A illustrates the preparation before the mentoring semester. They guided effective engagement with younger students during the training. Furthermore, faculty mentors delivered valuable training to mentors and mentees, imparting essential knowledge and skills.

Fig. 3. Selected pictures from the start of the semester to the end of the program

At our first meeting at the high school, Fig. 3B, we engaged in a dynamic speed mentoring exercise. This activity allowed us to connect with each student individually, enabling us to understand their backgrounds and gauge their existing knowledge in computer science. Following this initial interaction, the students and mentors could select their preferred partners based on their experiences during the speed mentoring rounds. This choice allowed for more personalized and meaningful mentor-mentee pairings.

Students and mentors collaborated to create a unique team identity as our mentorship progressed. Subsequent meetings were dedicated to familiarizing the students with the mBlock platform and the mBot robot, as shown in Fig. 3D. Based on the coding cards provided by mBlock, the presentations were designed to introduce new topics during each meeting. This approach facilitated a structured and engaging learning experience for the mentees, fostering their interest and proficiency in computer science.

4.3   Foundational Learning and Code Demonstration

Our mentorship approach, aimed at addressing our group of mentees' challenges, focused on enhancing their understanding of complex concepts and problem-solving skills through hands-on guidance and support. To achieve this objective, we adopted a multifaceted approach that included interactive presentations and code demonstrations as foundational elements to address initial difficulties with various subjects.

Initially, model code examples were generated to summarize the fundamental principles of complex subjects, guaranteeing a thorough comprehension. A detailed explanation of each component within these example codes was given throughout the presentations, offering clear insights into the purpose and functionality. The primary goal of this method was to equip the mentees with the essential knowledge required for customizing these examples to suit their coding projects.

4.4   Introduction to Robotics and AI Concepts

The engineering process for building an intelligent trash can robot involves several key stages. Initially, the project begins with conceptualization and problem definition, where each group identifies the need for the robot and the problem it aims to solve, such as automating waste collection or sorting. A detailed design and planning phase ensues, during which the robot's physical design and functionalities are planned. A prototype is then built to test the design's feasibility, and component selection is carried out, including the choice of the main functionalities and a set design.

The software development phase commences, crucial for integrating Machine Learning (ML) and Artificial Intelligence (AI) components into the robot's functionality. Programming is essential for tasks such as voice command recognition or computer vision for object identification. The students train different object recognition models using the AI/ML extension in the mBlock application. The mentees trained various objects that they deemed either recyclable or trash. Hardware integration follows, ensuring all components are assembled and interfaced with the software effectively. For this project, the groups were given servo motors to use on the robot; some used them for functionality, while others used them as a design factor.

For this project, the mBlock platform provided an accessible and user-friendly interface for controlling the robot. A user interface, a web application, was used to develop code for both the mBot and servos (thanks to an extension.) The students wrote code using the different code blocks, controlling elements of the mBot robot. Implementing ML and AI within the mBlock application requires data collection, preprocessing, and the selection of appropriate ML algorithms or pre-trained models. Training and integrating the ML model into mBlock's ecosystem allows the robot to interpret and respond to data effectively. This engineering process combines innovative design, hardware integration, and AI-powered capabilities to create an intelligent trashcan robot.

4.5 Interactive Problem-Solving

In addition to the presentation-based approach, we actively engaged in troubleshooting to address specific issues faced by the mentees. This process involved thorough code review and debugging sessions, where the mentors worked closely to identify and fix problems within their code. In cases where mentees struggled to progress with their project code, providing them with example code representing the desired solution serves as templates to aid in comprehending coding patterns and best practices.

Furthermore, the mentorship was personalized, addressing specific requests from individual mentees to meet their unique needs. This involved revisiting topics or concepts previously requested during the interactions, offering new perspectives and more profound insights as help. Tailored example code was developed to align with areas of interest or confusion identified by mentees during previous meetings.

4.6 Weekly Progress Reviews and Goal Setting

In addition to interactive mentorship methods, A weekly progress review was implemented. Mentees were encouraged to work on their assigned projects, experiment with learned concepts, and apply newfound knowledge. This proactive approach allowed them to gain practical experience and tackle many challenges.

At the end of each week, mentees showcased their accomplishments, creating a goal-setting exercise. This iterative learning, application, and presentation cycle fostered ownership and accountability among mentees, reinforcing their commitment to the learning process.

These showcases served as a focal point of the mentorship approach, motivating mentees to meet and exceed their self-set objectives and demonstrating progress. The feedback and interactions during these sessions enhanced the mentorship experience for mentors and mentees, allowing for tailored support based on individual progress and needs.

This mentorship approach enriched our mentees' grasp of complex subjects. The goal-oriented dynamic of the weekly showcases inspired mentees to apply their knowledge, share their accomplishments, and pursue continuous improvement. This feedback-driven approach played a pivotal role in amplifying our mentorship model's overall success and effectiveness.

5   Learning Outcomes

The high school mentees in this project learned valuable lessons and gained significant benefits from their interactions with their mentors. They found their mentors helpful in various aspects of their computer science learning, including design, programming, and troubleshooting.

Regarding design, the mentees appreciated that their mentors were readily available to assist them in figuring out their design concepts. The mentors' in-person guidance was particularly beneficial to the mentees, who could receive hands-on assistance and advice. This approach helped the mentees better understand and refine their design ideas. Regarding programming and troubleshooting, the mentors played a crucial role in improving the mentees' skills. They guided fixing code issues and offered insights on enhancing their mBot projects. The mentees found their mentors to be reliable problem-solvers, and they could rely on them for practical solutions to challenges they encountered.

They liked that the mentors completed tasks for them and actively showed them how to perform different tasks. This hands-on approach empowered the mentees to understand various computer science aspects better. Effective communication was another critical aspect of their mentorship experience. Mentees found it easy to communicate with their mentors, who answered their questions thoroughly and in a manner that helped them understand coding concepts better. The mentees highly valued the accessibility of mentors. Even when not meeting in person, mentors remained responsive and engaged. They promptly addressed questions and provided helpful resources, making the learning process smoother and more convenient.

Overall, the mentees felt comfortable discussing computer science-related concerns with their mentors. Through this project, the mentees not only improved their computer science skills but also gained confidence, effective communication, and problem-solving abilities, thanks to the support and guidance of their mentors.

5.1   Mentor Reflections

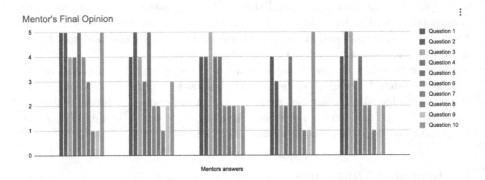

Fig. 4. Visualized data

The mentors in the cascade mentoring project were instrumental in shaping the experiences of their mentees. They played a pivotal role in fostering academic and personal growth through their dedication and guidance. Here are their responses to a series of questions, each rated on a scale from 1 to 5, with one representing "not at all" and five representing "very much" (Fig. 4):

Undergraduate mentors were initially concerned about the mentoring program but were eager to pass on their knowledge. They prepared for meetings differently, some addressing specific challenges others creating presentations. Building strong mentor-mentee relationships was a priority, achieved through open communication and tailored guidance. The mentors adjusted their approach based on the mentees' interests and abilities, such as relating concepts to their hobbies (Table 1).

Table 1. Survey topics and average ratings

Topic	Rating
Knowledge Enhancement	4.2
Boosting Confidence	4.4
Increasing Motivation	4.0
Social and Communication Skills	3.0
Leadership Skills	3.6
Challenging Activities	2.4
Lack of Knowledge	1.4
Lack of Social/Communication Skills	1.6
Lack of other Crucial Skills	1.6
Impact on Mentors	3.4

The mentees initially appeared uninterested but became highly engaged once they focused on their projects, which was inspirational and surprising. The students' projects met program expectations, showcasing their learning. Overall, the mentoring program successfully provided hands-on STEM education and personal growth, with mentors taking pride in their mentees' progress and their role as positive influences in their educational journeys.

5.2 Project Outcomes

At the end of the semester, we had five complete mBot projects. Each one has a unique design and set of functionalities. Each group worked with their mentor to complete the project, gaining valuable experience.

To track their progress, each group was required to write a reflection outlining the project's main goals, the desired design and functionalities for their robot, and the necessary supplies. This reflection also included daily records from the final week of working on their robots, describing their problems, the troubleshooting steps, and how they resolved them. The group reflections also contained their final thoughts on their semester-long project. This included their original project idea, the end product, how they overcame design challenges, what they would do differently if they continued or restarted the project, and their overall satisfaction with the robot upon completion.

These reflections provided insight into how the students assessed their progress and demonstrated the development of their skills throughout the semester (Fig. 5).

Fig. 5. Robots created by the mentee groups: (Left to right) Rosie the Robot, Gunter, R2-D2, PacMan, Atlas

Rosie The Robot: The team crafted their robot inspired by Rosie, the robotic maid from The Jetsons. It illuminates in red upon receiving trash and turns green when handling recycling. Respond to the command "Rosie" and play the Jetsons theme while approaching you. Additionally, it possesses the ability to identify various objects and categorize them as either trash or recycling. **Gunter:** The design for the robot is based on a character from Adventure Time, Gunter, the penguin. The robot can recognize different objects and tell whether they are trash or recycling based on the model they trained. Using the servo motors, they created doors that swing open based on what the model recognizes. **R2-D2:** The team made a line following the avoiding trash can robot. The robot used IR and Ultrasonic sensors to follow lines and avoid obstacles and a computer vision to categorize objects as trash or recyclables. The robot was modeled after R2-D2 from Star Wars. **PacMan:** They created the arcade machine design for the robot. The robot can recognize different objects and tell whether they are trash or recycling based on the model they trained. The robot has a single door that will open from the servo motor to put the items in. The robot has a single compartment separated into two, one section for the trash and the other for the recycling. **Atlas:** This robot's design draws inspiration from the Lakers' theme, incorporating their distinctive colors and logo. It distinguishes itself from the other four robots by taking a unique approach. This team created an efficient scoop mechanism rather than using the servo motors for traditional functions, opening and closing doors/flaps.

6   Discussion

Implementing Project-Based Learning (PBL) in AI and Robotics significantly boosts student engagement. Our trash can robot project demonstrates this, showing how PBL enhances involvement. PBL encourages active learning as students shape, design, and construct robots. This hands-on engagement nurtures ownership and motivation, making the learning more relevant.

Furthermore, the interdisciplinary nature of PBL, where students integrate multiple subjects such as engineering and technology, not only enriches their understanding but keeps them engaged by demonstrating the practical applications of their knowledge. The hands-on aspect of PBL allows students to see the direct impact of their efforts, creating a tangible connection between classroom learning and the real world. This connection, in turn, fuels their enthusiasm and engagement in the topics being taught.

Incorporating teamwork and collaboration further amplifies engagement. Students learn from mentors and peers, sharing ideas and creating a supportive learning atmosphere. PBL's success in AI and Robotics education suggests that high schools should shift from passive to experiential learning. PBL should be integrated into curricula, preparing students for future careers and life skills. Communication between mentors and mentees should be enhanced with regular check-ins and feedback for effective cascade mentoring. Open dialogue can address challenges, ensuring a productive mentorship experience.

7   Conclusion and Future Work

Artificial intelligence and robotics technologies are advancing at a rapid pace. They will inevitably play an even more significant role in society soon. The result is a world that relies on those technologies for essential day-to-day functions. To facilitate such a world, we must address the issue of AI and robotics education in young adult classrooms. Our study produced auspicious results which indicate that project-based learning will be a vital part of that education.

The pedagogical approach demonstrated in this study proved to be very effective at improving the high school students' knowledge of AI and Robotics. The cascade mentoring approach also enhanced mentees' and mentors' communication skills, confidence, and logical thinking. The mentees benefited from the accessibility and responsiveness of the mentors.

In the future, we hope to replicate this study at a larger scale to prove further the benefits for the students involved. Additionally, we intend to develop an exact classroom implementation procedure to increase the accessibility to this approach for educators. Finally, and critically, we hope to improve future generations' confidence and competence levels with these vital technologies.

Acknowledgements. We would like to thank NP High School for their great assistance and collaborative efforts during this project, as well as Dr. Shannon and her students for their valuable support and insightful contributions that have improved our research. This research was supported by the 2022 KSU EHHS Seed Award and the ATR lab research funding.

References

1. Aderibigbe, S.A., Alotaibi, E., Alzouebi, K.: Exploring the impact of peer mentoring on computer-supported collaborative learning among undergraduate students
2. Cervera, N., Diago, P.D., Orcos, L., Yáñez, D.F.: The acquisition of computational thinking through mentoring: an exploratory study
3. Chistyakov, A.A., Zhdanov, S.P., Avdeeva, E.L., Dyadichenko, E.A., Kunitsyna, M.L., Yagudina, R.I.: Exploring the characteristics and effectiveness of project-based learning for science and steam education
4. Clarke-Midura, J., Poole, F., Pantic, K., Hamilton, M., Sun, C., Allan, V.: How near peer mentoring affects middle school mentees
5. Coufal, P.: Project-based stem learning using educational robotics as the development of student problem-solving competence
6. Dolenc, N.R., Tai, R.H., Williams, D.: Excessive mentoring? An apprenticeship model on a robotics team
7. Jeon, M.: Developing middle schoolers' artificial intelligence literacy through project-based learning: investigating cognitive & affective dimensions of learning about AI. Ph.D. thesis
8. Kim, J.-H., Sharma, G., Prabakar, N., Iyengar, S.S.: Inspiring innovative aspirations among undergraduate students
9. Lee, I., Perret, B.: Preparing high school teachers to integrate AI methods into stem classrooms
10. Makeblock. Makeblock. https://www.makeblock.com/
11. mBlock. What is mblock5? https://support.makeblock.com/hc/en-us/articles/4408619146519-What-is-mBlock-5-
12. mBot. What is mBot? https://support.makeblock.com/hc/en-us/articles/360062396654-What-is-mBot-
13. Novak, E., Borgerding, L.: Perspectives of mentors and mentees in a cascading mentorship high school computer science robotics intervention
14. Sun, C., Clarke-Midura, J.: Testing the efficacy of a near-peer mentoring model for recruiting youth into computer science
15. Van Brummelen, J., Heng, T., Tabunshchyk, V.: Teaching tech to talk: K-12 conversational artificial intelligence literacy curriculum and development tools
16. Voldsund, K.H., Bragelien, J.J.: Student peer mentoring in an entrepreneurship course
17. Zadok, Y.: Project-based learning in robotics meets junior high school

Leveraging Human-Machine Interactions for Computer Vision Dataset Quality Enhancement

Esla Timothy Anzaku[1,2,3]([✉]) [ID], Hyesoo Hong[1] [ID], Jin-Woo Park[1] [ID], Wonjun Yang[1] [ID], Kangmin Kim[1] [ID], JongBum Won[1] [ID], Deshika Vinoshani Kumari Herath[5] [ID], Arnout Van Messem[4] [ID], and Wesley De Neve[1,2,3] [ID]

[1] Ghent University Global Campus, Incheon 21985, South Korea
eslatimothy.anzaku@ugent.be
[2] Center for Biosystems and Biotech Data Science, Ghent University Global Campus, Incheon 21985, South Korea
[3] IDLab, Ghent University, Technologiepark-Zwijnaarde 126, 9052 Ghent, Belgium
[4] University of Liège, 4000 Liège, Belgium
[5] Mediio, Seoul, South Korea

Abstract. Large-scale datasets for single-label multi-class classification, such as *ImageNet-1k*, have been instrumental in advancing deep learning and computer vision. However, a critical and often understudied aspect is the comprehensive quality assessment of these datasets, especially regarding potential multi-label annotation errors. In this paper, we introduce a lightweight, user-friendly, and scalable framework that synergizes human and machine intelligence for efficient dataset validation and quality enhancement. We term this novel framework *Multilabelfy*. Central to Multilabelfy is an adaptable web-based platform that systematically guides annotators through the re-evaluation process, effectively leveraging human-machine interactions to enhance dataset quality. By using Multilabelfy on the ImageNetV2 dataset, we found that approximately 47.88% of the images contained at least two labels, underscoring the need for more rigorous assessments of such influential datasets. Furthermore, our analysis showed a negative correlation between the number of potential labels per image and model top-1 accuracy, illuminating a crucial factor in model evaluation and selection. Our open-source framework, Multilabelfy, offers a convenient, lightweight solution for dataset enhancement, emphasizing multi-label proportions. This study tackles major challenges in dataset integrity and provides key insights into model performance evaluation. Moreover, it underscores the advantages of integrating human expertise with machine capabilities to produce more robust models and trustworthy data development.

Keywords: Computer Vision · Dataset Quality Enhancement · Dataset Validation · Human-Computer Interaction · Multi-label Annotation

B. J. Choi et al. (Eds.): IHCI 2023, LNCS 14531, pp. 295–309, 2024.
https://doi.org/10.1007/978-3-031-53827-8_27

1 Introduction

Deep learning, the engine behind advanced computer vision, has been largely propelled by training on expansive resources like the ImageNet Large Scale Visual Recognition Challenge (ILSVRC) dataset [1], commonly known as ImageNet-1k. However, recent performance trends in deep neural network (DNN) models trained on these datasets have shown top-1 and top-5 accuracy stagnation across diverse DNN architectures and training techniques, irrespective of model complexity and dataset size [2,3]. This performance plateau suggests that we may be nearing the limits of model accuracy with the current ImageNet-1k dataset using the top-1 accuracy.

A potentially overlooked factor contributing to the observed stagnation may be attributed to the inherent multi-label nature of the dataset in question. It is plausible that a substantial proportion of the images in the dataset are related to more than a single ground truth label. However, the dataset only provides labels for a singular ground truth, which may impose limitations [4–6]. This single-label ground truth constraint could inadvertently lead to underestimating the performance of DNN models, particularly when utilizing the top-1 accuracy metric.

Furthermore, the performance of models significantly degrades when assessed on newer but similar datasets, such as ImageNetV2 [7]. Despite being developed following a similar protocol to the original ImageNet-1k dataset, ImageNetV2 exhibits unexplained accuracy degradation across various models, regardless of model architecture, training dataset size, or other training configurations. While efforts are being made to investigate this degradation [8,9], we found only one work that partially studied this problem [10].

Prior work has acknowledged the need for more accurate dataset labels and has published reassessed labels that reflect the multi-label nature of the ImageNet-1k validation set [4]. However, label reassessment is not a trivial task. It requires considerable resources and expertise, presenting a substantial challenge for smaller research groups. Given the vital role of the validation and test sets in DNN model selection and benchmarking, meticulous analysis of the ImageNet-1k validation set and its replicates remains indispensable. This critical importance highlights the necessity for accessible and effective frameworks to scrutinize and tackle the multi-label nature of computer vision single-label classification datasets. To address this, we propose an accessible and scalable framework, termed Multilabelfy, that combines human and machine intelligence to efficiently validate and improve the quality of computer vision multi-class classification datasets. Multilabelfy comprises four stages: (i) label proposal generation, (ii) human multi-label annotation, (iii) annotation disagreement analysis, and (iv) human annotation refinement. It is designed with two primary objectives: to strategically harness the capabilities of a diverse pool of annotators and to seamlessly blend human expertise with machine intelligence to improve the quality of a dataset. These objectives are made accessible through a user-friendly interface.

This research effort enriches existing literature by offering Multilabelfy for improving the quality of computer vision multi-class classification datasets. Utilizing Multilabelfy, we reassessed the labels for ImageNetV2, revealing that 47.88% of the images in this dataset could have more than one valid label. We also identified other noteworthy dataset issues. Our work accentuates the importance of recognizing and addressing the multi-label nature of ImageNet-1k and its replicates. Our ultimate goal is to contribute towards developing robust DNN models that can effectively generalize beyond their training data.

2   Related Work

2.1   Label Errors

Label errors have been identified within the test sets of numerous commonly used datasets, including a 6% error rate in the ImageNet-1k validation set [11]. The importance of tackling the issue of label errors in test partitions of datasets was further emphasized. It was found that high-capacity models are prone to mirroring these systematic errors in their predictions, potentially leading to a misrepresentation of real-world performance and distortion in model comparisons. In another work, an extensive examination of 13,450 images across 269 categories in the ImageNet-1k validation set, which predominantly includes wild animal species, was conducted [12]. Through collaboration with ecologists, it was found that many classes were ambiguous or overlapping. An error rate of 12% in image labeling was reported, with some classes being erroneously labeled more than 90% of the time. Our work further accentuates the critical role of addressing label errors in datasets used for model evaluation. It underscores the need for more precise and thorough dataset construction and assessment methodologies.

2.2   From Single-Label to Multi-Label

Single-label evaluation has traditionally served as the standard for assessing models on the ImageNet-1k dataset. However, a reassessment of the ImageNet-1k validation ground truth labels revealed that a good proportion of the images could have multiple valid labels, prompting the creation of Reassessed Labels (ReaL) [4], incorporating these multi-labels.

In a related study [6], the remaining errors that models made on the ImageNet-1k dataset were examined, focusing on the multi-label subset of *ReaL*. Nearly half of the perceived errors were identified as alternative valid labels, confirming the multi-label nature of the dataset. However, it was also observed that even the most advanced models still exhibited about 40% of errors readily identifiable by human reviewers.

2.3   ImageNet-1k Replicates

When tested on replication datasets like ImageNetV2, DNN models have been observed to demonstrate a significant, yet unexplained, drop in accuracy [7].

Despite these replication datasets, including ImageNetV2, being created by following the original datasets' creation protocols closely, the performance decline raises significant questions about the models' generalization capabilities or the integrity of the datasets. The significant performance drop on ImageNetV2, between 11% to 14% [7], was based on the conventional approach of evaluating model accuracy using all data points in the test datasets.

However, it has been argued that the conventional evaluation approach may not fully capture the behavior of DNN models and may set unrealistic expectations about their accuracy [8,9]. A more statistically detailed exploration into this unexpected performance degradation on ImageNetV2 found that standard dataset replication approaches can introduce statistical bias [8]. After correcting for this bias and remeasuring selection frequencies, the unexplained part of the accuracy drop was reduced to an estimated $3.6\% \pm 1.5\%$, significantly less than the original $11.7\% \pm 1.0\%$ earlier reported in [7]. An alternative evaluation protocol that leverages subsets of data points based on different criteria, including uncertainty-related information, provides an alternative perspective [9]. Through comprehensive evaluation leveraging the predictive uncertainty of models, the authors found that the degradation in accuracy on ImageNetV2 was not as steep as initially reported, suggesting possible differences in the characteristics of the datasets that warrant further investigation. A closely related research work studies various aspects of the ImageNet-1K and ImageNetV2 datasets using human annotators. Using a sample of $1,000$ images from both datasets, the proportion of images with multiple labels was estimated to be 30.0% and 34.4% for the two datasets, respectively. This information is detailed in Section B.2 of the supplementary material in [10]. The cited work suggested that the difference in the multi-label composition between the two datasets could be a possible explanation for the accuracy degradation.

2.4   Key Modifications to Existing Approaches

Our research expands upon a previous work [4] with several essential modifications:

Model Selection for Candidate Label Proposals. In contrast to the original study's use of a hand-annotated sample of 256 images from a 50,000-image dataset to guide the selection of an optimal model ensemble, we built upon their work, utilizing their generated multi-labels and proposed *ReaL accuracy*. We selected the best-performing pre-trained model utilizing the ReaL accuracy metric, designed to evaluate multi-class classification DNN models on a multi-label test dataset. Further details on this process are provided in Sect. 3.2.

Image Pre-selection for Multi-label Annotation. The original study only utilized an ensemble of pre-trained single-label models to generate eight candidate labels. In contrast, our approach extended the candidate proposals to 20, thereby decreasing the risk of omitting valid labels and increasing selection accuracy (Sect. 3.2).

Annotation Refinement. We introduced an additional stage, wherein the top twenty model-proposed labels, alongside all human-selected labels, are presented to an additional pool of experienced annotators for further refinement (Sect. 3.5).

Open-Source Platform. Recognizing that platforms like Mechanical Turk might be inaccessible or not affordable for some research labs, we developed Multilabelfy, an open-source alternative. This platform allows in-house dataset quality improvement while maintaining a user-friendly interface.

Section 3 provides more comprehensive information regarding these contributions.

3 Proposed Framework

3.1 Overview

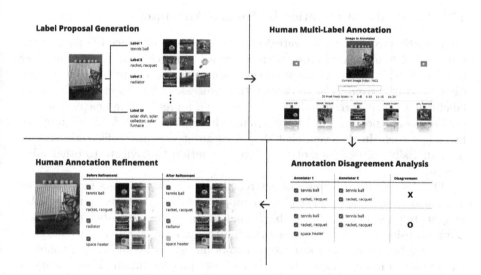

Fig. 1. Overview of the proposed framework for enhancing computer vision datasets from single-label to multi-label, enabling a more comprehensive capture of their descriptions.

The proposed multi-label dataset enhancement framework (Fig. 1) comprises four key stages: (i) label proposal generation, (ii) human multi-label annotation, (iii) annotation disagreement analysis, and (iv) human annotation refinement. The label proposal generation and annotation disagreement analysis can be automated using the appropriate algorithms while the human multi-label annotation and human annotation refinement require the involvement of human annotators.

3.2 Label Proposal Generation

Our qualitative analysis shows that pre-trained models, originally trained on single-label computer vision datasets, can effectively rank the predicted probability vector. This capability is corroborated by the near-perfect top-5 accuracies of state-of-the-art DNN classification models reaching approximately 99% [2]. For model selection, we utilize the *ReaL accuracy* metric, specifically designed to assess the performance of single-label pre-trained DNN models in multi-label scenarios. *Under this metric, an image prediction is considered correct if the prediction belongs to the set of ground truth labels assigned to the image.* The selected model is then used to generate the top-20 label proposals, an increase from the eight proposals presented in previous work, to ensure broader coverage of valid labels. Given the potential for information overload with many label proposals, we designed the annotation user interface to mitigate this concern. Additional details regarding the role of human annotators and the annotation interface are discussed in Sect. 3.3.

3.3 Multi-label Annotation by Human Annotators

Multilabelfy incorporates a strategically designed web interface to alleviate the workload of human annotators, with a screenshot provided in Fig. 2. This user interface is characterized by several key features engineered to enhance the efficiency and effectiveness of the annotation process. It facilitates the display of label names, their corresponding synonyms, and representative images from the pool of twenty potential labels systematically organized into four subgroups of five labels each. In the event that the initial group does not sufficiently encompass all visible objects, annotators have the option to navigate to other label groups.

The design also incorporates a streamlined selection process facilitated by a singular checkbox assigned to each proposed label. Moreover, ten exemplar images are presented in a scrollable format for each proposed label, providing a comprehensive view without overwhelming the annotator. Further attention to detail is reflected in the feature that allows images to be clicked on, enabling annotators to inspect these images at their original resolution. These elements combined optimize the multi-label annotation process, yielding higher accuracy and efficiency.

3.4 Annotation Disagreement Analysis

Single-label multi-class classification computer vision datasets often comprise images featuring multiple objects. However, a prior research work [5] estimated that about 80% of the ImageNet-1k images contain a single object. We also expect some images with multiple labels to pose no challenges to the annotators. Considering the aforementioned observations, our framework seeks to effectively exclude such images from the pool intended for further refinement. We target

Fig. 2. The user interface of the annotation platform. It showcases key features like label presentation in groups of five, a single checkbox per proposed label, scrollable sample images, and click-to-enlarge functionality for detailed inspection of images. These features are designed to streamline the annotation process and efficiently accommodate multi-label data annotation.

images that require additional human annotation refinement during the *annotation disagreement analysis* stage, as depicted in Fig. 1. Images are selected for further annotation refinement if the labels generated by human annotators, as discussed in Sect. 3.3, fail to meet a predefined annotation agreement condition. This annotation agreement condition requires: *complete consistency across all labels identified by human annotators for a particular image and the inclusion of the originally provided ground truth label within the array of labels selected by the annotators.* This strategic condition facilitates focused refinement of annotations for the subset of images that pose more significant challenges to annotators. As a result, we minimize the misuse of annotators' time and provide an avenue for a more detailed examination of the more complex images, ultimately fostering a more thoroughly annotated dataset.

3.5 Refinement of Human Annotation

This stage follows the process described in Sect. 3.3 but with some critical distinctions. In the stage described in Sect. 3.3, annotators with varying degrees of experience with the dataset contribute to the labeling. However, the refinement stage exclusively engages more experienced annotators. These experienced annotators are provided with the labels previously selected, which are pre-checked for the annotators to review: uncheck (to correct) or check additional missing labels. Furthermore, the annotators are instructed to document any changes they make to the labels using the comments section of the web interface. This provision ensures that a clear record is maintained for each correction, which can be invaluable in resolving potential discrepancies in the annotations. It is important to note that these annotators have undergone several tutorial sessions on the label issues of the ImageNet-1k dataset. Additionally, they reviewed and summarized related literature to ensure that they are aware of the nuanced issues that are encountered when annotating images into 1,000 categories, especially within the fine-grained categories.

4   Results

4.1   Experimental Setup

Our goal is to re-assess the labels for the ImageNetV2 dataset to accommodate and account for its multi-label nature. The four stages in Sect. 3 were carefully followed. In the label proposal generation stage, the EVA-02 [13] model was used to generate the proposal. It is one of the top performing models (90.05% top-1 accuracy [2]) on the ImageNet-1k dataset; additional details of the model can be found in the cited paper. Subsequently, in the human multi-label annotation stage, the 10,000 images of the ImageNetV2 dataset were partitioned into seven batches, and each batch was assigned to two human annotators. Fourteen human annotators having varying experience levels with the ImageNet-1k dataset and computer vision in general, participated during this stage. Upon the annotation disagreement analysis (detailed in Sect. 3.4), the annotations for 6,425 of the 10,000 images fulfilled our disagreement criteria and were selected for subsequent refinement by five more experienced annotators, four of whom were previously referenced among the group of fourteen. Each annotator refined the annotations for 1,285 images. The refined annotations were then used to generate the results presented and discussed in the following sections.

4.2   The Extent of ImageNetV2 Multi-Labeledness

Here, we provide visual statistics summarizing the multi-label nature of the labels we generated for the ImageNetV2 dataset. Specifically, we show what percentage of the dataset contains which label count, i.e., the number of ground truth labels assigned to an image. As shown in the pie chart of Fig. 3, the annotation process could not find labels for 1.29% of the images. Moreover, 50.83% of the images contain one label, 23.85% contain two labels, and 24.03% contain more than two labels.

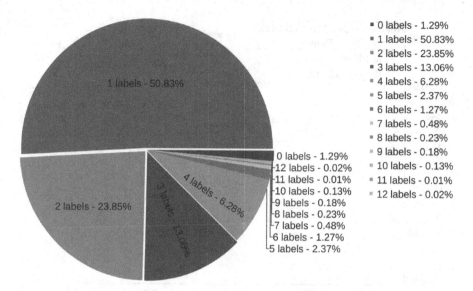

Fig. 3. The distribution of images based on the number of labels assigned to them during our annotation process.

4.3   Re-evaluation of Models on ImageNetV2 Improved Labels

Top-1 Accuracy Versus ReaL Accuracy. We provide a Scatterplot to understand the relationship between ReaL and top-1 accuracy on our generated labels (Fig. 4). Each dot in the plot represents a pre-trained model, and 57 models were evaluated on the ImageNet-1k validation set and ImageNetV2. These models are sourced from a publicly available GitHub repository [2] and represent state-of-the-art models pre-trained either exclusively on the ImageNet-1k dataset, or on additional external data. Details of these models can be found together with the paper's code at https://github.com/esla/Multilabelfy The regression analysis indicates a significant correlation between the two metrics. Specifically, for every percentage point increase in top-1 accuracy, the ReaL accuracy rises by approximately 0.5788 percentage points. The coefficient of determination, R^2, is 75.69%, suggesting that 75.69% of the variation in ReaL accuracy is explained by its linear relationship with top-1 accuracy. This result reflects a consistent positive relationship: as the top-1 accuracy of models improves, there is a proportional increase in ReaL accuracy. It is worth noting that four models visibly diverge from the regression line; these models merit additional scrutiny to identify potential model-specific quirks or underlying reasons for their divergence. A detailed investigation of these models will be addressed in future work.

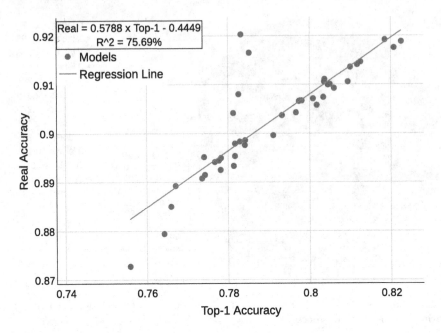

Fig. 4. Scatterplot of ReaL accuracy versus top-1 accuracy for 57 top-performing DNN models, pre-trained either exclusively on the ImageNet-1k dataset or additionally on external datasets.

Visual Statistics of Top-1 Accuracy Versus Image Count.

We investigate the relationship between top-1 accuracy and the variability in image label assignments using heatmaps (Fig. 5). While we presented the results for 57 models in Sect. 4.2, for visual brevity, we randomly selected 5 models for the heatmaps. We determine top-1 accuracy using ground truth labels from the ImageNetV2 dataset, comparing them with our multi-label annotations. To this end, we employ a heatmap (Fig. 5, top) that presents top-1 accuracies for each evaluated model across different *label count* categories. While this heatmap is informative, it does not factor in the variability stemming from different sample sizes across label counts. For instance, images with a single label may be more prevalent than those with multiple labels, potentially leading to biases in accuracy measurements.

To enhance our understanding of accuracy computations and account for inherent uncertainties, we incorporate a secondary heatmap as shown in Fig. 5, bottom. The margin of error related to the top-1 accuracy is denoted as $U(i, j)$ and is determined using the following formula: $U(i, j) = 1.96 \times \sigma(i, j)/\sqrt{n}$. Here, $\sigma(i, j)$ stands for the standard deviation stemming from the binary outcomes of individual predictions for a specific *model* and *label count*. This standard deviation for a binary variable is expressed as $\sigma(i, j) = \sqrt{p(1-p)/n}$, where p represents the proportion of correct predictions. The variable n symbolizes the number of observations for the considered model-label count pairing. This margin

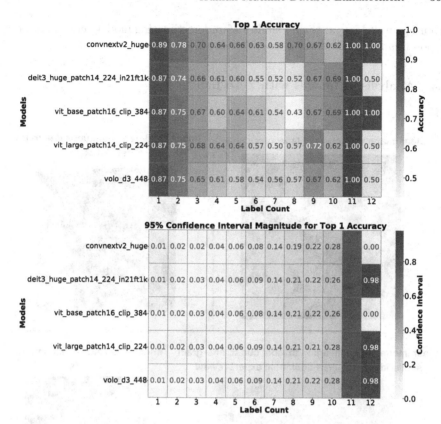

Fig. 5. Heatmaps displaying top-1 accuracy (top) for five randomly selected models evaluated on our multi-labeled ImageNetV2 dataset, and the half-width of the 95% confidence interval (bottom) associated with these accuracies. Red cells without numbers represent NaN values due to sets with one or no images for a given label count. (Color figure online)

of error, corresponding to half the width of the 95% confidence interval, offers a gauge of uncertainty for each model-label count combination. Differences in sample sizes across subsets can lead to variations in the width of the confidence interval. This variance emphasizes the significance of jointly considering both accuracy and its associated uncertainty when interpreting model performance across different label counts.

In our analysis, while results for only five models are presented for clarity, the observations are representative of numerous other models evaluated. A notable observation is that models consistently exhibit higher top-1 accuracy for images associated with a single label. However, as the number of potential labels expands, a discernible decrease in accuracy is evident. This pattern potentially indicates that models might be predicting alternative valid labels, and the top-1 accuracy metric penalizes them for such predictions. Such a negative correlation warrants attention, as it hints at the possibility of underestimating model performance due to potentially skewed dataset assumptions.

4.4   Analysis of Images with Zero Labels

Clear image with no valid label proposals

Image rendered ambiguous due to low resolution

Clear image but challenging to label due to fine-grain similarities; Requires additional expert review

Image showcasing uncommon objects

Fig. 6. Example images where annotators did not find matching labels from the 20 proposed labels. The images are categorized based on possible explanations for not finding matching labels in the labels proposed (see Sect. 4.4).

During our dataset annotation process, despite the meticulous efforts of fifteen annotators, 1.29% (129 images) had no labels assigned to them at the completion of human annotation refinement. Consequently, two of the experienced annotators further scrutinized these images. They classified the images without valid annotations into (i) clear images with no valid label proposals (21.79%), (ii) images rendered ambiguous due to low resolution (10.26%), (iii) clear images but challenging to label due to fine-grain similarities, thereby requiring additional expert review (38.46%), and (iv) images showcasing uncommon objects

Fig. 7. Example images where annotators did not find matching labels from the 20 proposed labels. The images are categorized based on whether or not our two annotators agree with the provided ImageNetV2 ground truth label (see Sect. 4.4).

or atypical viewpoints (29.49%). One example from each of these categories is shown in Fig. 6.

While our finalized annotations did not provide labels for these images, ground truth labels from the creators of the ImageNetV2 dataset existed for reference. Using these, the annotators further categorized the images based on their alignment with the ImageNetV2 ground truth as (i) those they agree with (26.92%), (ii) those they disagree with (19.23%), and (iii) those they remain uncertain about and highly doubt (53.85%). Examples of this type of categorization are provided in Fig. 7.

5   Conclusions

Single-label multi-class classification datasets like ImageNet-1k are crucial for advancing deep learning in computer vision. However, as the demand for reliable DNN models grows, it is vital to examine these datasets for biases that could impede progress. We provide a practical framework for smaller research groups to enhance the quality of multi-class classification datasets, especially those that could contain multi-labeled images. Furthermore, we introduce new labels for the ImageNetV2 dataset to account for its multi-label nature. The purpose of our dataset enhancement platform and the provided multi-labels for ImageNetV2 is to facilitate research on the performance degradation of ImageNet-1k-trained

DNN models on the ImageNetV2 dataset. Interestingly, only about half of the 10,000 images in the ImageNetV2 dataset can be confidently categorized as having a single label, thereby underscoring the need for further investigation into the impact of the multi-labeled images on ImageNet-based benchmarks and their potential implications for downstream utilization. Such research endeavors will help us better understand how models perform on complex vision datasets.

Acknowledgment. This research was supported by Ghent University Global Campus (GUGC) in Korea. This research was also supported under the National Research Foundation of Korea (NRF), (2020K1A3A1A68093469), funded by the Korean Ministry of Science and ICT (MSIT). We want to specifically thank the following people for their contribution to the annotation process: Gayoung Lee, Gyubin Lee, Herim Lee, Hyesoo Hong, Jihyung Yoo, Jin-Woo Park, Kangmin Kim, Jihyung Yoo, Jongbum Won, Sohee Lee, Sohn Yerim, Taeyoung Choi, Younghyun Kim, Yujin Cho, and Wonjun Yang.

References

1. Krizhevsky, A., Sutskever, I., Hinton, G.: ImageNet classification with deep convolutional neural networks. Commun. ACM **60**(6), 84–90 (2017). https://doi.org/10.1145/3065386
2. Wightman, R.: Pytorch image models. GitHub (2019). https://github.com/huggingface/pytorch-image-models/blob/main/results/results-imagenet.csv
3. Ozbulak, U., et al.: Know your self-supervised learning: a survey on image-based generative and discriminative training. Trans. Mach. Learn. Res. (2023). https://openreview.net/forum?id=Ma25S4ludQ
4. Beyer, L., Hénaff, O., Kolesnikov, A., Zhai, X., Oord, A.: Are we done with ImageNet? arXiv preprint (2020). http://arxiv.org/abs/2006.07159
5. Tsipras, D., Santurkar, S., Engstrom, L., Ilyas, A., Madry, A.: From ImageNet to image classification: contextualizing progress on benchmarks. In: 37th International Conference on Machine Learning, Article no. 896, pp. 9625–9635 (2020). https://dl.acm.org/doi/10.5555/3524938.3525830
6. Vasudevan, V., Caine, B., Gontijo-Lopes, R., Fridovich-Keil, S., Roelofs, R.: When does dough become a bagel? Analyzing the remaining mistakes on ImageNet. In: NeurIPS (2022). https://openreview.net/pdf?id=mowt1WNhTC7
7. Recht, B., Roelofs, R., Schmidt, L., Shankar, V.: Do ImageNet classifiers generalize to ImageNet? In: 36th International Conference on Machine Learning (2019). http://proceedings.mlr.press/v97/recht19a/recht19a.pdf
8. Engstrom, L., Ilyas, A., Santurkar, S., Tsipras, D., Steinhardt, J., Madry, A.: Identifying statistical bias in dataset replication. In: 37th International Conference on Machine Learning (2020). http://proceedings.mlr.press/v119/engstrom20a/engstrom20a.pdf
9. Anzaku, E., Wang, H., Van Messem, A., De Neve, W.: A principled evaluation protocol for comparative investigation of the effectiveness of DNN classification models on similar-but-non-identical datasets. arXiv preprint (2022). http://arxiv.org/abs/2209.01848
10. Shankar, V., Roelofs, R., Mania, H., Fang, A., Recht, B., Schmidt, L.: Evaluating machine accuracy on ImageNet. In: 37th International Conference on Machine Learning, vol. 119, pp. 8634–8644 (2020). https://proceedings.mlr.press/v119/shankar20c.html

11. Northcutt, C., Athalye, A., Mueller, J.: Pervasive label errors in test sets destabilize machine learning benchmarks. In: Proceedings of the Neural Information Processing Systems Track on Datasets and Benchmarks (2021). https://openreview.net/pdf?id=XccDXrDNLek

12. Luccioni, A., Rolnick, D.: Bugs in the data: how imagenet misrepresents biodiversity. In: Proceedings of the Thirty-Fifth Conference on Innovative Applications of Artificial Intelligence and Thirteenth Symposium on Educational Advances in Artificial Intelligence, Article no. 1613, pp. 14382–14390 (2023). https://dl.acm.org/doi/10.1609/aaai.v37i12.26682

13. Fang, Y., Sun, Q., Wang, X., Huang, T., Wang, X., Cao, Y.: Eva-02: a visual representation for neon genesis. arXiv preprint (2023). https://doi.org/10.48550/arXiv.2303.11331

A Comprehensive Survey on AgriTech to Pioneer the HCI-Based Future of Farming

Ashutosh Mishra[ID] and Shiho Kim[✉][ID]

School of Integrated Technology, Yonsei University, Incheon, South Korea
shiho@yonsei.ac.kr

Abstract. The burgeoning global population constricted arable land availability, exacerbated farming input expenditures, and a dwindling labor workforce underscore the imperative for pioneering advancements within the realm of agriculture and cultivation. AgriTech represents the synergistic integration of cutting-edge technologies and human-computer interaction (HCI) into traditional agricultural methodologies, poised to usher in a transformative era in farming practices. It heralds a promising frontier for the implementation of intelligent farming techniques. Within this scholarly exposition, we delve into the profound challenges that beset the domain of intelligent agriculture and advanced agricultural technologies. We proffer an in-depth exploration of historical developments, leveraging innovative applications of artificial intelligence (AI), automation, robotics, and the Internet of Things (IoT) to propel the paradigm of intelligent farming forward. Furthermore, we elucidate the impediments intrinsic to these technologies and proffer potential remedies encapsulated within the purview of AgriTech. This research not only serves as a comprehensive elucidation of the multifaceted intricacies within the domain of AgriTech but also serves as a launching pad, fostering fertile grounds for the cultivation of future AgriTech innovations.

Keywords: Autonomous Vehicles · AgriTech · Future Farming · Human-Computer Interaction (HCI) · Intelligent Systems · Smart Farming

1 Introduction

In light of the burgeoning global population and the simultaneous degradation of environmental conditions, the agricultural sector confronts a persistent imperative: to augment its productivity while safeguarding long-term sustainability [1, 2]. Traditional farming methodologies, heavily reliant on manual labor, exhibit inherent limitations in crop monitoring capabilities and rely on antiquated business management and production practices. Confronted with a scarcity of skilled labor and escalating managerial costs, these conventional farming approaches are conspicuously ill-suited to attain optimal productivity. Additionally, many agricultural tasks necessitate operators to meticulously control vehicles at precise intervals, with minimal margin for error. This onerous responsibility demands real-time modifications to machine settings while simultaneously ensuring constant vigilance over the trajectory of travel.

© The Author(s), under exclusive license to Springer Nature Switzerland AG 2024
B. J. Choi et al. (Eds.): IHCI 2023, LNCS 14531, pp. 310–325, 2024.
https://doi.org/10.1007/978-3-031-53827-8_28

To surmount these formidable challenges, contemporary agriculture has ushered in a transformative era through the integration of agricultural robots and intelligent machinery, gradually diminishing human involvement in the trajectory of agricultural progress [1]. Cutting-edge technologies such as artificial intelligence (AI) and human-computer interaction (HCI) imbued into the agricultural domain to heighten its intelligence collectively fall under the purview of "AgriTech." Agricultural robots, as exemplars of this revolution, amalgamate state-of-the-art technologies such as artificial intelligence (AI), geographic information systems, advanced navigation systems, sensor arrays, and communication protocols [2].

Recent times have borne witness to a pronounced upswing in interest in agricultural robotic systems, propelling the evolution of increasingly autonomous and intelligent agricultural vehicles. These self-directed agricultural robots and vehicles hold profound potential for elevating agricultural production efficiency while concurrently reducing resource consumption. Across a diverse spectrum of agricultural applications, these robotic marvels are entrusted with tasks encompassing tillage, transplanting, pruning, weeding, harvesting, seeding, spraying, fertilization, and a myriad of other functions [1, 2].

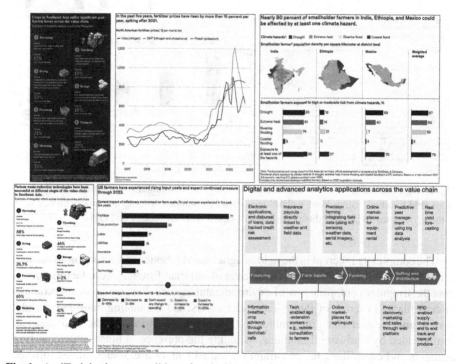

Fig. 1. AgriTech horizons: condition of agriculture, market potentials in smart farming, and the possibilities of innovations in AgriTech. (Figure adapted from: Goyal et al. [7], Bland et al. [8], Frost et al. [10]).

The bedrock prerequisite for the effective execution of these multifarious tasks is the achievement of autonomous navigation in agricultural domains [2]. A dependable navigation system hinges upon the adept guidance of agricultural robots and vehicles along predetermined routes, leveraging data from a constellation of sensors to comprehensively interpret the environment and achieve precise localization. Nevertheless, autonomous navigation within agricultural contexts is rife with multifaceted challenges, primarily owing to the intricate and unstructured nature of agricultural environments, characterized by a profusion of potential sources of interference and disturbances.

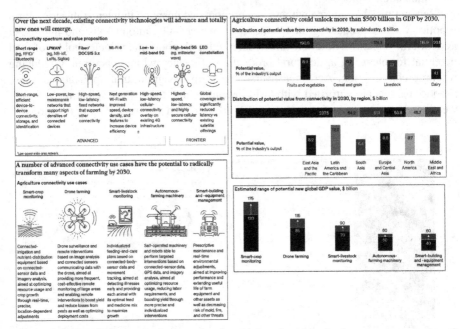

Fig. 2. AgriTech advancements in smart farming: high-end communication, smart monitoring of crops, and livestock using drone technologies, and autonomous farming machinery unlock surge in the agriculture and food industries. (Figure adapted from Goedde et al. [11]).

The complexity inherent in the field of agricultural robotics and autonomous vehicles is multifaceted, stemming from a confluence of intricate factors. These factors encompass the irregular configurations of environmental layouts, the vast array of variations in crop plants in terms of their shapes and colors, the undulating and unpredictable terrain, and the ever-changing ambient light conditions. Moreover, the challenges are compounded by hardware limitations, including the presence of noise emanating from actuators and controllers, the constraint of limited battery life, issues related to wheel slippage, and the inherent imprecision in sensor measurements [2]. The paramount challenge in this domain lies in the conceptualization and design of a navigation system that not only achieves precision but also demonstrates resilience in adapting to intricate and dynamic agricultural settings. The automation of agricultural robots and vehicles hinges on successfully addressing these complex issues.

The agricultural workflow encompasses a sequence of critical steps, commencing with soil preparation, followed by sowing, the application of manure and fertilizers, irrigation, harvesting, and ultimately, storage [3].

Numerous surveys have delved into the state of agriculture, exploring the market potential of smart farming and examining the scope for innovations in AgriTech. Figure 1 provides a glimpse of select surveys and emerging trends. On a global scale, post-harvest wastage constitutes a substantial loss in the food supply chain, contributing significantly to the prevailing food crisis [4]. A substantial portion of this loss is experienced in Asian nations, where post-harvest wastage remains a pervasive issue [5, 6]. The inefficiencies in commodity supply chains between the point of harvest and the consumer are chiefly responsible for this loss. Goyal et al. have meticulously documented the extent of post-harvest loss in Southeast Asian crops in their comprehensive report [7]. Furthermore, Bland et al. have cataloged the evolving trends in agricultural automation [8]. Their research underscores the escalating input costs in farming, including labor, fertilizers, crop protection chemicals, seeds, and more, as a pressing challenge confronting the food and agriculture industries. A recent survey conducted by McKinsey & Company Inc. reflects the concerns of American farmers, who identify rising input costs as the most significant risk to their profitability. They point to the escalating prices of fertilizers and crop protection chemicals as primary culprits contributing to this financial strain [9].

Furthermore, the deleterious carbon footprint and resultant global warming have instigated a capricious climate fluctuation that has exerted substantial repercussions on the vulnerable smallholder agricultural sector. Notably, climate-related adversities, including prolonged droughts, extreme heatwaves, riverine inundations, and coastal deluges, have precipitated significant hardships for approximately 80% of smallholder farmers in regions such as India, Ethiopia, and Mexico, as evidenced by research [10].

To ameliorate these predicaments, the field of AgriTech has emerged as a vanguard, proffering multifaceted solutions that harness cutting-edge technologies. These solutions encompass an array of next-generation innovations, including but not limited to digital advancements, enhanced sensor technologies, robotic systems, and the integration of modern Information and Communications Technologies (ICTs) such as machine learning (ML), AI, and the Internet of Things (IoT) [11]. Figure 2 in our discourse serves as a graphical representation, elucidating the pivotal role played by AgriTech in bolstering the economic dynamics of food and agriculture-based industries. Nevertheless, the integration of advanced AgriTech technologies is not devoid of challenges. These challenges necessitate dedicated research and innovation endeavors to formulate effective resolutions.

2 The Key Contributions

This research endeavor is principally concerned with the meticulous identification of prevailing research gaps within the realm of state-of-the-art AgriTech as applied to the domain of smart farming. The pivotal contributions of this comprehensive survey encompass:

- An exhaustive and intricately detailed examination of the landscape of smart farming, meticulously scrutinizing the most recent advancements in AgriTech, thereby affording a holistic perspective on the subject matter.
- An in-depth exploration of the cutting-edge AI methodologies and HCI techniques that are extensively employed within the domain of smart farming, encompassing a rigorous evaluation of their recent developments, offering a profound insight into their evolving applications.
- A meticulous survey delineates the extant challenges that impede the seamless integration and adoption of smart farming, and HCI in AgriTech, while also providing a critical analysis of potential strategies and solutions to circumvent these impediments.
- An in-depth discourse on smart farming practices that extends beyond a mere surface-level examination, delving into the intricate challenges faced by practitioners, and offering a forward-looking perspective that charts the course for future research endeavors in the realm of smart farming.

This survey serves as an illuminating beacon, casting a more intense and analytical spotlight on various facets of contemporary smart farming practices, especially in the context of harnessing advanced AI methodologies and pertinent analytical technologies.

3 A Few Related Works

The routine agricultural operations encompass a spectrum of activities, including but not limited to crop monitoring, harvesting, seeding, precision spraying, and fertilization. In response to the growing complexities of modern agriculture, numerous scholars and experts have proposed the integration of advanced Agricultural Technology (AgriTech) within the framework of smart farming. To provide a comprehensive overview of AgriTech applications within these common agricultural tasks, a detailed summary has been meticulously curated and is presented in Table 1. This tabulated representation serves as a valuable reference for understanding the technological advancements that are reshaping contemporary agriculture.

Table 1. A summary of AgriTech involved in agricultural tasks ([1–3, 12–22]).

Farming tasks	AgriTech	Description
Crop monitoring	AI-assisted techniques, Physical sensors, Vison sensors, Multispectral/Hyperspectral sensors, IoT technologies, wireless communication protocols	Two-dimensional visual and three-dimensional stereo visual imaging, advanced artificial intelligence methods, photodiodes, tensiometers, ion-selective electrodes, field-effect transistors, the global positioning system, and wireless sensor networks, as well as technologies like Wi-Fi, WiMAX, LoRa, Bluetooth, etc., were all involved in this study. Significant linear correlations were observed between the normalized difference vegetation index (NDVI) and factors such as plant nitrogen and aerial biomass. This discovery holds promise for shedding light on effective management practices and strategies
Irrigation	AGROBOTs, Deep learning, Moisture sensing, IoT sensors, semi-automatic methods, image-processing techniques on single-band images, maps of Vegetation Indices, and Digital Elevation Models to detect crop canopy, remote sensors, etc.	Using the UAV (Drone) and Robot, Fog, and cloud computing, AI/ML techniques: K-means clustering, Minimum Distance to the Mean classifier, Moisture sensors, Fuzzy Logic Controller, etc.
Spraying and Weeding	A smart sprayer, Herbicide Spray, High-power lasers for intra-row weeding, Rotatory hoe/Mechanical removal, Robotic arms, Electrical Discharge, Chemical spraying, Wireless Sensor Networks, Gyroscope and Accelerometer sensors, Bradley Method, Filter-paper ratio assured methods for spraying droplets, etc.	Machine Vision algorithm incorporating mathematical morphology, color, and texture-based techniques. It employs these methods for tasks such as identifying greenness, employing a fuzzy real-time classifier, and utilizing a Pulse Width Modulation (PWM) controller for precise spraying by UAVs. The system also integrates Centrifugal Energy Sprayers and Kinetic Energy Sprayers. For weed control, a mechanized weeding actuator is employed, which comprises a servo motor integrated with a computer vision system. This setup detects plant locations and guides the weeding actuator to perform mechanical weeding operations without causing damage to crops. In the context of filter-paper ratio-based methods, the algorithm measures spray droplets using filter paper-assumed titration. It establishes a functional relationship between the diameter of colored spots and the diameter of liquid drops through regression analysis
Ploughing and Harvesting	Autonomous tractor/ harvester, Robot-based, AI-assisted intelligent machines, etc.	AI algorithms such as random forest, XGBoost, Cubist, MLP, SVR, Gaussian Process, k-NN, and Multivariate Adaptive Regression Splines, to detect the yield gaps in a farm

In pursuit of overcoming the multifaceted challenges within the realm of smart farming, researchers have strategically harnessed a diverse array of advanced intelligent techniques, integrating cutting-edge AI algorithms, sophisticated robotics, and state-of-the-art smart sensing technologies. Presented herewith is a comprehensive exploration of the latest advancements in the field of Agricultural Technology (AgriTech).

In their work documented in [2], Bai et al. conduct an exhaustive examination of cutting-edge developments in vision sensors and autonomous navigation systems tailored specifically for agricultural robotics and autonomous vehicles (AVs). These vision sensors represent a cornerstone technology in the realm of agricultural robotics, facilitating contactless assessment of agricultural environments. Their categorization hinges on the fundamental principles and capabilities of imaging technology, resulting in two primary categories: 2D vision imaging and 3D stereo vision imaging.

2D vision imaging excels in providing critical insights into the geometrical aspects of agricultural scenes, including the shapes of trees, crops, obstacles, and various elements present within the agricultural context. In stark contrast, 3D vision sensors operate on an entirely different plane, quite literally. They generate three-dimensional maps of the agricultural surroundings, yielding precise spatial coordinates for both the autonomous robot and the objects in its vicinity. The spectrum of vision sensors available for agricultural automation spans monocular sensors, stereo sensors, RGB-D sensors, and thermal imaging sensors.

The applications of vision-based assistance in agriculture are extensive, encompassing key tasks such as planting, picking, fertilizing, spraying, weeding, and harvesting. However, the integration of vision-based automation in this domain isn't without its formidable challenges. These challenges include the vagaries of illumination stemming from factors like natural daylight variations, unpredictable weather conditions, and diverse viewing angles. Furthermore, the uneven topography of farms, densely vegetated areas, and the presence of tall grasses can all conspire to impede the seamless operation of vision-based automation systems.

To surmount these challenges, the research community has proposed a multifaceted approach. This approach includes the fusion of data from multiple sensors, the creation of highly accurate 3D maps, collaborative automation through the coordination of multiple robots, and the implementation of efficient AI frameworks.

The core tasks within precision agriculture revolve around the identification, segmentation, and tracking of fruits and vegetables. These tasks serve as the foundation for critical applications such as robotic harvesting and yield estimation. However, modern CV methods deployed for these tasks demand substantial volumes of data, which can pose significant logistical and financial challenges, particularly for small-scale agricultural businesses.

In 2022, Ciarfuglia et al. proposed an innovative solution rooted in the concept of weak supervision to address this data bottleneck [12]. Their approach aims to achieve state-of-the-art fruit and vegetable detection and segmentation in precision agriculture while utilizing a reduced dataset. In a subsequent article published in 2023, they expanded upon their previous work, focusing on the intricate task of tracking fruits within orchards [13]. This endeavor encompasses several neural networks, including the Source Detector Network (SDet) and the Source Segmentation Network (SSeg). While SDet might not be perfectly calibrated for the target environment, it demonstrates strong performance on selected frames, known as keyframes, extracted from the video input. These keyframes are strategically spaced to maintain computational simplicity. Initial bounding boxes are generated based on these keyframes, and a stringent confidence threshold is applied to minimize false positives. To refine detection results for the remaining frames, the video data undergoes processing through a Geometric Consistency block (GC block), which extracts features from each frame and establishes associations between them. This geometric information, combined with the initial bounding boxes, aids in estimating bounding boxes for the remaining frames. These estimated labels, termed pseudo-labels, subsequently serve as training data for the Target Detector (TDet). The proposed system comprises two main components: Detection Pseudo-Label Generator and Segmentation Pseudo-Label Generator. Both subsystems necessitate an initial coarse detection or segmentation learning algorithm to provide initial label estimates.

Recent advancements in smart agriculture constitute a paradigm shift from traditional farming practices, driven by the infusion of equipment, technology, machinery, and a plethora of devices. The emerging field of Agritech encompasses a diverse array of technologies, with sensors, information technology, and CV at its core. The future of agriculture hinges on the seamless integration of advanced tools and technologies such as robots, moisture sensors, aerial imagery, and precise localization systems.

The establishment of an intelligent agricultural ecosystem necessitates the integration of multiple elements to enhance crop production, streamline food distribution, foster crowd farming initiatives, and more. This integration hinges on the real-time collection and processing of data. The deployment of decision support systems in real-time can significantly boost productivity, optimize resource allocation, adapt to changing climatic conditions, enhance food supply chain management, and facilitate the early identification of crop diseases.

Within the realm of CV, recent advancements have ushered in a transformative era for the automated diagnosis of crop diseases. However, this remains a formidable challenge, given the myriad of non-infectious and infectious agents that can produce similar symptoms across various parts of plants. Detecting and classifying crop diseases through continuous monitoring now constitutes a primary focus for researchers and agronomists.

Jackulin et al. have undertaken an extensive review of plant disease detection methodologies employing AI techniques [14]. In the domain of CV, segmentation stands out as a fundamental technique for identifying plant diseases. Four prominent segmentation approaches, namely clustering, edge detection, regional methods, and thresholding, are widely employed for this purpose. Each of these methods is subject to individual evaluation, highlighting their respective advantages and drawbacks. Thresholding segmentation, for instance, excels by not requiring prior knowledge of plant images and offering relatively swift and computationally efficient outcomes. Nonetheless, it grapples with the identification of connected regions due to the absence of spatial information. Furthermore, the choice of threshold assumes paramount significance, and this method proves highly sensitive to noise. Edge detection, while proficient in scenarios with high image contrast, falters when confronted with images replete with numerous edges. Clustering-based approaches deliver homogeneous regions and expedited computation but thrive best when dealing with clusters of similar sizes. Region-based segmentation, although supporting interactive and automated image segmentation techniques, demands greater computational time and memory resources due to sequential processing.

The authors have also conducted a comprehensive survey of various ML techniques for plant disease detection, including Support Vector Machines (SVM), K-Nearest Neighbors (KNN), decision trees, and Random Forest classifiers. Metre et al. have introduced swarm intelligence-based techniques to address real-time optimization challenges in smart farming [15].

Jerhamre et al. have undertaken a thorough investigation into the vulnerabilities and opportunities within the realm of intelligent agriculture [16]. Their primary focus lies in strategies for data collection in the context of smart farming and the integration of advanced technologies within the agricultural sector. Their study encompasses both remote sensing-based and IoT-based data collection approaches.

Remote sensing entails the identification and monitoring of physical attributes of the Earth's surface from a distance, typically facilitated by satellites and drones. Critical characteristics of remote sensing data encompass spatial, spectral, and temporal resolutions. Spatial resolution pertains to the size of individual pixels in an image and significantly impacts the capacity to detect objects through visual information. Spectral resolution relates to the size and number of spectral intervals utilized for sensor-based detection in different electromagnetic regions. Temporal resolution refers to the frequency of data acquisition.

In the realm of data resolution, Meier et al. contend that precision-oriented smart farming hinges on high-resolution data, as the detection of irregularities becomes impractical or ineffective when employing excessively large pixel dimensions [17]. The requisite level of resolution varies according to the specific analytical goals of the data. For instance, accurately predicting crop yields in a field might be achievable with relatively coarse resolution data, while the precise detection of plant diseases through hyperspectral imaging demands a much more detailed level of resolution.

IoT-based data collection, on the other hand, leverages electronic components and connectivity to enable remote control and data exchange. In the agricultural context, IoT is predominantly employed to gather data through a diverse array of sensors. Subsequent analysis of this data yields valuable insights that aid in decision-making.

Kamienski et al. have delineated the challenges associated with IoT-based intelligent agriculture, emphasizing the need for adaptability, efficiency, configurability, reliability, and scalability in IoT systems deployed for smart farming [18]. The intricate nature of IoT systems presents an additional challenge, further compounded by security concerns in IoT-based smart farming.

Kleinschmidt et al. have stressed the critical importance of end-to-end encrypted communication, spanning from sensors to application interfaces [19]. In practical terms, this entails implementing a comprehensive security strategy for the entire IoT sensor network, covering cloud databases and potential fog computing networks. Robust security measures enhance the trustworthiness of IoT systems among farmers.

Autonomous robots designed for deployment in smart farming scenarios are engineered to operate at reduced velocities along pathways characterized by irregularities. They must navigate regions with moist terrains while surmounting obstacles such as small stones, soil clumps, and depressions. One strategic approach involves the use of D-type (intelligent-sensory) rolling robots equipped with caterpillar-shaped gears. These gears enhance traction, promote optimal weight distribution, and reduce the risk of immobilization [20].

Boukens et al. [21] have introduced a real-time self-tuning motion controller specifically tailored for mobile robot systems, incorporating a mechanism for adaptive adjustments during operation. Jiang et al. have developed an intelligent control system dedicated to manipulating color ratios using LED lights within a greenhouse [22]. They have also introduced a testbed that provides an accessible plant growth system enriched with IoT-enabled supervisory and control functionalities.

Durmus et al. [23] have presented a mobile robot with autonomous capabilities, designed to collect data from agricultural fields and greenhouse environments, subsequently transmitting this data to a web application for processing and analysis. Rosero-Montalvo et al. have demonstrated a quadruped robot designed for monitoring rose crops, employing supervised learning algorithms to enhance its capabilities [20]. Furthermore, numerous researchers have made significant strides in optimizing harvesting processes through innovative data acquisition schemes within the context of intelligent farming. Notably, some of these endeavors have focused on devising methodologies for collision avoidance, addressing the intricate task of coordinating robot movements within greenhouse confines. They have introduced a concept for a smart farming robot primed for the detection of environmental conditions within greenhouse settings, thereby presenting a novel application of robotic technology to the sphere of agricultural monitoring and management.

4 Role of HCI in Future Farming

HCI plays a significant role in shaping the future of farming. In the context of future farming, HCI contributes in several ways:

- User-Friendly Interfaces: HCI helps in creating user-friendly interfaces for the various technologies and systems used in agriculture. Farmers and agricultural professionals need intuitive interfaces to interact with drones, sensors, automated machinery, and data analytics tools.
- Data Visualization: HCI principles are essential for designing data visualization tools that enable farmers to understand complex data and make informed decisions. Visualizations can present real-time data about soil conditions, weather, crop health, and yield predictions.
- Mobile Apps and Dashboards: Mobile applications and web-based dashboards are increasingly used in agriculture to provide real-time information and control over farm operations. HCI ensures that these apps are easy to use and provide a seamless experience.
- Wearable Technology: HCI is essential for designing wearable devices like smart glasses, which can assist farmers in identifying plant diseases, monitoring crop health, and offering guided navigation within fields.
- Voice and Gesture Control: Hands-free interfaces that use voice commands or gestures can be valuable in farming, where users often have their hands full. HCI design is crucial for creating efficient and accurate voice and gesture recognition systems.

- Remote Monitoring and Control: HCI enables farmers to remotely monitor and control equipment, such as irrigation systems, without physically being present on the field. This is particularly important for large-scale and automated farms.
- Robotics and Automation: HCI is essential for developing user interfaces for autonomous farming robots and drones. Farmers need to interact with these machines, set tasks, and monitor their progress.
- Decision Support Systems: HCI plays a vital role in designing decision support systems that provide farmers with insights and recommendations based on data analytics. These systems help farmers optimize planting, harvesting, and resource allocation.

Therefore, HCI meets the demands of future farming by providing user-friendly interfaces, efficient, and effective in assisting farmers with the adoption of advanced agricultural technologies. This ultimately leads to increased productivity, reduced resource waste, and more sustainable farming practices.

5 Challenges and Opportunities

Weather pattern fluctuations represent just a single facet of the myriad obstacles confronting the agricultural sector. The future is being shaped by three fundamental technologies: Electrification, the evolution from Automation to Autonomy, and the integration of AI. In general, farming faces a variety of challenges that can impact agricultural productivity, sustainability, and food security. Some of these challenges include:

- Climate Change: Changing weather patterns, extreme weather events, and shifts in temperature and precipitation can disrupt growing seasons, impact crop yields, and increase the prevalence of pests and diseases.
- Resource Scarcity: Water scarcity, soil degradation, and dwindling arable land availability can limit agricultural production. Irrigation practices and soil management techniques must be optimized for sustainable resource use.
- Pests and Diseases: Insects, fungi, bacteria, viruses, and other pathogens can devastate crops, leading to significant yield losses. The development of resistant varieties and integrated pest management strategies is crucial.
- Market Access: Farmers often struggle to access fair markets for their produce, leading to low income and economic instability. Lack of infrastructure, transportation, and market information can hinder their ability to reach consumers.
- Labor Shortages: Many regions face labor shortages due to rural-to-urban migration and changing demographics. This can result in reduced farm productivity and increased reliance on mechanization.
- Costs: Increasing costs of seeds, fertilizers, pesticides, and machinery including the fluctuations in global commodity prices also affect profitability for the farmers.
- Technology Adoption: While technological advancements have the potential to improve efficiency and yield, some farmers may lack access to or knowledge about modern farming technologies and practices.
- Land Tenure and Ownership: Issues related to land tenure, land rights, and land fragmentation can create uncertainty and hinder investment in sustainable agricultural practices.

- Environmental Impact: Intensive farming practices usually lead to soil erosion, water pollution, and habitat destruction. Transitioning to more sustainable practices that promote biodiversity and soil health is essential.
- Regulatory and Policy Challenges: Inconsistent or outdated agricultural policies, regulations, and trade barriers also hinder the development of the agricultural sector and limit farmers' ability to adapt to changing conditions.
- Food Safety and Quality: Ensuring food safety and quality standards throughout the supply chain is crucial to protect consumer health and maintain market access.
- Access to Finance: Limited access to credit and financing options hinders farmers' ability to invest in modern equipment, technologies, and inputs.
- Rural Development: Insufficient rural infrastructure, healthcare, education, and basic services impact the overall quality of life in rural areas and contribute to migration to urban centers.

In order to surmount these pressing challenges, contemporary agriculture is diligently exploring innovative technological breakthroughs and comprehensive solutions. Nevertheless, the domain of modern agriculture is concurrently grappling with a multitude of intricate challenges concerning the assimilation and harmonization of cutting-edge technologies.

Fig. 3. Automated technologies in smart farming: the innovations in agricultural technologies to boost the yields in farming and utilize the resources intelligently. (Figure adapted from various web sources).

As elucidated in Fig. 3, the landscape of smart farming is intricately woven with an array of automated technologies. These AgriTech innovations play a pivotal role in augmenting agricultural productivity by leveraging existing resources with astute precision. Remarkably, they serve as a conduit for mitigating the depletion of human labor resources within the agricultural sector, thereby substantially bolstering the economic viability of the food and agricultural industries. These challenges vary depending on the region, scale of operation, and the specific crops involved. Some of the key challenges include:

- High Initial Costs: Implementing modern technologies like precision agriculture, robotic systems, and sensor networks often requires significant upfront investments. This represents a barrier for small and medium-sized farmers to afford these technologies.
- Technical Knowledge and Training: Many modern farming technologies require specialized technical knowledge to set up, operate, and maintain. Farmers need to be trained to effectively use these technologies, which is challenging, especially for older generations as they are mostly unfamiliar with digital tools.
- Data Management and Privacy: Advanced farming technologies generate large amounts of data, including information about soil conditions, weather patterns, crop health, and more. Managing, analyzing, and protecting this data from potential breaches or misuse is essential though challenging.
- Interoperability and Compatibility: As the agricultural technology landscape evolves, various solutions and tools are developed by different companies. Ensuring that these technologies can seamlessly communicate and integrate with each other is another challenge.
- Access to Reliable Connectivity: Many modern farming technologies rely on real-time data transmission, which requires reliable internet connectivity. However, rural and remote farming areas often lack robust internet infrastructure, limiting the effectiveness of these technologies.
- Energy Supply: Advanced technologies, such as precision irrigation systems and automated machinery, require a stable and sufficient energy supply. In areas with unreliable or limited access to electricity, using these technologies poses difficulty.
- Adaptation to Local Conditions: Locality adaptation is also a challenge to the advanced Agritech. Not all modern technologies are universally applicable. Some technologies might work well in certain climates or soil types but may be less effective in others. Customizing these technologies to suit local conditions is crucial for their success.
- Environmental Concerns: While modern technologies enhance productivity, they also raise environmental concerns. For instance, increased mechanization contributes to soil compaction, and excessive use of fertilizers or pesticides can harm ecosystems.
- Regulatory Hurdles: The rapid advancement of agricultural technologies sometimes outpace regulatory frameworks. Unclear or restrictive regulations are used to impede the adoption of certain technologies or create confusion for farmers.
- Cultural and Social Factors: Farming often has deep-rooted cultural and social aspects that influence the adoption of new technologies. Traditional practices and beliefs usually clash with modern approaches, creating resistance to change.
- Risk Management: Investing in new technologies carries inherent risks. If a particular technology fails to deliver the expected outcomes, farmers may face financial losses. Balancing the potential benefits with the associated risks is a challenge.

- Labor Displacement: Automation and robotics have the potential to reduce the need for manual labor on farms. While this improves efficiency, it also leads to concerns about job displacement in rural communities.

6 Conclusions and Future Scope

In this comprehensive survey, a multifaceted array of challenges plaguing the realm of farming and agriculture has been meticulously uncovered. The burgeoning global population, dwindling workforce in the agricultural sector, alarming rates of agricultural land depletion, the scourge of post-harvest losses, and the capricious shifts in weather patterns collectively exert monumental pressure upon the shoulders of small-scale farmers and stakeholders. This unrelenting pressure inevitably contributes to the escalating devastation within the agricultural domain. The arsenal of solutions offered by the dynamic field of AgriTech and the concept of smart farming does not emerge unscathed from this relentless onslaught of challenges. These innovations encountered a myriad of obstacles in their quest to revolutionize agriculture. Thus, it becomes evident that a deeper exploration of these complexities through sustained research efforts is an imperative endeavor, crucial for surmounting the impending hurdles and bottlenecks entwining AgriTech and smart farming.

In the domain of prospective research, a pressing need arises for the refinement and advancement of AI techniques. These advancements are indispensable for enhancing machine vision capabilities, particularly in adverse environmental conditions such as blurred or low-light settings. The aim is to bolster the efficacy of crop monitoring, disease detection, and operations like spraying, sowing, weeding, ploughing, and harvesting. To facilitate these advancements, the development of intelligent sensors takes center stage. These sensors must furnish precise and real-time data on critical parameters, including moisture levels, soil attributes, humidity dynamics, pH levels, fertilizer compositions, pigment information, and manure content, amongst others. Moreover, it is paramount to recognize the crucial role of economic AgriTech in smart farming, particularly in its support of small-scale farmers and stakeholders. Given their sheer number, these individuals collectively constitute the backbone of the agricultural sector. Consequently, addressing their needs becomes of paramount importance. The primary thrust of these endeavors should be geared toward the reduction of post-harvest losses and the provision of cost-effective AgriTech solutions. These actions are pivotal in forging a sustainable and prosperous future for the agricultural domain.

Acknowledgments. This work was partially supported by the Brain Pool Program through the National Research Foundation of Korea (NRF) funded by the Ministry of Science and ICT (NRF-2019H1D3A1A01071115), and partially supported by Korea Evaluation Institute of Industrial Technology (KEIT) grant funded by the Korea government (MOTIE). (No. 1415181754, 3D semantic camera module development capable of material and property recognition).

Conflicts of Interest. The authors declare no conflict of interest.

References

1. Sharma, V., Tripathi, A.K., Mittal, H.: Technological revolutions in smart farming: current trends, challenges & future directions. Comput. Electron. Agric. **201**, 107217 (2022)
2. Bai, Y., Zhang, B., Xu, N., Zhou, J., Shi, J., Diao, Z.: Vision-based navigation and guidance for agricultural autonomous vehicles and robots: a review. Comput. Electron. Agric. **205**, 107584 (2023)
3. Sinha, B.B., Dhanalakshmi, R.: Recent advancements and challenges of Internet of Things in smart agriculture: a survey. Future Gener. Comput. Syst. **126**, 169–184 (2022)
4. Depta, L.: Global food waste and its environmental impact. Reset (2018). https://en.reset.org/global-food-waste-and-its-environmental-impact-09122018/. Accessed 03 Sept 2023
5. Bradford, K.J., et al.: The dry chain: Reducing postharvest losses and improving food safety in humid climates. Trends Food Sci. Technol. **71**, 84–93 (2018)
6. Gunasekera, D., Parsons, H., Smith, M.: Post-harvest loss reduction in Asia-Pacific developing economies. J. Agribus. Dev. Emerg. Econ. **7**(3), 303–317 (2017)
7. Goyal, A., Lock, E., Moorthy, D., Perera, R.: Saving Southeast Asia's crops: Four key steps toward food security. McKinsey & Company Inc. (2023). https://www.mckinsey.com/industries/agriculture/our-insights/saving-southeast-asias-crops-four-key-steps-toward-food-security. Accessed 03 Sept 2023
8. Bland, R., Ganesan, V., Hong, E., Kalanik, J.: Trends driving automation on the farm. McKinsey & Company Inc. (2023). https://www.mckinsey.com/industries/agriculture/our-insights/trends-driving-automation-on-the-farm. Accessed 04 Sept 2023
9. Ferreira, N., Fiocco, D., Ganesan, V., de la Serrana Lozano, M.G., Mokodsi, A.L., Gryschek, O.: Global Farmer Insights. McKinsey & Company Inc. (2022). https://globalfarmerinsights2022.mckinsey.com/#autores. Accessed 04 Sept 2023
10. Frost, C., Jayaram, J., Pai, G.: What climate-smart agriculture means for smallholder farmers. McKinsey & Company Inc. (2023). https://www.mckinsey.com/industries/agriculture/our-insights/what-climate-smart-agriculture-means-for-smallholder-farmers. Accessed 04 Sept 2023
11. Goedde, L., Katz, J., Menard, A., Revellat, J.: Agriculture's connected future: how technology can yield new growth. McKinsey & Company Inc. (2020). https://www.mckinsey.com/industries/agriculture/our-insights/agricultures-connected-future-how-technology-can-yield-new-growth. Accessed 04 Sept 2023
12. Ciarfuglia, T.A., Marian Motoi, I., Saraceni, L., Nardi, D.: Pseudo-label generation for agricultural robotics applications. In: 2022 IEEE/CVF Conference on Computer Vision and Pattern Recognition Workshops (CVPRW), pp. 1685–1693. IEEE, New Orleans (2022)
13. Ciarfuglia, T.A., Motoi, I.M., Saraceni, L., Fawakherji, M., Sanfeliu, A., Nardi, D.: Weakly and semi-supervised detection, segmentation and tracking of table grapes with limited and noisy data. Comput. Electron. Agric. **205**, 107624 (2023)
14. Jackulin, C., Murugavalli, S.: A comprehensive review on detection of plant disease using machine learning and deep learning approaches. Meas.: Sens. **24**, 100441 (2022)
15. Metre, V.A.: Research review on plant leaf disease detection utilizing swarm intelligence. Turk. J. Comput. Math. Educ. (TURCOMAT) **12**(10), 177–185 (2021)
16. Jerhamre, E., Carlberg, C.J.C., van Zoest, V.: Exploring the susceptibility of smart farming: identified opportunities and challenges. Smart Agric. Technol. **2**, 100026 (2022)
17. Meier, J., Mauser, W., Hank, T., Bach, H.: Assessments on the impact of high-resolution-sensor pixel sizes for common agricultural policy and smart farming services in European regions. Comput. Electron. Agric. **169**, 105205 (2020)
18. Kamienski, C., et al.: Smart water management platform: IoT-based precision irrigation for agriculture. Sensors **19**(2), 276 (2019)

19. Kleinschmidt, J.H., Kamienski, C., Prati, R.C., Kolehmainen, K., Aguzzi, C.: End-to-end security in the IoT computing continuum: perspectives in the SWAMP project. In: 9th Latin-American Symposium on Dependable Computing (LADC), pp. 1–2 IEEE (2019)
20. Rosero-Montalvo, P.D., Gordillo-Gordillo, C.A., Hernandez, W.: Smart farming robot for detecting environmental conditions in a greenhouse. IEEE Access 11, 57843–57853 (2023)
21. Boukens, M., Boukabou, A., Chadli, M.: A real time self-tuning motion controller for mobile robot systems. IEEE/CAA J. Autom. Sinica 6(1), 84–96 (2019)
22. Jiang, J., Moallem, M.: Development of an intelligent LED lighting control testbed for IoT-based smart greenhouses. In: IECON 2020 The 46th Annual Conference of the IEEE Industrial Electronics Society, pp. 5226–5231. IEEE, Singapore (2020)
23. Durmus, H., and Günes, E.O.: Integration of the mobile robot and Internet of Things to collect data from the agricultural fields. In: 8th International Conference on Agro-Geoinformatics (Agro-Geoinformatics), pp. 1–5. IEEE, Istanbul (2019)

RoboRecycle Buddy: Enhancing Early Childhood Green Education and Recycling Habits Through Playful Interaction with a Social Robot

Saifuddin Mahmud[1], Zina Kamel[2], Aditi Singh[3], and Jong-Hoon Kim[2(✉)]

[1] Computer Science and Information Systems, Bradley University, Peoria, IL, USA
[2] ATR Lab, Computer Science, Kent State University, Kent, OH, USA
jkim72@kent.edu
[3] Computer Science, Cleveland State University, Cleveland, OH, USA

Abstract. The escalating environmental challenges faced by our world today make it imperative to instill eco-friendly habits and environmental awareness in young children, who will be at the forefront of addressing these issues in the future. In this paper, we present the "RoboRecycle Buddy", a voice and chat GPT-integrated social robot designed to enhance early childhood green education and foster positive recycling habits through playful interaction. By incorporating engaging, voice feedback-based educational content, the RoboRecycle Buddy aims to make learning about recycling enjoyable, accessible, and relevant to children, while simultaneously sparking their interest in the rapidly growing field of robotics. We discuss the design, development, and implementation of the RoboRecycle Buddy, highlighting its innovative features such as object recognition, face recognition, and natural voice interaction which encourages children to practice responsible waste disposal. The robot leverages advanced voice recognition technology and the GPT language model, enabling it to engage in contextually relevant conversations with children, both through voice and text, further enhancing their learning experience. The robot is equipped with an image recognition module based on a Convolutional Neural Network, enabling it to detect and classify waste materials in a manner similar to a child, i.e., by simply looking at them. We emphasize the importance of providing a supportive learning environment that encourages children to explore, question, and develop a deeper understanding of the recycling process. The paper concludes with a discussion on future research and development directions, including potential improvements and adaptations to the RoboRecycle Buddy, as well as broader implications for the field of Child-Robot Interaction.

Keywords: Social robot · green education · recognition · face recognition · voice interaction · Convolutional Neural Network · Child-Robot Interaction · ChatGPT

B. J. Choi et al. (Eds.): IHCI 2023, LNCS 14531, pp. 326–343, 2024.
https://doi.org/10.1007/978-3-031-53827-8_29

1   Introduction

In the past decade, pollution has become an issue of paramount importance, as humans, animals, and the environment have all suffered due to pollution-related environmental calamities [1]. These calamities include floods, rising global temperatures, a slew of illnesses, and other natural disasters [2]. As its effects on our surrounding ecosystem have become increasingly clear, scientists have sought to evaluate its leading causes. Waste is generally created either by human activities or by means of natural ecological processes, and has been identified to be a major contributing factor to pollution. Waste comes in many forms and in order to properly manage it, each form must be processed using slightly different methods. Waste that may be handled for reuse is referred to as recyclable waste. Plastics, bottles, glass, iron, and a number of other materials fall into this category. Even though many communities provide distinct bins to realize these recyclable disposal methods, many individuals still fail to properly dispose of their waste. This is a major contributing factor to the environmental issues previously outlined. To alleviate this problem, it is of critical importance to educate the future generations on proper waste management practices [3]. In fact, in 2018, Gustavo Teixera and his col- leagues [4] indicated that one of the reasons the Earth is facing a multitude of environmental problems due to the insufficient emphasis on environmental education offered to society in today's age. Clearly, educating students on waste management would go a long way towards solving the waste management issue. However, finding strategies to teach green education in school could be an issue. One of the strategies being implemented by educators in many fields is to take a hands-on approach with students [5–7]. Studies have found that exposing students to science, technology, engineering and mathematics (STEM) principles using this approach can improve their understanding [8,9]. This same principle could be applied to the education of environmental topics to young students. Additionally, it can be found that hands-on activities can stimulate creativity and passion within children [10–13]. These same strategies could be applied to not only teach about but also instill a passion for sustainability in the next generation of students. Previous studies have investigated the use of robotics for this purpose and found it to be an effective approach [14,15]. The studies suggest the use of robotics as one of many potential hands-on methods to teach and impassion students about waste management. However, robots are especially valuable for this purpose given the ever-increasing prevalence of robotic systems in our world. Robots are already serving crucial roles in many industries, such as food service, healthcare, and agriculture, and the number of industries that use robots is expected to see massive growth in the near future [16]. With robots becoming so ubiquitous, future generations must be well-informed on their requirements and the mechanisms through which they operate. Moreover, it is our responsibility as the current generation to ensure that they are equipped to meet the demands of future robotics systems. To accomplish this, students must be introduced to robots and the concepts that surround them at a young age. Research shows that children begin forming opinions on jobs as early as three years old [17].

Therefore, it is critical to start forming positive opinions in order to direct children towards the field of robotics. However, the interaction between children and robots imposes a number of requirements that must be met to achieve positive educational outcomes.

Such requirements include a larger user interface, repetition in actions, pictorial representations of actions, and communicating with children [18–20]. These are some of the aspects considered within the field of Child-Robot Interaction (CRI), the study of the interaction between children and robots. One study found that a robot that has been equipped with empathetic capabilities is more likely to hold a child's attention, which is an important aspect of education [21]. Overall, any robotics system intended to educate children must incorporate many of the principles of CRI if it is to achieve a desirable educational outcome. This paper proposes RoboRecycle, a smart trashcan robot created to meet the aforementioned demands in both environmental and robotics education, and designed in consideration of CRI and its principles. RoboRecycle employs a wide array of innovative technologies to reach these goals. First, designed to be very appealing to children, the RoboRecycle has a user interface that uses large font and pictorial representations. Furthermore, the robot is equipped with two different sections in the bin, each color-coded and labeled with a picture symbolizing its corresponding type of waste. Moreover, by employing facial recognition, it specializes its communication with each child, thereby increasing its interactability and the child's sense of engagement. Through the deployment of these features, the RoboRecycle seamlessly familiarizes children with robotics while also promoting sustainable waste habits. Finally, the incorporation of an analysis utility powered by ChatGPT into the robot allows for the generation of comprehensive descriptions and responses to queries. By leveraging this augmented capability, users can not only obtain the classification outcome, but also gain insight into the underlying methodology, facilitating their ability to apply the same approach in subsequent scenarios.

The RoboRecycle Buddy is a sophisticated extension of our earlier smart trashcan model [22], designed with a focus on affordability and retaining a high degree of functionality. By utilizing the cost-effective and widely available Jetson Nano AI processing board, along with a custom-built Android application, the system effectively employs artificial intelligence to achieve outstanding performance. The open hardware approach of the RoboRecycle Buddy ensures that it remains both economically viable and easily accessible to a broad range of users. This artificial intelligence affords the robot with its facial recognition abilities it employs to greet a student by name upon their approach. Additionally, the robot uses a recognition module based on a Convolutional Neural Network (CNN), allowing it to categorize various wastes as recyclable or not. Upon categorization, the application communicates with the Arduino microcontroller to open the associated waste bin.

2   Background

2.1   Interaction Based Learning

The utilization of interaction like games as a tool in children's education has been a successful and captivating approach. This strategy in teaching can effectively deliver information while simultaneously boosting motivation, engagement, and overall performance. Research has shown the use of interaction-based games in various teaching environments including language, multiple school subjects as well as other everyday skills [23–25]. This evidence has also validated the interactive techniques' effectiveness in improving the children's acquired skills and understanding. In particular, integrating these interactive techniques into the educational process provides children the opportunity to learn by doing, which has proved to be more effective than traditional methods [26].

The effects of this can also be seen on a wider scale when reflecting on their results across global problems such as pollution and waste disposal. A study done by Gaggi O. et al. [27] has shown the effectiveness of using an interactive game to teach waste matching with different kinds of trash cans. Another game is "Fox the Recycler" which was developed to teach students how to recycle. Upon the first launch of the game in 2018, the recycling rate of biowaste has increased by 21% to reach 97% [28]. Kinect recycling game is yet another example that helps students learn to recycle using gesture interfaces. The research shows that participating students found this educational method more entertaining and engaging than traditionally used methods [29].

Prior research substantiates the efficacy of incorporating interactive techniques into educational systems, yielding benefits such as increased motivation, scalability, and effectiveness. These methodologies facilitate experiential learning, enabling children to acquire knowledge through hands-on application, and fostering an engaging, immersive educational experience. In this paper, we introduce an innovative approach to integrating advanced interactive techniques into green education, employing social robots and cutting-edge technology, such as waste recognition and the ChatGPT Model. This approach not only enhances the educational experience but also extends its reach to a larger number of students simultaneously, reducing the reliance on human resources during the learning process. Although traditional methods may face constraints in terms of interaction capabilities when scaling up, our proposed solution ensures a higher degree of interaction, enabling a more personalized and dynamic learning experience for young learners in the realm of environmental sustainability and recycling habits.

2.2   Social Robots

With the increasing advancements in technology, robots are becoming increasingly common in the lives of children, as they are being used for educational and entertainment purposes. From robot pets to interactive learning aids, robots have the potential to engage and stimulate children's curiosity and learning. One particularly interesting type of robot is the social robot. Social robots are advanced machines designed to interact with humans in a way that feels natural and intuitive. They are typically equipped with sensors, cameras, and microphones that allow them to perceive their environment and communicate with people through speech, gestures, and facial expressions. Since social robots are designed to interact with humans in a way that feels natural and intuitive, this makes them ideal companions for children. They can be used to teach children new skills, such as coding or foreign languages, and can even provide emotional support and companionship [30, 31]. In the context of recycling, social robots have been designed to teach different types of recycling to children. One such example is PeppeRecycle, which has been developed by Giovanna et al. [14]. PeppeRecycle utilizes Pepper, which is a social humanoid robot designed by SoftBank Robotics. Pepper is programmed to recognize and respond to human emotions and engage in natural conversations through voice recognition and natural language processing. It can also interpret body language, facial expressions, and tone of voice to adapt its behavior and responses accordingly. Because of its friendly and engaging interactions with humans, Pepper is used in various applications where social interaction and assistance are needed. In PeppeRecycle, pepper has been used to teach children recycling in the form of a serious game (games whose primary purpose is not entertainment or fun but rather have educational objectives). The game is divided into three parts, with each paper having 3 attempts. There is also a human judge who oversees the game and verifies the correctness. Throughout the game, the child and pepper exchange turns to guess the recycling category of different items. After every guess, the human judge verifies its accuracy and the game proceeds. After every turn, Pepper gives additional information about the item. The results of the study further support its effectiveness. High success percentages in choosing the right bin for plastic, glass, and paper were registered. This paper discusses the approach we took to also teach recycling through a serious game form but with a different type of robot. We also touch upon gaps in previous implementations such as those in PeppeRecycle to enhance the trustability, effectiveness and scalability of the idea.

2.3 AI Voice Interaction

Teaching children is a complex process that involves a lot of factors to ensure their learning and retention. Among these factors, interaction has proven to be a crucial component in helping children to understand and retain new concepts effectively. Children learn best when they are actively engaged in the learning process, and interaction provides an excellent platform for that. Moreover, the use of voice interaction [32], where children can interact with their teachers or a digital assistant, has been shown to have numerous benefits in enhancing children's learning experience. Voice interaction helps children to develop language skills, improves their cognitive abilities, and fosters their independence and self-confidence. In this way, interaction and voice interaction play an essential role in creating a conducive environment for children to learn and grow. According to a study done by researchers at Temple University, the University of Washington, and the University of Delaware, responsiveness and interactions are key to their learning, even if they are coming from a screen [33]. In the study, two-year-olds were assigned to learn new verbs through three different methods: live training, video chat, and prerecorded video instruction. The results showed that children only learned new words through conversing with a person or through live video chat, which both involved responsive interactions. The authors explain that "back-and-forth" social interactions, such as live instruction and video chats, are crucial for learning words. Hence, when developing an educational robot to be used by children, it is important to ensure enough interactions and responsiveness in the application. RoboRecycle employs the mentioned factors through various ways through the application. During the first few interactions with the robot, RoboRecycle initiates communication through a personalized greeting. Studies have shown the effectivenss of greeting someone by their name especially during educational settings. One study found that within the first 10 min of instruction time, the effects of a personal greeting can be observed [34]. Several other studies have shown that when students are acknowledged by their teacher before class, they tend to engage more quickly in academic instruction and display on-task behavior [34,35]. This underscores the significance of greeting students by name in establishing positive rapport and enhancing learning outcomes. Another central form of interaction throughout the application is voice interaction during information communication. This can be seen when the robot provides a description of the classified object and when it answers the questions posed by the user. This helps in enhancing effective communication between the user and the robot, as well as provides a scalable and efficient approach to teaching children how to recycle.

2.4 ChatGPT

ChatGPT is a leading state-of-the-art artificial intelligence language model developed by OpenAI. It is based upon the transformer-based language processing model as well as deep learning algorithms to generate human-like text responses. ChatGPT is developed on top of OpenAI's [36] GPT-3.5 and GPT-4 large language models (LLMs) and has been fine-tuned using both supervised and reinforcement learning techniques. It has access to a large database of information and can provide informative answers to a large number of questions. Since its release in November 2022, ChatGPT has been utilised in a wide range of applications, including chatbots, virtual assistants, and customer service support systems, due to its ability to understand natural language. This paper employs ChatGPT's advanced generative and comprehensive capabilities to offer the child further analysis after categorizing the waste object as either recyclable or not.

In the context of educational developments like RoboRecycle, one must ensure that the child attains the learned information and applies it independently. The ability to think critically, problem-solve, and develop self-sufficiency is vital for a child's academic and personal success. When children learn how to find answers independently, they develop a sense of confidence and accomplishment that can help them overcome future challenges. Therefore, when developing interactive educational robots, it is essential to not only provide children with answers but also to teach them the skills and knowledge needed to arrive at the answers themselves. With the use of ChatGPT, users can comprehend the rationale behind categorizing their waste into the appropriate group. As the robot is repeatedly employed, children can acquire the knowledge required to accurately classify their waste, ultimately allowing them to independently recycle the items.

3   System Design

The RoboRecycle Buddy is an interactive and versatile robotic recycling assistant, meticulously designed to educate and engage children in the recycling process, fostering environmentally responsible habits from an early age. By utilizing a user-friendly and interactive robotic recycling assistant, children can develop a better understanding of the importance of recycling and its impact on the environment.

The RoboRecycle Buddy system encompasses a suite of interconnected components orchestrated to cultivate environmentally responsible recycling habits in children. Robotic Trash Can, Interactive Mobile Application, Face Recognition System, Trash Recognition, Voice Interaction, and Chat GPT Integration; these components work together to offer an interactive learning environment, ensuring an effective and impressive experience for young learners. Figure 1 presents an illustration of the RoboRecycle Buddy's architecture, which consists of four integral modules: robot hardware, navigation, processing, and stakeholder (user) engagement.

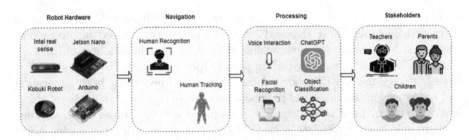

Fig. 1. System Architecture of RoboRecycle Buddy

3.1   Hardware Configuration

The RoboRecycle Buddy has been designed as a versatile robotic recycling assistant with carefully chosen hardware components that optimize its performance, enhance user interaction, and ensure smooth operation. The main hardware components depicted in Fig. 2 including a Kobuki robot base as its foundation that provides a robust, reliable, and mobile platform for excellent stability and maneuverability during recycling tasks. An ultrasonic sensor is incorporated for collision avoidance, emitting ultrasonic waves and measuring their time of return to calculate distances to nearby objects. The RealSense depth camera facilitates human tracking by capturing depth data and enabling the robot to detect and track human presence, ensuring safe operation and adaptability. A Jetson Nano serves as the processing unit for garbage classification, employing machine learning and computer vision algorithms to accurately identify and classify recyclable materials.

A tablet is integrated for face recognition and user interaction, with a custom build application, its camera, and display allowing the robot to recognize and track human faces while providing a user interface for monitoring and accessing information. Lastly, an Arduino Mega and servo motor control the opening and closing mechanism of the garbage can, with the Arduino directing the servo motors to actuate the specific lid, enabling efficient depositing of classified materials into the appropriate parts of recycling bins.

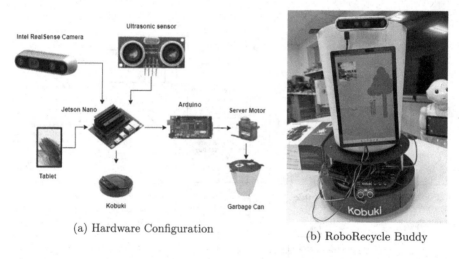

(a) Hardware Configuration

(b) RoboRecycle Buddy

Fig. 2. From Interconnected Components to a Complete RoboRecycle Buddy

3.2 Workflow Diagram

The workflow of the RoboRecycle Buddy, as illustrated in Fig. 3, illustrates a seamless and interactive process designed for an engaging user experience. The system initiates its operation by detecting the presence of a human within its proximity using human tracking through realsense camera. Upon identifying an individual, the robot strategically approaches them, maintaining a certain distance of 0.5 m. It then greets the person using their name, which is made possible by a custom-built application installed on an accompanying tablet. To achieve this personalized interaction, the robot scans the individual's face, recognizing them if they have interacted before and greeting them accordingly. In the event the person is a new user, the robot courteously inquires about their name, records it in its database, and offers a warm greeting.

Following the introduction, the system prompts the user to present the waste item they intend to dispose of. The robot skillfully captures an image of the object using its high-resolution camera, employing an object classification algorithm to determine the item's category and name. To provide users with insight into the object's classification, the robot utilizes the object's name to activate a ChatGPT script that gathers relevant information. This data is then conveyed back to the user through clear voice feedback. Subsequently, the robot opens the appropriate trash can lid (recyclable or landfill) based on the item's classification and inquires if the user has any further questions. Should the user pose additional queries, the robot encourages them to ask and runs the ChatGPT script once more to generate a comprehensive response. Once all questions have been addressed, the robot bids the user farewell and resumes its cycle by scanning its surroundings for other individuals in need of assistance.

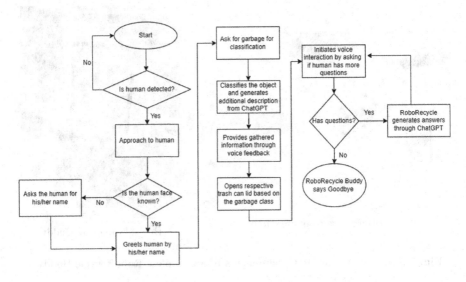

Fig. 3. System workflow diagram of RoboRecycle Buddy

3.3 Software Architecture

The software architecture of the RoboRecycle Buddy is a harmonious blend of several components in a custom-built app. The components are human tracking, facial recognition, Waste Recognition, voice interaction and ChatGPT which creates a seamless and intuitive experience for users.

3.3.1   Human Tracking

In the RoboRecycle Buddy, human tracking is implemented by integrating the Intel RealSense camera and the Histogram of Oriented Gradients (HOG) human detection method. The RealSense camera, equipped with depth-sensing capabilities, captures a three-dimensional view of the environment, enabling the robot to accurately perceive distances between objects and people. The HOG human detection algorithm extracts features from the captured images by analyzing the distribution of gradient directions in localized portions of the image. This process enables the identification of human shapes within the images, regardless of variations in clothing, lighting, and pose. By combining the depth information provided by the RealSense camera with the HOG human detection method, the RoboRecycle Buddy can accurately track and locate individuals within its proximity. This integration ensures smooth and safe interactions with users while maintaining optimal distances and enabling efficient navigation within its environment.

3.3.2   Face Recognition

The face recognition technique employed by the RoboRecycle Buddy system plays a crucial role in creating a personalized and engaging learning experience for children. By recognizing individual children, the social robot can address them by their names and tailor its interactions to suit their unique needs and learning styles.

To achieve this, the RoboRecycle Buddy uses the Python Face Recognition library to enable its facial recognition capabilities. This library is built upon the deep learning-based library, Dlib, which employs state-of-the-art techniques to efficiently and accurately detect and recognize faces in images captured by the tablet's camera and extracts unique facial features to create face encoding. The system maintains a database of face encodings associated with the names of children who have previously interacted with the robot.

When a child approaches the RoboRecycle Buddy, the system captures their image, extracts the face encoding, and searches its database for a matching entry. If a match is found, the robot recognizes the child and addresses them by their name, fostering a sense of familiarity and rapport. If no match is found, the robot gently prompts the child to provide their name and adds their face encoding and name to the database for future reference.

3.3.3   Waste Recognition

To enable RoboRecycle Buddy to accurately identify waste materials during the interaction, we designed a computer vision module that leverages the power of Convolutional Neural Networks (CNNs).

More specifically, we utilized the YOLOv7 neural network customized for Jetson Nano, which has been pre-trained on the extensive COCO dataset [37], to achieve our goals. YOLOv7, an advanced convolutional neural network, has been developed explicitly for object detection and classification tasks, making it well-suited for this application.

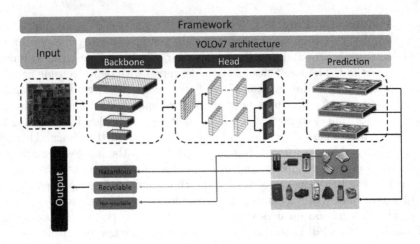

Fig. 4. Framework of the waste classification

The original architecture of YOLOv7 and its various components are depicted in Fig. 4, providing a clear overview of its structure. To adapt the YOLOv7 network to our specific use case, we employed a transfer learning technique based on fine-tuning. This approach allowed us to harness the existing knowledge within the pre-trained network and adjust it to suit the particular requirements of waste classification. We trained the YOLOv7 network on the Garbage Classification Dataset [38], which is publicly accessible on Kaggle. This comprehensive dataset consists of 21,984 waste-related images, meticulously arranged into 10 distinct categories as illustrated in Fig. 5. For our training purposes, we have selected a subset of 400 images from each class. Within the dataset, each category is represented by an integer value ranging from 0 to 9. When the robot encounters an object, its primary task is to identify the corresponding class to which the object belongs. Subsequently, we have consolidated these ten classes into three overarching categories. During the disposal process, the object is placed into one of the two compartments of the trash can based on its category, as illustrated in Table 1.

The distribution of images for each class is structured in a balanced manner, ensuring that the neural network can effectively learn the unique features and characteristics of each waste material type. The split of the data into training, testing, and validation sets was 70-10-20 ratio. By training our YOLOv7-based computer vision module on this dataset, we aimed to equip Pepper with the capability to accurately recognize and classify waste materials during the interaction, thereby enhancing its educational value and effectiveness in promoting proper waste disposal practices.

3.3.4   Voice Interaction and ChatGPT

The RoboRecycle Buddy system employs the Google Voice Recognition API and ChatGPT to facilitate seamless and natural voice interactions between the

Fig. 5. Garbage images count by class

Table 1. Different waste categories and disposal locations.

Garbage class	Garbage Category	Disposal Location
Cardboard	Recyclable	Recyclable
Glass	Recyclable	Recyclable
Trash	Non-recyclable	Landfill
Metal	Recyclable	Recyclable
Paper	Recyclable	Recyclable
Plastic	Recyclable	Recyclable
Biological	Non-recyclable	Landfill
Shoes	Recyclable	Recyclable
Clothes	Recyclable	Recyclable
Battery	Hazardous	Recyclable

social robot and the children. These technologies work together to enable the robot to understand and respond to spoken language, creating an engaging and immersive educational experience. The Google Voice Recognition API is responsible for converting the children's speech into text. When a child speaks to the RoboRecycle Buddy, the system captures the audio and sends it to the Google Voice Recognition API, which processes the audio and returns a text transcript. This API is known for its high accuracy and ability to recognize different languages and accents, making it an ideal choice for RoboRecycle Buddy's diverse audience. Once the speech has been converted to text, ChatGPT comes into play. ChatGPT generates contextually relevant and meaningful responses based on the input text. The model understands and generates human-like responses, which are then sent back to the RoboRecycle Buddy.

The robot's text-to-speech system then converts ChatGPT's generated response into audible speech, allowing the RoboRecycle Buddy to engage with children in a conversational manner. This combination of Google Voice Recognition API and ChatGPT makes voice interactions with the RoboRecycle Buddy feel natural and intuitive, ultimately enhancing the learning experience for young children and fostering positive recycling habits.

3.3.5   Playful Interaction with "EcoQuest" Reward Games

One of the standout features of the RoboRecycle Buddy is its ability to create an engaging learning environment through playful interactions. Dubbed "Eco-Quest", this rewards-based game system is specifically designed to enhance the learning experience while promoting positive recycling habits among children.

The Game Mechanics: "EcoQuest" involves a series of interactive games and challenges centered around waste sorting and recycling. Each game session begins with the RoboRecycle Buddy presenting a recycling-related scenario or challenge to the child. For instance, a game could involve sorting waste items into the correct bins within a given timeframe or identifying recyclable items from a mixed set of waste. As the child successfully completes each challenge, they earn virtual badges or points, which accumulate over time.

Reward System: The accumulated points can be exchanged for rewards, which might include virtual stickers, eco-friendly tales narrated by the RoboRecycle Buddy, or unlocking new game levels. These rewards not only serve as motivation for children to actively participate but also reinforce the lessons learned during the gameplay.

Adaptive Learning: The RoboRecycle Buddy's underlying AI system constantly monitors the child's performance and feedback during the games. Based on this data, the games adapt in difficulty, ensuring that each child is appropriately challenged based on their understanding and skill level. For instance, if a child consistently misclassifies a particular type of waste, subsequent games will focus on that category to reinforce the correct behavior.

Feedback Mechanism: After each game session, the RoboRecycle Buddy offers feedback on the child's performance. This includes not only the points earned but also insightful comments on areas of improvement. The robot's natural voice interaction, powered by Google Voice Recognition API and ChatGPT, ensures that the feedback is delivered in a positive, constructive, and child-friendly manner.

Benefits: By combining learning with play, the "EcoQuest" system addresses the unique needs of young learners, making the educational process fun and engaging. This playful approach is vital in capturing children's attention and retaining their interest over prolonged periods. Furthermore, by receiving instant rewards and feedback, children can immediately see the impact of their actions, leading to better retention and understanding of recycling principles.

4   System Evaluation

In the results and discussion section of the RoboRecycle Buddy study, we assess the performance of the waste classification system, a critical component in the overall effectiveness of the social robot in fostering eco-friendly habits and environmental awareness among young children. The waste recognition model was trained to identify and categorize a wide array of waste materials, covering ten distinct waste classes. Utilizing a desktop computer equipped with an RTX A6000 GPU and the YOLOv7 model, the training process spanned 300 epochs. To evaluate the model's performance, we considered various metrics, including precision, recall, mAP, F1 score, and confusion matrix.

Table 2. Model performance by class

Class	P	R	mAP@.5	mAP@.5:.95
All	0.553	0.587	0.559	0.242
Battery	0.559	0.651	0.587	0.336
Biological	0.397	0.210	0.268	0.0822
Cardboard	0.675	0.617	0.611	0.249
Clothes	0.709	0.827	0.839	0.371
Glass	0.590	0.675	0.641	0.271
Metal	0.563	0.654	0.593	0.211
Paper	0.498	0.642	0.458	0.178
Plastic	0.439	0.426	0.428	0.183
Shoes	0.624	0.535	0.629	0.336
Trash	0.477	0.636	0.534	0.208

As indicated by Table 2, the model achieved varying levels of accuracy across the 10 different waste categories. The overall mAP@.5 score was 0.559, while the mAP@.5:.95 was 0.242, suggesting that there is room for improvement in terms of classification accuracy. The F1 curve revealed a score of 0.56 at a confidence threshold of 0.111, indicating a balance between precision and recall. Analyzing the individual waste classes, we observed that the model performed well in certain categories like clothes (mAP@.5: 0.839) and cardboard (mAP@.5: 0.611). However, other categories like biological (mAP@.5: 0.268) and plastic (mAP@.5: 0.428) had lower performance scores. The confusion matrix diagonal values were 0.46, 0.05, 0.18, 0.50, 0.41, 0.29, 0.10, 0.21, 0.35, and 0.37, further emphasizing the disparities in classification accuracy across different waste classes.

While the present results exhibit satisfactory performance, there is still potential for enhancement. To improve the waste classification system's performance, various approaches can be considered, such as data augmentation to enrich the diversity of the training data, fine-tuning the model's parameters, leveraging transfer learning, and so on (Fig. 6).

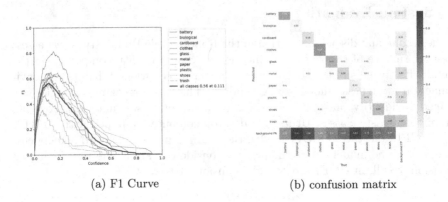

(a) F1 Curve   (b) confusion matrix

Fig. 6. Performance Evaluation of the Waste Classification Model

5   Conclusion

The RoboRecycle Buddy has demonstrated its potential to enhance early childhood green education and foster positive recycling habits through playful interaction with a social robot. By leveraging advanced technologies such as object recognition, face recognition, and natural voice interaction, it provides a unique and engaging educational experience for children. The introduction of the "EcoQuest" reward games further accentuates this playful aspect, intertwining education with gamified elements to foster both enthusiasm and retention. Children are not just learning about recycling; they're incentivized to actively participate and engage, making the educational journey more rewarding and memorable.

However, there are some limitations and areas for future improvement. First, the current waste recognition module may not be able to accurately classify certain waste materials with complex or ambiguous features, necessitating further refinement of the image recognition algorithm. Additionally, the GPT language model could be fine-tuned to better cater to the specific needs and language proficiency of young children, ensuring more effective communication and learning. Moreover, ensuring the accuracy of AI-generated content is vital for the effectiveness of the RoboRecycle Buddy in fostering eco-friendly habits and environmental awareness among children. As we move forward, our plan includes gathering information from multiple sources and implementing a comparison-based approach to verify and enhance the reliability of the content provided by the robot.

Another limitation is that the current RoboRecycle Buddy system primarily focuses on recycling education, while there are numerous other environmental topics that could benefit from a similar interactive approach, now enriched by gamified experiences such as the "EcoQuest" reward games. Future work could involve expanding the system's scope to cover a wider range of environmental subjects, such as energy conservation, pollution reduction, and sustainable living. Moreover, the inclusion of additional sensors, such as olfactory sensors for detecting hazardous waste, could further enhance the system's capabilities and safety features.

As the field of Child-Robot Interaction continues to evolve, ongoing research and development efforts should be directed towards improving the RoboRecycle Buddy's artificial intelligence, allowing it to better adapt to individual learning styles and offer more personalized educational experiences. In conclusion, the RoboRecycle Buddy, enhanced by the "EcoQuest" games, represents a promising avenue for cultivating eco-friendly habits and environmental awareness in young children. With further refinement and expansion of its capabilities, it has the potential to make a significant impact on early childhood education in the realm of environmental sustainability.

References

1. Manisalidis, I., Stavropoulou, E., Stavropoulos, A., Bezirtzoglou, E.: Environmental and health impacts of air pollution: a review. Front. Public Health 14 (2020)
2. Cavallo, E., Noy, I.: Natural disasters and the economy - a survey. Int. Rev. Environ. Resour. Econ. 5(1), 63–102 (2011)
3. Debrah, J.K., Vidal, D.G., Dinis, M.A.P.: Raising awareness on solid waste management through formal education for sustainability: a developing countries evidence review. Recycling 6(1), 6 (2021)
4. Teixeira, G., Bremm, L., dos Santos Roque, A.: Educational robotics insertion in high schools to promote environmental awareness about e-waste. In: 2018 Latin American Robotic Symposium, 2018 Brazilian Symposium on Robotics (SBR) and 2018 Workshop on Robotics in Education (WRE), pp. 591–597 (2018)
5. DeCoito, I., Myszkal, P.: Connecting science instruction and teachers' self-efficacy and beliefs in stem education. J. Sci. Teach. Educ. 29(6), 485–503 (2018)
6. Saunders, R., McFarland-Piazza, L., Jacobvitz, D., Hazen-Swann, N., Burton, R.: Maternal knowledge and behaviors regarding discipline: the effectiveness of a hands-on education program in positive guidance. J. Child Fam. Stud. 22(3), 322–334 (2013)
7. Ma, J., Tucker, C.S., Kremer, G.E.O., Jackson, K.L.: Exposure to digital and hands-on delivery modes in engineering design education and their impact on task completion efficiency. J. Integrated Des. Process Sci. 21(2), 61–78 (2017)
8. Ekwueme, C.O., Ekon, E.E., Ezenwa-Nebife, D.C.: The impact of hands-on-approach on student academic performance in basic science and mathematics. High. Educ. Stud. 5(6), 47–51 (2015)
9. Rothe, I.: Organization of a Lego-robots contest offered to high school kids by engineering students within a project based learning environment. In: 2014 IEEE Global Engineering Education Conference (EDUCON), pp. 36–39 (2014)
10. Ball, C.L.: Sparking passion: engaging student voice through project-based learning in learning communities. Learn. Commun. Res. Pract. 4(1), 9 (2016)
11. Albo-Canals, J., et al.: A pilot study of the KIBO robot in children with severe ASD. Int. J. Soc. Robot. 10, 371–383 (2018)
12. Alves-Oliveira, P., Arriaga, P., Paiva, A., Hoffman, G.: Guide to build YOLO, a creativity-stimulating robot for children. HardwareX 6, e00074 (2019)
13. Kasibhatla, R., Mahmud, S., Sourave, R.H., Arnett, M., Kim, J.H.: Design of a smart puppet theatre system for computational thinking education. In: Kim, J.H.,

Singh, M., Khan, J., Tiwary, U.S., Sur, M., Singh, D. (eds.) IHCI 2021. LNCS, vol. 13184, pp. 301–312. Springer, Cham (2022). https://doi.org/10.1007/978-3-030-98404-5_29

14. Castellano, G., De Carolis, B., D'Errico, F., Macchiarulo, N., Rossano, V.: PeppeReCycle: improving children's attitude toward recycling by playing with a social robot. Int. J. Soc. Robot. **13**, 97–111 (2021)

15. Vega, J., Cañas, J.M.: PiBot: an open low-cost robotic platform with camera for stem education. Electronics **7**(12) (2018)

16. Shukla, M., Shukla, A.N.: Growth of robotics industry early in 21st century. Int. J. Comput. Eng. Res. **2**(5), 1554–1558 (2012)

17. Gottfredson, L.S.: Circumscription and compromise: a developmental theory of occupational aspirations. J. Couns. Psychol. **28**(6), 545 (1981)

18. Alcorn, A.M., et al.: Educators' views on using humanoid robots with autistic learners in special education settings in England. Front. Robot. AI **6** (2019)

19. Laban, G., Cross, E.S., Henschel, A.: What makes a robot social? A review of social robots from science fiction to a home or hospital near you. Curr Robot Rep **2**, 9–19 (2021)

20. van Straten, C.L., Peter, J., Kühne, R.: Child-robot relationship formation: a narrative review of empirical research. Int. J. Soc. Robot. **12**, 325–344 (2020)

21. Serholt, S., Barendregt, W.: Robots tutoring children: longitudinal evaluation of social engagement in child-robot interaction. In: Proceedings of the 9th Nordic Conference on Human-Computer Interaction, NordiCHI 2016. Association for Computing Machinery, New York (2016)

22. Arnett, M., Mahmud, S., Sourave, R.H., Kim, J.H.: Smart trashcan brothers: early childhood environmental education through green robotics. In: Kim, J.H., Singh, M., Khan, J., Tiwary, U.S., Sur, M., Singh, D. (eds.) IHCI 2021. LNCS, vol. 13184, pp. 313–324. Springer, Cham (2022). https://doi.org/10.1007/978-3-030-98404-5_30

23. Al Neyadi, O.S.: The effects of using games to reinforce vocabulary learning (2007)

24. Russell, J.V.: Using games to teach chemistry: an annotated bibliography. J. Chem. Educ. **76**(4), 481 (1999)

25. Farrell, D., et al.: Computer games to teach hygiene: an evaluation of the e-bug junior game. J. Antimicrob. Chemother. **66**(suppl_5), v39–v44 (2011)

26. Vanbecelaere, S., Van den Berghe, K., Cornillie, F., Sasanguie, D., Reynvoet, B., Depaepe, F.: The effectiveness of adaptive versus non-adaptive learning with digital educational games. J. Comput. Assist. Learn. **36**(4), 502–513 (2020)

27. Gaggi, O., Meneghello, F., Palazzi, C.E., Pante, G.: Learning how to recycle waste using a game. In: Proceedings of the 6th EAI International Conference on Smart Objects and Technologies for Social Good, pp. 144–149 (2020)

28. Santti, U., Happonen, A., Auvinen, H.: Digitalization boosted recycling: gamification as an inspiration for young adults to do enhanced waste sorting. In: AIP Conference Proceedings, vol. 2233, p. 050014. AIP Publishing LLC (2020)

29. de Jesús Luis González Ibánez, J., Wang, A.I.: Learning recycling from playing a Kinect game. Int. J. Game-Based Learn. (IJGBL) **5**(3), 25–44 (2015)

30. Johal, W., Castellano, G., Tanaka, F., Okita, S.: Robots for learning (2018)

31. Jeong, S., et al.: A social robot to mitigate stress, anxiety, and pain in hospital pediatric care. In: Proceedings of the Tenth Annual ACM/IEEE International Conference on Human-Robot Interaction Extended Abstracts, pp. 103–104 (2015)

32. Mahmud, S., Sourave, R.H., Islam, M., Lin, X., Kim, J.-H.: A vision based voice controlled indoor assistant robot for visually impaired people. In: 2020 IEEE Inter-

national IOT, Electronics and Mechatronics Conference (IEMTRONICS), pp. 1–6. IEEE (2020)

33. Roseberry, S., Hirsh-Pasek, K., Golinkoff, R.M.: Skype me! socially contingent interactions help toddlers learn language. Child Dev. **85**(3), 956–970 (2014)

34. Allan Allday, R., Pakurar, K.: Effects of teacher greetings on student on-task behavior. J. Appl. Behav. Anal. **40**(2), 317–320 (2007)

35. Allan Allday, R., Bush, M., Ticknor, N., Walker, L.: Using teacher greetings to increase speed to task engagement. J. Appl. Behav. Anal. **44**(2), 393–396 (2011)

36. OpenAI - OpenAI.com. https://openai.com/. Accessed 16 Apr 2023

37. Lin, T.Y., et al.: Microsoft COCO: common objects in context. In: Fleet, D., Pajdla, T., Schiele, B., Tuytelaars, T. (eds.) Computer Vision ECCV 2014. Lecture Notes in Computer Science, vol. 8693, pp. 740–755. Springer, Cham (2014). https://doi.org/10.1007/978-3-319-10602-1_48

38. Garbage Classification. https://www.kaggle.com/datasets/sumn2u/garbage-classification-v2. Accessed 14 Apr 2023

Optimizing Interface and Interaction Design for Non-immersive VR Firefighting Games: A User Experience Approach

Linjing Sun[1], Boon Giin Lee[1(✉)] [iD], Matthew Pike[1], and Wan-Young Chung[2]

[1] School of Computer Science, University of Nottingham Ningbo China,
Ningbo 315100, China
{linjing.sun,boon-giin.lee,matthew.pike}@nottingham.edu.cn

[2] Department of Electronic Engineering, Pukyong National University,
Busan 48513, Korea
wychung@pknu.ac.kr

Abstract. In light of the pressing imperative for enhanced fire safety education within the Chinese context, this research delves into the intricacies of interface and interaction design for non-immersive Virtual Reality (VR) firefighting simulations anchored on computer platforms. Utilizing a bespoke VR prototype developed within the Unity environment, the study offers a systematic exploration of variables such as the spatial orientation of informational panels, luminance calibrations, and the dynamics of keyboard-mediated spray interactions. Empirical findings underscore a prevailing inclination among users towards luminous fire scenario visualizations, with a manifest predilection for the strategic alignment of task-centric panels to the left and ancillary hint panels to the right. Interactionally, a sustained keypress modality for simulating spraying actions emerged as the favored paradigm. While the concealment functionality of panels garnered predominantly affirmative reception, the hover-induced display modality was met with notable reservation. The derived insights proffer pivotal heuristics for the optimization of interface design in VR-centric fire safety pedagogical tools, encapsulating prospects for enriched user engagement and augmented instructional efficacy.

Keywords: Firefighting · User Interface Design · Interaction Design · Virtual Reality

1 Introduction

1.1 Background

According to a comprehensive report by the Chinese Ministry of Emergency Management [1], between the years 2012 and 2021, the nation experienced a

This work was supported in part by the Zhejiang Provincial Natural Science Foundation of China under Grant LQ21F020024, and in part by the Ningbo Science and Technology (S&T) Bureau through the Major S&T Program under Grant 2021Z037 and 2022Z080.

B. J. Choi et al. (Eds.): IHCI 2023, LNCS 14531, pp. 344–352, 2024.
https://doi.org/10.1007/978-3-031-53827-8_30

staggering 1.324 million residential fires. These incidents led to 11,634 fatalities, 6,738 casualties, and a consequential property loss approximated at 7.77 billion yuan. Given these figures, there emerges an unequivocal imperative for robust fire safety education.

While conventional methods, such as slide-based lectures, knowledge manuals, and videos, have their merits [2], they are predominantly passive in nature. They do, however, offer flexibility, cost-efficiency, and reusability. Conversely, fire drills, albeit effective, introduce potential safety liabilities for participants and can invoke substantial socio-economic repercussions [3].

In the realm of innovative pedagogical approaches, Virtual Reality (VR) stands out, primarily due to its cost-efficacy and the immersive learning environment it engenders. This immersive modality facilitates a deeper engagement, prompting users to actively participate and introspect on the instructional content. Furthermore, given its safe instructional environment, VR is postulated to serve as a potential surrogate for hazardous real-life drills. VR's categorization, based on immersion levels—fully immersive, semi-immersive, and non—immersive-offers varied experiential dimensions. Notably, non-immersive VR, given its minimal hardware prerequisites, has garnered substantial acclaim [4].

Numerous scholarly investigations integrating VR into fire safety pedagogy have enlisted computers as essential research instruments [5–8]. This paper, in its essence, aims to delve into the non-immersive VR interaction design anchored in computer-based platforms. Addressing the prevalent monotony in fire safety educational tasks, gamification emerges as a pivotal mechanism, stimulating motivation and augmenting learner engagement [9]. With an emphasis on a user-centric design ethos, gamification in fire safety instruction augments user experience, catalyzing knowledge acquisition and thereby elevating efficiency [10]. The inherent immersive capabilities of Extended Reality (XR), coupled with its dynamic 3D visualization, offer unparalleled enhancements in interface aesthetics and overall user experience [11]. Notably, a study conducted by Chittaro and Ranon [12] illuminated the superior efficacy of gamified VR in safety knowledge dissemination and retention, especially when juxtaposed against traditional instructional methodologies.

This paper underscores the paramountcy of user interface and interaction design as the linchpin mediating users and the virtual milieu. As users navigate the virtual ecosystem, their interaction with the digital interface becomes their most tangible point of contact. A meticulously crafted interface, encompassing design elements such as layout, color schemes, and interactive buttons, can significantly augment comprehension, task completion rates, and user engagement [13]. Contrariwise, suboptimal design paradigms can engender user dissonance, truncating efficiency and satisfaction [14]. The crux of our study orbits around the intricate nuances of interface layout and interactive methodologies specific to computer-based semi-immersive firefighting VR simulations. Through this investigative lens, we aspire to elucidate an optimal, intuitive, and user-centric design framework for VR firefighting simulations, thereby ensuring superior learning outcomes and enriched user experiences.

1.2 Related Work

Layout and interaction design in games are pivotal components in crafting a successful user experience. The Head-Up Display (HUD) plays a crucial role in many games, predominantly presenting relevant information as an overlay on the screen [15]. While some studies suggest that such additional displays might disrupt the immersion of the game world, they are vital in assisting novice players, especially those unfamiliar with using firefighting equipment, to make progress within the game [15]. The nine types of computer game screen layouts proposed by Rollings and Adams [16] serve as a cornerstone in this domain, with their ubiquity and significance in game interface design validated by various research endeavors. A study by Rho *et al.* [17] investigated the disparity between actual layout in MMORPG models and user preferences. The results revealed that while many users believed status and information windows were best placed in the top-left, there was also a preference for chat windows in that location. In contrast, main menu placement preferences were scattered between the bottom-right and top-right. This discrepancy underscores the importance of tailored interaction and layout explorations for specific applications, such as fire simulations. Ultimately, the success of the user experience hinges on how information is elegantly presented. Overloading or inappropriate overlaying of information might engender user aversion to the UI. This elegance in presentation often depends more on individual user preferences [18].

Although numerous studies [19–24] have focused on VR computer games for fire safety education or training and examined their efficacy and utility, specific research on in-game UI design seems relatively scant. From these works, we can discern several recurring UI layout patterns. For instance, the status bar is frequently positioned in the top-left corner of the screen [19,23], potentially owing to the majority of players' left-to-right and top-to-bottom reading habits. Concurrently, informational prompts, such as task indicators or feedback, often appear in the bottom-right, perhaps to prevent clashing with primary game content [19,20]. However, there are exceptions, with some studies showing the status bar on the screen's top-right [21,22] or bottom-left corners [20,24]. Moreover, some games utilize specific NPCs as means to convey informational prompts, offering players a more immersive experience [20]. As for timers, they often directly correlate with game missions or objectives, and as such, their positioning on the screen lacks a standardized norm. While we can distill certain UI design patterns from these observations, existing research does not explicitly provide the rationale behind choosing these particular layouts. Perhaps it's grounded in intuitiveness, usability, or other player experience considerations, but these assumptions necessitate further empirical studies for validation. This highlights that there remain underexplored questions in the domain of UI design for VR computer fire training games, warranting further attention and in-depth exploration by researchers.

2   Methodology

2.1   System Design

Our firefighting training system, fundamentally user-centric in its design philosophy, revolves around a robust application framework tailored to address critical pedagogical objectives in fire safety education. The essence is to not just familiarize individuals with fire scenarios but also empower them with the ability to judiciously select and proficiently deploy the correct firefighting equipment.

Within this immersive virtual environment, users navigate through four distinct fire scenarios: Computer Fire, Shelf Fire, Room Fire, and Warehouse Fire. Each scene is meticulously crafted to challenge users in terms of fire identification, equipment location, and deployment. This ensures that learners are not passive observers but active participants, making informed decisions and learning from real-time feedback. The interactive framework ensures users are continuously engaged, using cues like glowing arrows for guidance and diminishing the need for extensive textual instructions.

The gameplay mechanics, pivotal to the application framework, are embedded with features that enhance realism and foster learning. Time constraints introduce an element of urgency, compelling users to make swift yet informed decisions. The system's dynamic simulation of fire propagation, extinguishing, and consequences for incorrect equipment choices further deepens the immersive experience. For instance, selecting inappropriate equipment could lead to an explosion, emphasizing the importance of proper equipment knowledge.

Interface Design. The simulation includes four distinct fire scenarios: Computer Fire, Shelf Fire, Room Fire, and Warehouse Fire. The Computer and Shelf fires are categorized as small fires, suitable for control with fire extinguishers. In contrast, the Room and Warehouse fires are classified as big fires, requiring fire hoses for effective containment. As shown in Fig. 1, the Bookshelf and Room fires are presented in darker settings, while the others are in brighter environments. Figure 1 also highlights the strategic placement of information panels in each scenario. The functionalities of 'hide' and 'hover' were tested specifically in the Warehouse and Room fire scenarios. Regarding spraying methods, the Computer and Room fires utilized a tap-to-spray approach, while the Bookshelf and Warehouse fires adopted a hold-to-spray technique.

Brightness. Brightness stands as a cardinal visual parameter within interface design, commanding influence over user engagement and content discernibility [25]. Elevated brightness levels possess the potential to induce visual stimuli, magnetizing user attention and enhancing information clarity [26]. Conversely, diminished brightness settings might mitigate visual fatigue, promoting prolonged user engagement [27]. It's pivotal to clarify that the study's brightness discourse is centralized around the ambient luminance within the game's fire scenarios. Consequently, two distinct fire scenarios were delineated: minor conflagrations addressable via fire extinguishers, and major infernos necessitating

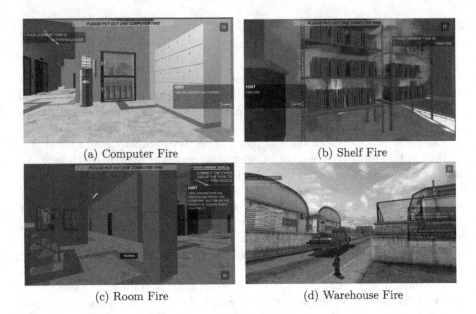

(a) Computer Fire   (b) Shelf Fire

(c) Room Fire   (d) Warehouse Fire

Fig. 1. Fire Senarios: (a) Computer Fire (Small Fire), (b) Shelf Fire (Small Fire), (c) Room Fire (Big Fire), (d) Warehouse Fire (Big Fire)

water jet interventions. Each scenario was meticulously calibrated with dichotomous luminance settings—bright and dim.

Information Panel Setting. In the landscape of UI design, the strategic placement and visualization dynamics of informational panels emerge as paramount. Within the context of firefighting simulations, elements such as task directives and procedural cues are indispensable. To extrapolate optimal panel placement paradigms, we initiated experimental deployments with panels anchored to both lateral extremities across diverse fire scenarios. Rho *et al.* [17] posits that the interactivity of game windows, such as the ability to close or move them, is a vital facet of interface interacation. Hence, augmenting this, an array of user-centric interactions was devised, encompassing mechanisms for discretionary panel concealment and hover-induced display, facilitating an empirical assessment of user proclivity.

Keyboard Interaction. Recognizing the indispensable nature of extinguishing agent deployment within fire education simulations, this research extends its purview to encompass keyboard-mediated spray dynamics. Specifically, we juxtaposed two disparate interaction modalities: a toggled spray initiation and termination versus a sustained keypress mechanism. This examination is geared toward deciphering the intuitive alignment of these modalities vis-à-vis user preference.

3   Result and Discussion

3.1   Participants

The experiment recruited 44 participants (20 females, 24 males), with ages ranging from 16 to 49 (mean = 24.3, standard deviation = 8.7). This study received approval from the university's ethics committee. Before the commencement of the research, all participants received an information sheet detailing the study's content and subsequently signed a consent form.

3.2   Brightness Results

Table 1. Brightness Results

Condition	Light	Dark	Same for Me	Not Sure
Small fire	33	6	5	0
Big fire	38	5	1	0

Referencing Table 1, data elucidates that in the context of small indoor fires, a notable 75% of respondents exhibit a predilection for brighter environmental settings. This inclination amplifies in large-scale fire scenarios with 86.4% advocating for increased luminance. Consequently, a holistic assessment infers a dominant consensus favoring brighter fire incident environments. An array of feedback encapsulates sentiments associating increased brightness with elevated mood and enhanced clarity in data discernment. Contrarily, a subset of participants exhibited an affinity for darker environments, asserting that subdued lighting intensifies the realistic urgency concomitant with fires, thereby augmenting the pedagogical efficacy of the simulation.

Table 2. Panel Position Results

Condition	Left	Right	Same for Me	Not Sure
Task panel	21	13	8	2
Hint Panel	13	24	7	0

3.3   Information Panel Results

Delving deeper into the nuances of interface design, as delineated in Table 2, a substantial 47.7% of respondents vouch for the strategic emplacement of task panels on the left. This tendency potentially mirrors entrenched visual scanning

Table 3. Panel Interaction Results

Condition	Like	Dislike	Not Sure
Click can hide panel	22	20	2
Hover show panel	11	20	13

patterns, typically from left to right in many cultures. Conversely, 54.5% of participants advocate for positioning hint panels on the right, citing their ancillary nature, as shown in Table 3. There's a shared sentiment suggesting an overt left placement might obfuscate primary information, inducing potential in-game misconceptions. Feedback pertaining to the 'hide information panel' feature yielded a relatively symmetrical distribution, albeit with a marginal proclivity towards its endorsement. Additionally, many participants registered reservations about the hover-to-display modality. Predominantly, the unfamiliarity and subtlety of this feature inadvertently nudged participants towards its oversight. Constructive feedback underscored the desirability of concise panel content, emphasizing the integration of intuitive icons, such as arrows, to bolster user immersion.

3.4   Key Interaction Results

Table 4. Key Interaction Results

Hold-to-Operate	Single-Press	Same for Me	Not Sure
24	9	10	1

Pivoting to the data presented in Table 4, a discernible trend emerges: a majority showcases an inclination for the 'Hold-to-Operate' key interaction over its 'Single-Press' counterpart. Proponents of the hold-down method underscore its immersive nature, postulating a closer simulation of real-life extinguishing operations. Conversely, a faction gravitating towards the single keypress mechanism deemed the hold-to-operate method as somewhat cumbersome, citing streamlined operations as a priority.

4   Conclusion

This research meticulously interrogated the interface and interaction design paradigms within non-immersive VR firefighting simulations, grounded in computer architectures. Empirical data underscored the superior user experience and enhanced clarity associated with luminous interface tonalities. Furthermore, results advocated for a strategic placement of task-oriented panels on the left and hint panels on the right, reflecting dominant visual and cognitive user tendencies. While the concealment feature of information panels received moderate endorsement, the hover-to-display interaction was less favored. Notably, a sustained keypress interaction for simulating spray actions emerged as the preferred modality. Collectively, these findings enrich the discourse on VR interface design. However, it's imperative to emphasize the evolutionary nature of this field, warranting further refined studies that validate and expand upon these initial discoveries.

References

1. MEM. 1011634, 03
2. All, A., Plovie, B., Castellar, E.P.N., Van Looy, J.: Pre-test influences on the effectiveness of digital-game based learning: a case study of a fire safety game. Comput. Educ. **114**, 24–37 (2017)
3. Cha, M., Han, S., Lee, J., Choi, B.: A virtual reality based fire training simulator integrated with fire dynamics data. Fire Saf. J. **50**, 12–24 (2012)
4. Gilson, S., Glennerster, A.: High fidelity immersive virtual reality. Virtual Reality - Hum. Comput. Interact. (2012)
5. Bode, N.W.F., Kemloh Wagoum, A.U., Codling, E.A.: Human responses to multiple sources of directional information in virtual crowd evacuations. J. Roy. Soc. Interface **11**, 20130904 (2014)
6. Zhang, K., Suo, J., Chen, J., Liu, X., Gao, L.: Design and implementation of fire safety education system on campus based on virtual reality technology (2017)
7. Padgett, L.S., Strickland, D., Coles, C.D.: Case study: using a virtual reality computer game to teach fire safety skills to children diagnosed with fetal alcohol syndrome. J. Pediatr. Psychol. **31**, 65–70 (2005)
8. Rahouti, A., Lovreglio, R., Datoussaïd, S., Descamps, T.: Prototyping and validating a non-immersive virtual reality serious game for healthcare fire safety training. Fire Technol. (2021)
9. Tori, A.A., Tori, R., de Lourdes dos Santos Nunes, F.: Serious game design in health education: a systematic review. IEEE Trans. Learn. Technol. **15**, 827–846 (2022). https://ieeexplore.ieee.org/abstract/document/9867976
10. What is gamification (2017). https://yukaichou.com/gamification-examples/what-is-gamification/
11. Falah, J., et al.: Identifying the characteristics of virtual reality gamification for complex educational topics. Multimodal Technol. Interact. **5**, 53 (2021)
12. Chittaro, L., Ranon, R.: Serious games for training occupants of a building in personal fire safety skills (2009)
13. Pamudyaningrum, F.E., Rante, H., Zainuddin, M.A., Lund, M.: UI/UX design for Metora: a gamification of learning journalism interviewing method. In: E3S Web of Conferences, vol. 188, p. 00008 (2020)
14. Johnson, D., Wiles, J.: Effective affective user interface design in games. Ergonomics **46**, 1332–1345 (2003)
15. Wahlroos, V.: Design and implementation of a gameplay UI for a MOBA mobile game. Master's thesis, Aalto University. School of Science (2018)
16. Rollings, A., Adams, E.: Andrew Rollings and Ernest Adams on Game Design. New Riders (2003)

17. Rho, J.-H., Chun, Y.-D.: Analysis on MMORPG UI layout and research on effective UI demanded by heavy user. Arch. Design Res. **20**(1), 263–272 (2007)

18. Llanos, S.C., Jørgensen, K.: Do players prefer integrated user interfaces? A qualitative study of game UI design issues. In: Proceedings of DiGRA 2011 Conference: Think Design Play, pp. 1–12 (2011)

19. Rahouti, A., Lovreglio, R., Datoussaïd, S., Descamps, T.: Prototyping and validating a non-immersive virtual reality serious game for healthcare fire safety training. Fire Technol. 1–38 (2021)

20. All, A., Plovie, B., Castellar, E.P.N., Van Looy, J.: Pre-test influences on the effectiveness of digital-game based learning: a case study of a fire safety game. Comput. Educ. **114**, 24–37 (2017)

21. Mystakidis, S., et al.: Design, development, and evaluation of a virtual reality serious game for school fire preparedness training. Educ. Sci. **12**(4), 281 (2022)

22. ETC Simulation - Advanced Disaster Management Simulator – etcsimulation.com. https://www.etcsimulation.com/

23. de Carvalho, P.V.R., Ranauro, D.O., de Abreu Mol, A.C., Jatoba, A., de Siqueira, A.P.L.: Using serious game in public schools for training fire evacuation procedures. Int. J. Serious Games **9**(3), 125–139 (2022)

24. Backlund, P., Engstrom, H., Hammar, C., Johannesson, M., Lebram, M.: Sidh - a game based firefighter training simulation. In: 2007 11th International Conference Information Visualization (IV 2007), pp. 899–907 (2007)

25. Zuo, H., Niu, Y., Tian, J., Yang, W., Xue, C.: Study on the brightness and graphical display object directions of the single-gaze-gesture user interface. Displays **80**, 102537 (2023)

26. Yan, Z., Hu, L., Chen, H., Lu, F.: Computer vision syndrome: a widely spreading but largely unknown epidemic among computer users. Comput. Hum. Behav. **24**, 2026–2042 (2008)

27. Xie, X., Song, F., Liu, Y., Wang, S., Dong, Yu.: Study on the effects of display color mode and luminance contrast on visual fatigue. IEEE Access **9**, 35915–35923 (2021)

User Centred Design

Enhancing E-Rickshaw Driving Experiences: Insights from User-Centric EV Dashboards in India

Lipsa Routray, Abhishek Shrivastava$^{(\boxtimes)}$, and Priyankoo Sarmah

Indian Institute of Technology, Guwahati, Guwahati, India
{lroutray,shri,priyankoo}@iitg.ac.in

Abstract. Three-wheeler E-rickshaws (E3W or E-rickshaws) are one of India's most widely adopted electric vehicles (EVs) for short-distance transportation. These E-rickshaws have two different types of dashboards (Analog and Digital dashboards), which provide essential information needed to operate the vehicle like battery status, speed, and odometer. Hence, the design of the dashboards emerges as one of the critical factors influencing driving experiences and overall EV adoption. This study aims to identify the needs, barriers, and facilitators among the drivers in using different E-rickshaw dashboards in their everyday work. Using a qualitative approach involving observation and semi-structured interviews, 15 male drivers with varying E-rickshaw experience (mean age 42.93 years) were studied. Seven analog and eight digital dashboards from different models were analyzed. The findings reveal the significance of integrating analog and digital dashboards into E-rickshaws. Drivers emphasized accurate battery percentage information and timely low-battery alerts as crucial needs. Language barriers, inadequate visibility, technical complexity, lack of standardized dashboard layout, and absence of feedback were identified as design barriers. Accurate battery status and speed information were facilitators for digital dashboards, while familiar interface design and easy battery monitoring techniques were facilitators for analog dashboards. This study pioneers research in the Indian context, providing guidelines to enhance E-rickshaw dashboard designs for a more user-friendly experience. The insights gained offer valuable contributions to improving E-rickshaw technology, aligning it more closely with user needs and preferences.

Keywords: Electric Vehicles (EVs) · E-rickshaws (E3W) · Dashboard design · User experience · Human-computer Interaction (HCI) · EV Interface · EV adoption · Driver preferences · User-centric Study

1 Introduction

The electric three-wheeler (E-rickshaws or E3W) market in India is expected to surpass USD 1 billion by the fiscal year 2023 [15]. This expansion is due to the increasing need to manage air pollution levels and different government initiatives, plans, and subsidies. EV manufacturers' increased investment in developing more innovative, efficient, and economical E3Ws is expected to fuel growth in the Indian E-rickshaw market in the coming years. E-rickshaws have gained immense popularity owing to the ease of recharging, reduced fuel costs, and minimal human labor, making them a compelling alternative to traditional gasoline, diesel, and CNG vehicles [15].

1.1 Types of Three Wheelers

According to [15], there are mainly three types of E3Ws:

- **Plug-in Electric Vehicle (PEV):** Similar to conventional hybrids, PEVs incorporate both electric motors and internal combustion engines. However, what sets them apart is the ability to recharge their batteries from external sources like wall sockets, utilizing stored electricity to power the electric motors.
- **Battery Electric Vehicle (BEV):** They operate solely on electricity from onboard batteries, charged either at dedicated stations or regular outlets, completely eliminating the need for fuel tanks. Propelled by electric motors, the three-wheelers rely entirely on battery power for their operation.
- **Solar-powered Battery:** Solar rickshaws are like regular electric rickshaws, but their batteries are charged with sunlight. When the batteries run out, they are swapped with charged ones to keep the rickshaws running smoothly.

The dashboard serves as the primary interface between the driver and the vehicle's vital information, playing a pivotal role in conveying real-time data crucial for safe and efficient driving [11]. While the importance of dashboards is widely acknowledged, there exists a paucity of research that specifically delves into how different dashboard designs impact user interactions and battery management within the context of E-rickshaws. This study seeks to bridge this gap by conducting an in-depth study of E-rickshaw drivers' experiences with various dashboard designs.

The research questions for the study are:

- **RQ1. How do different design attributes of analog and digital dashboards impact E-rickshaw drivers' battery management, range estimation, and driving behaviors concerning energy efficiency and overall usability?**
- **RQ2. What are the specific user needs, barriers, and facilitators associated with analog and digital dashboard designs among E-rickshaw drivers?**
- **RQ3. What improvements are required to be implemented to the dashboard design to improve the user experience of E-rickshaw drivers?**

In the subsequent sections, this paper presents the Literature Review in Sect. 2. Section 3 describes the data collection and analysis in Methodology, Sect. 4 discusses the results derived from driver interactions with different dashboard designs, a comprehensive discussion of the findings is presented in Sect. 5, and Sect. 6 concludes by highlighting the broader significance of the study. Finally, Sect. 7 includes the limitation and future works.

2   Literature Review

2.1   E-Rickshaws in India

Battery-operated e-rickshaws have become popular in India's public transport due to their comfort and affordability. A case study in West Bengal found them to be the most efficient passenger vehicle with an average specific energy consumption of 53.76 kJ/passenger-km, but challenges hinder their widespread adoption in the public transport sector [9]. Urban areas, responsible for 70% of global greenhouse gas emissions with a third from transportation, necessitate decarbonizing urban mobility. This study assesses India's progress in electric vehicle adoption, covering availability, infrastructure, schemes, and recommends a roadmap to address challenges for effective adoption [15]. This study develops a real-world driving cycle for auto-rickshaws, highlighting their unique emission patterns, and suggests that a 5% modal shift to electric auto-rickshaws in India by 2030 could reduce emissions by 6.30%, providing valuable insights for pollution reduction policies [2]. Soleckshaw, a solar electric rickshaw, as an example of how connecting existing knowledge and technology can create socially acceptable, environmentally friendly innovations that drive policy interventions and new markets [3].

2.2   User Experience

A user experience is a worthwhile episode related to an interaction between a user and a product, situated in context, and involves all subjective aspects the user thought, felt, and did while expecting, experiencing, and/or remembering the interaction [7] (Refer Fig. 1). Positive user experiences emerge when product interaction results in fulfilling users' needs and when it fulfills or even exceeds users' expectations. In the context of driving, a driver could describe his interaction with the driving mode interface as a positive experience [6].

2.3   Human-Computer Interaction (HCI) Studies in EVs

The interaction between humans and EVs concerns the activities and procedures that users perform while operating an EV. Cognition and specific tasks influence user behavior [12]. This means the design of the dashboard and the information available on the dashboard interface influence the driving pattern of the E-rickshaw driver. The first-ever study on EV interfaces was done by

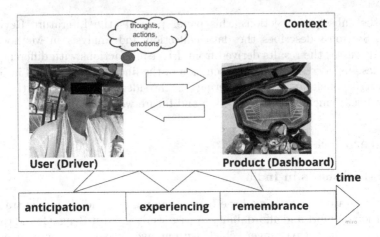

Fig. 1. Aspects of user experience

Stromberg et al. [16]. He surveyed what information is relevant to the driver of an EV and how that information should be presented to the driver innovatively and familiarly. The new iconography of EV dashboards often leads to confusion. So, A Lundstrom and F Hellstrom [8] used a mobile application on the regular ICE vehicle to familiarize drivers with the new EV attributes. A usability study of the user interface has been done for EVs on a simulator using various display modalities [14]. However, it has been observed that the domain of driver-dashboard interaction in the case of EVs remains understudied in the field of HCI [1].

3   Methodology

3.1   Participants

The E-rickshaw drivers were conveniently sampled from the population available at College Square Chouk, North Guwahati, Assam. All participants were male, with ages ranging from 22 to 60 years and a mean age of 42.93 years. The driving experience of participants varies from six months to two years.

3.2   Data Collection Procedure

In the study, two researchers accompanied the selected driver on a ride in this study. During the ride, one researcher observed the driver's behavior while the other took notes. This direct observation was followed by a semi-structured interview, which provided in-depth insights. The detailed questionnaire used in this study is listed in Table 1. The data collection tools included a GoPro camera for video recording the study. Additionally, a mobile phone was used for audio recording the interviews. Prior to data collection, verbal consent was obtained from all participants, ensuring their willingness to share their experiences and insights. Confidentiality and anonymity were maintained throughout the study, with pseudonyms assigned to participants to protect their identities.

Table 1. Questionnaire

Sl. No	Context	Questions
A	Participant's demographic information	1. Full Name
		2. Age
		3. Gender
		4. E-Rickshaw Model Name
B	Dashboard and Charging	1. What type of dashboard does your e-rickshaw have? (Analog/Digital)
		2. On average, how long does it take to charge your e-rickshaw fully?
		3. How many hours do you typically spend driving your e-rickshaw per day?
C	Dashboard Information and Interaction	1. What specific information do you typically check when you glance at the dashboard?
		2. Can you explain the detailed information shown on the dashboard of your e-rickshaw?
		3. Is there any additional information you need for driving that is not currently available on the dashboard?
D	Challenges and Low Battery Situations	1. Do you experience any challenges or difficulties while driving the e-rickshaw?
		2. If yes, can you please elaborate on them?
		3. How do you typically notice when the battery is running low while you're driving?
E	Responses to Low Battery	1. In a worst-case scenario of low battery, what steps do you take?
		2. Could you share any specific actions you take to manage your driving when the battery is critically low?
F	Financial Aspects	1. How much do you typically earn per day through your e-rickshaw operations?
		2. Could you briefly describe your daily expenses related to e-rickshaw usage?
		3. How much do you spend on electricity to charge your e-rickshaw?

3.3 Data Analysis Procedure

The video and audio recordings, along with the field notes, were transcribed to generate the interview transcript in MS word. The interview transcripts were meticulously reviewed and coded to identify recurring themes and patterns. These themes were then organized to draw meaningful insights regarding dashboard preferences, battery monitoring practices, and usability.

4 Results

The study captured a total of fifteen E-rickshaw drivers as participants. The participants used to operate various E-rickshaw models. Those include seven analog dashboards (equipped with analog gauges) and eight digital dashboards (equipped with LED screen). The details of the dashboard are provided in Table 2.

From the interview analysis, we learned that the drivers typically charge their E-rickshaws for about 8–10 h overnight. Once fully charged, they can operate the E-rickshaws for 8–9 h during the day, covering approximately 70–80 km. However, the limited charging infrastructure confines their operations within a narrow area (within 4–5 km of their homes). Some drivers recharge during lunch breaks at home. Currently, they can only charge at home, resulting in monthly electricity bills ranging from Rs. 1200 to Rs. 2000. Despite this cost, drivers earn around Rs. 500 to Rs. 700 per day, showcasing a substantial income. The maximum speed the E-rickshaw can achieve is 30–35 km/hr.

Table 2. Details of dashboards

Type of dashboards	E-rickshaw Models	Numbers
Analog (7 models)	Sarthi	3
	Kumar	2
	Galaxy	1
	Morni	1
Digital (8 models)	Anant	4
	Ele	2
	GK	2

During the study, we observed the E-rickshaw dashboards provide information on the Speedometer, Battery, and Odometer to the driver. In analog dashboards, the battery level represented was a circular meter, whereas digital dashboards use mobile-like vertical/horizontal battery icons. The dashboards also feature directional indicators (Left/Right) and provide auditory feedback for horn and reverse gear.

In the following subsections, we present themes obtained from the interview and the participant's quotes supporting the theme.

4.1 Driver Preferences and Experiences: Analog vs. Digital Dashboards in E-Rickshaws

Driver's Preferences in Analog Dashboard Features: Participant_KD, operating a Saarthi model (analog dashboard), found the analog design intuitive for monitoring battery levels, appreciating its quick glance accessibility:

"I find the analog design quite intuitive, especially for monitoring the battery levels. It's like a quick glance tells me how much charge is left." -Participant_KD

Participant_TB (Galaxy model), interacting with an analog battery meter, highlighted the need for clearer battery monitoring. He underscored the importance of accurately estimating battery levels to prevent running out of charge during trips.

"I wish the battery monitoring was clearer. It's important for me to know how much charge I have left, especially on longer trips."-Participant_TB

Participant_DS (Saarthi model) reported challenges with dim LEDs in the battery meter, demonstrating the impact of visual clarity on effective battery management.

"The battery meter LEDs aren't very bright, and that makes it a bit hard to see in certain conditions. It could be clearer."-Participant_DS

Driver's Preferences in Digital Dashboard Features: Participant_AH, driving a Kumar E-rickshaw (analog dashboard), preferred digital displays for accurate real-time battery readings, emphasizing the importance of real-time information:

"Honestly, I'd prefer a digital dashboard. Sometimes, it's hard to know the exact battery percentage without accelerating, and that affects my planning." - Participant_AH

In contrast, Participant_ND driving Ele model (digital dashboard) praised the clear presentation of battery percentage and speed on the digital dashboard:

"The digital display is pretty clear. I like how I can see the battery percentage and speed easily. It helps me focus more on driving." -Participant_ND

These insights show how E-rickshaw drivers feel about analog and digital dashboards and what are their preferences. Analog and digital dashboards have their merits and demerits; it's mostly about information and how clearly they are presented to the drivers.

4.2 Usability and Battery Monitoring

The interface between E-rickshaw drivers and their dashboards is where technology meets everyday reality, shaping the driver's usability experience in profound ways.

Participant_RK (Anant model), relying on battery bars and distance traveled, expressed discomfort with the lack of precise battery readings during trips.

"It would be great to have a more accurate way of seeing the battery level. Right now, I rely on the bars and distance traveled, but that's not always very precise."-Participant_RK

Participant_LD (G K Rickshaw model) encountered visibility issues on the analog dashboard, highlighting the importance of ambient lighting.

"The analog display is a bit hard to see, especially in certain lighting conditions. It's like the display isn't as clear as it could be."-Participant_LD

Participant_VD (Morni model) emphasized the necessity of accurate battery estimation while navigating uphill terrain, as two out of four cores often appeared depleted, creating a misleading illusion of a full charge.

"When I'm going uphill, sometimes the battery seems to drop faster. It's like I wish I had a clearer idea of how much charge is left." -Participant_VD

Participant_MD (Anant model) proposed auditory feedback for gear changes, suggesting it could enhance energy efficiency. This feedback mechanism, he believed, would optimize gear shifts and, subsequently, battery consumption.

"I think if there were some sound feedback when changing gears, it would help me save battery. It's like a little nudge to be more efficient."-Participant_MD

4.3 Preferences for Battery Management Practices

Drivers' experiences with different dashboard designs extended to their battery management practices:

Participant_AC (Anant model) effectively gauged battery capacity with a clear understanding of the icons present on the digital dashboard. His practice of charging the vehicle at home was informed by precisely estimating the remaining battery.

"I can usually tell how much battery is left based on the icons on the digital display. It helps me plan when to charge the vehicle."-Participant_AC

Participant_BD (Kumar model) noted the discrepancy between the initial battery percentage display and the subsequent actual readings during vehicle operation. This compelled him to monitor battery slots closely to avoid depleting the charge unexpectedly.

"The battery level sometimes changes quite a bit after I start driving. I have to be careful and watch the battery slots to make sure I don't run out of batteries."-Participant_BD

Participant_IS (Kumar model) highlighted the significance of accurate battery readings, recounting an incident where faulty batteries necessitated replacements, emphasizing the need for timely low-battery alerts.

"I had to replace the batteries because they weren't holding a good charge. It would be helpful to get a warning when the battery is low." -Participant_IS

5 Discussion and Proposed Guidelines

From the results we obtained from the analysis, the needs, barriers, and facilitators among the drivers in using different E-rickshaw dashboards in their everyday driving tasks. The details are described in the Table 3.

Based on the findings of our study, we derived guidelines that can enhance E-rickshaw dashboard design and improve the driver's user experience. These proposed guidelines are detailed in Subsect. 5.1.

5.1 Proposed Guidelines

- **Clarity and Visibility are Keys:**
  Consider enhancing the brightness of LEDs or exploring high-contrast color schemes to improve visibility, especially under different lighting conditions.
- **Real-Time Battery Monitoring:**
  Design dashboards should provide real-time and accurate battery percentage information.
- **User-Centered Information:**
  Simplify dashboard information to essentials like battery percentage and speed. Irrelevant or less useful features, such as voltage indicators, might clutter the display and should be reconsidered.
- **Auditory Feedback for Efficiency:**
  Incorporate auditory feedback for gear changes to promote energy-efficient driving. This could help drivers optimize gear shifts, leading to improved battery utilization.
- **Hill-Specific Information:**
  Consider including features that provide more detailed battery information when navigating uphill terrains.
- **Accurate Low-Battery Warnings:**
  Implement accurate and timely low-battery warnings to prevent unexpected vehicle shutdowns.

Table 3. Needs, Barriers and Facilitators

Category	Main themes	Subthemes
Needs	Accurate Battery Percentage Information	*It's important to have a clear idea of the battery level, especially for longer trips* [5]
		Knowing how much charge is left would help drivers plan to charge better [10]
		Consistent and accurate battery percentage is essential to prevent discrepancies between initial and real-time battery readings during driving
	Timely Low-Battery Alerts	*Ensuring timely low-battery alerts is crucial. Accurate battery readings and prompt warnings prevent unexpected depletions, ensuring smooth operations* [13]
Barriers	Language barriers	*Difficulty in reading jargons present on the dashboard (Low literate users)* [8]
		Information primarily presented in English is not familiar to the user, causing difficulties in understanding
	Inadequate visibility	*Dim lighting, glare, or reflections on the dashboard hinder clear visibility, especially during varying weather conditions.*
	Technological Complexity	*Overly complicated digital interfaces that require extensive training or technical knowledge, posing challenges for drivers who may not be tech-savvy* [4]
	Lack of Standardized dashboard Layout	*Inconsistencies in dashboard layouts and information presentation across different E-rickshaw models, leading to confusion for drivers operating multiple vehicles* [16]
	Inadequate Feedback	*Insufficient or unclear feedback mechanisms, making it challenging for drivers to assess how their driving behaviors affect battery consumption* [14]
Facilitators	Digital Dashboards	*Accurate battery percentage and speed information* [17]
	Analog Dashboards	*Intuitive for monitoring battery levels for its simplicity and ease of understanding (familiarity), allowing quick monitoring of battery levels during trips* [14]

- **Customized Charging Information:**
  Provide accurate estimations of the remaining range based on the current battery level. This information would be invaluable for E-rickshaw drivers planning their routes and charging stops.
- **User Training and Familiarity:**
  Offer user training or guides for new E-rickshaw drivers to help them understand and utilize dashboard features effectively, particularly for digital displays with additional functionalities.
- **Integration of Charging Station Data:**
  Explore integrating information about nearby charging stations into the dashboard. This could alleviate driver anxiety about running out of battery and help them plan their routes more efficiently.

– **Design Flexibility for Different Models:**
Recognize the diversity of E-rickshaw models and their unique user requirements. Design dashboards that can accommodate various vehicle specifications and adapt to different driving conditions.

6   Conclusion

This study sheds light on the drivers' preferences regarding the features of different analog and digital E-rickshaw dashboards. Along with this, the study
reveals the needs, barriers, and facilitators of various E-rickshaw dashboard
types, shedding light on the crucial role of dashboard design. The diverse group of
E-rickshaw drivers featured in this study shared invaluable insights that underscore the significance of user-centered design. The analog and digital dashboard
designs elicited distinct responses, with drivers articulating nuanced perspectives on each. Analog displays were valued for their simplicity, allowing for a
quick glance and providing essential information, facilitating efficient driving.
On the other hand, digital displays offered accurate and detailed insights into
battery status. The implications of these insights extend beyond individual user
preferences. In light of these findings, it becomes evident that dashboard design
is a crucial interaction point between drivers and their E-rickshaws. Tailoring
designs to accommodate diverse user needs and providing clear, relevant, and
easily accessible information are paramount. As the EV landscape continues to
evolve, incorporating user feedback into design iterations is imperative for creating sustainable and user-friendly transportation solutions. In this scenario,
our analysis and guidelines can be helpful to dashboard designers, E-rickshaw
manufacturers, and researchers in this field.

7   Limitation and Future Work

This study, while valuable in its insights, is not without its limitations. First and
foremost, the sample size was limited to a specific number of participants, potentially introducing biases that might not accurately represent the broader population of E-rickshaw drivers. Furthermore, the study was geographically con-
fined, capturing the preferences and experiences of drivers in a specific area. Cultural and social factors inherent to individual preferences could have influenced
the participants' experiences, although these elements were not exhaustively
explored. Lastly, external factors such as weather conditions, traffic, and road
quality, which were beyond the study's control, could have unpredictably influenced participants' experiences. Addressing these limitations in future research
endeavors will be pivotal in ensuring a more comprehensive and nuanced understanding of E-rickshaw drivers' preferences and experiences.

Future advancements in dashboard design might prioritize enhancing battery monitoring precision, streamlining vital information, and accommodating
diverse user needs. By doing so, E-rickshaw manufacturers can contribute to
more efficient and user-friendly electric vehicle operations.

Acknowledgments. We extend heartfelt gratitude to all the E-rickshaw drivers who generously shared their experiences and insights, enabling this study to provide valuable contributions to the field of electric vehicle dashboard design.

References

1. Ba, T., Li, S., Gao, Y., Wang, S.: Design of a human-computer interaction method for intelligent electric vehicles. World Electr. Veh. J. **13**(10), 179 (2022)
2. Bagul, T.R., Kumar, R., Kumar, R.: Real-world emission and impact of three wheeler electric auto-rickshaw in India. Environ. Sci. Pollut. Res. **28**, 68188–68211 (2021)
3. Chandran, N., Brahmachari, S.K.: Technology, knowledge and markets: connecting the dots-electric rickshaw in India as a case study. J. Frugal Innov. **1**(1), 1–10 (2015)
4. Jamson, S.L., Hibberd, D.L., Jamson, A.H.: Drivers' ability to learn eco-driving skills; effects on fuel efficient and safe driving behaviour. Transp. Res. Part C: Emerg. Technol. **58**, 657–668 (2015)
5. Jung, M.F., Sirkin, D., Gür, T.M., Steinert, M.: Displayed uncertainty improves driving experience and behavior: the case of range anxiety in an electric car. In: Proceedings of the 33rd Annual ACM Conference on Human Factors in Computing Systems, pp. 2201–2210 (2015)
6. Jung, T., Kaß, C., Schramm, T., Zapf, D.: So what really is user experience? An experimental study of user needs and emotional responses as underlying constructs. Ergonomics **60**(12), 1601–1620 (2017)
7. Lallemand, C., Gronier, G., Koenig, V.: User experience: a concept without consensus? Exploring practitioners' perspectives through an international survey. Comput. Hum. Behav. **43**, 35–48 (2015)
8. Lundström, A., Hellström, F.: Getting to know electric cars through an app. In: Proceedings of the 7th International Conference on Automotive User Interfaces and Interactive Vehicular Applications, pp. 289–296 (2015)
9. Majumdar, D., Jash, T.: Merits and challenges of e-rickshaw as an alternative form of public road transport system: a case study in the state of west Bengal in India. Energy Procedia **79**, 307–314 (2015)
10. Neumann, I., Krems, J.F.: Battery electric vehicles-implications for the driver interface. Ergonomics **59**(3), 331–343 (2016)
11. Patil, R.M., Chethan, K.P., Ramaprasad, R., Nithin, H.K., Rangayyan, S.: Infotainment system using CAN protocol and system on module with Qt application for formula-style electric vehicles. In: Gupta, D., Khanna, A., Bhattacharyya, S., Hassanien, A.E., Anand, S., Jaiswal, A. (eds.) International Conference on Innovative Computing and Communications. AISC, vol. 1165, pp. 215–226. Springer, Singapore (2021). https://doi.org/10.1007/978-981-15-5113-0_16
12. Pikhart, M.: Human-computer interaction in foreign language learning applications: applied linguistics viewpoint of mobile learning. Procedia Comput. Sci. **184**, 92–98 (2021)
13. Rolim, C., Baptista, P., Duarte, G., Farias, T., Shiftan, Y.: Quantification of the impacts of eco-driving training and real-time feedback on urban buses driver's behaviour. Transp. Res. Procedia **3**, 70–79 (2014)
14. Salmanzadeh, H., Pishnamazzadeh, M., Mirmajlesi, S., Samimi, Y., Landau, K.: Investigating the usability of a user interface for display of an electro vehicle in a driving simulation. J Ergon. Res. 1: 2. **13**, 2 (2018)

15. Singh, S., et al.: Electric vehicles for low-emission urban mobility: current status and policy review for India. Int. J. Sustain. Energ. **41**(9), 1323–1359 (2022)

16. Strömberg, H., Andersson, P., Almgren, S., Ericsson, J., Karlsson, M., Nåbo, A.: Driver interfaces for electric vehicles. In: Proceedings of the 3rd International Conference on Automotive User Interfaces and Interactive Vehicular Applications, pp. 177–184 (2011)

17. Wellings, T., Binnersley, J., Robertson, D., Khan, T.: Human machine interfaces in low carbon vehicles: market trends and user issues. Low Carbon Vehicle Technology Project, Document No. HMI 2, pp. 1–42 (2011)

Interactive Learning Tutor Service Platform Based on Artificial Intelligence in a Virtual Reality Environment

Chang-Ok Yun, Sang-Joong Jung, and Tae-Soo Yun[✉]

Dongseo University, 47 Jurye-ro, Sasang-gu, Busan 47011, Korea
{coyun,sjjung,tsyun}@dongseo.ac.kr

Abstract. In recent years, the COVID-19 pandemic has led to the rise of the "EduTech" market, which integrates technology with education. In this paper, the use of information and communication technologies (ICT) in education has led to the widespread adoption of cutting-edge technologies such as Virtual Reality (VR), Augmented Reality (AR), and Artificial Intelligence (AI). This paper proposes a conversational AI-powered VR training service for elementary English language learning. The service is designed to maximize the effectiveness of learning by providing realistic and immersive English language learning experiences through VR technology and personalized tutoring services through AI. The service will use a learning management system (LMS) to analyze and track individual learning progress. Through voice recognition and AI response processing, the service will provide tailored tutoring services that are tailored to each learner's needs. This service has the potential to address the educational divide that exists in many countries, especially for young learners. By providing realistic and immersive English language learning experiences, the service can help learners from all backgrounds develop the skills they need to succeed.

Keywords: VR · Interactive Learning · UI/UX · AI Tutoring · Gamification

1 Introduction

The COVID-19 pandemic has accelerated the digitalization of education, leading to a growing interest in the EduTech market. EduTech is a technology that incorporates ICT into education to improve existing services or provide new services and new value. It is expected that the global EduTech market will exceed 400 trillion won in five years. In recent years, there have been increasing efforts to use AR/VR technology for educational services [1–3]. These studies have explored the overview, technological advantages, application areas, and future prospects of education services utilizing AR/MR technology. Furthermore, research [4, 5] systematically reviews the learning effects of education services using AR/VR technology, showing positive impacts in terms of learning interest, immersion, comprehension, and learning achievements. In particular, the COVID-19 pandemic has led Korean English education companies to expand their business areas into new forms of distance learning services and develop language services that incorporate various ICT technologies, including AI.

B. J. Choi et al. (Eds.): IHCI 2023, LNCS 14531, pp. 367–373, 2024.
https://doi.org/10.1007/978-3-031-53827-8_32

The current state of VR-based ELL (English language learning) content for elementary school students in Korea is limited. SpeakIT is a VR app that allows students to experience real-world English conversations in a variety of settings, such as a restaurant, a museum, and a park. EnglishTown VR is a VR app that allows students to take virtual field trips to English-speaking countries. However, most of the currently available VR English language learning services are designed for easy device accessibility and are tailored to adult content, such as business travel. As a result, elementary school students, particularly those in lower grades, often struggle to experience a sense of realism and immersion, and there is a shortage of situation-appropriate English learning content for them.

In March 2020, Korea successfully launched an online school opening for all elementary, middle, and high school students (5.4 million), which surprised the world. However, as time has passed, the evaluation of online education in Korea has been somewhat lukewarm. Although education could be continued thanks to the government-led EBS Online Class and e-learning centers, the learning engagement of students was low, and many students were neglected due to lack of individual assistance from parents or teachers, which further exacerbated the education gap. Therefore, this paper aims to develop a training service that can maximize the learning effect by providing elementary school situational English in a conversational UI/UX using artificial intelligence and immersive content VR technology. Through this, it can provide various learning contents that can increase engagement through a tutoring service platform that can access learning more easily and enjoyably.

2 Design of the Service Platform Structure

The goal of this paper is to provide a training service that maximizes the effectiveness of English learning for elementary school students by utilizing artificial intelligence and VR technology to deliver situational English in an interactive UI/UX (See Table 1).

Table 1. The proposed service's structure.

Category	Description
Length	• Approximately 5 to 10 min
System	• Android-based mobile device (tab, mobile phone) • Resolution: 2048 × 1090 (optimal), HMD-type headset
Display	• Supports headsets such as Google Cardboard and Samsung Gear
Player target age	• For lower elementary school students

The overview of the service platform is shown in Fig. 1. In the general user mode, the user progresses to learning English step by step through unit-by-unit learning in a game format that maximizes user access. At this time, a ranking system that incorporates gamification elements rather than simple game-type learning content is applied to provide a more motivating learning environment. The information learned in this way provides artificial intelligence tutoring services for individual learning through analysis of learning data through learning history management through individual diagnosis. To this end, data is accumulated on a cloud server and learning history can be managed in real time through the LMS server.

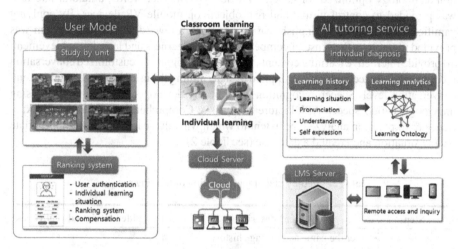

Fig. 1. Service platform overview.

This paper was conducted based on English textbooks linked to an existing renowned domestic English textbook publisher (KEYMAP, a domestic English publisher). Based on the basic structure of the book without departing from the framework of existing English textbooks, this English learning content consists of a total of three sections. It consists of ① Listen & Repeat ② Picture Analysis & Comprehension ③ Personal experience & Expression, and is ultimately designed in a 3+1 format so that the degree of learning can be determined through an overall analysis of the learning situation. Artificial intelligence deep learning technology is used to score and express the accuracy of sentences in three stages. The first stage is Listen & Repeat, which consists of listening to and repeating the key keywords and sentences of each unit. After completing the 3rd stage of learning, each individual's learning history and achievement status was presented in the form of learning situation, pronunciation, comprehension, and self-expression to check learning achievement and encourage learning.

3   Implementation of the Service Platform

3.1   Development of AI Interactive VR English Learning Content

This service platform was developed as a friendly UI/UX system targeting elementary school students. To this end, in order to produce a VR realistic video, the world and terrain were created and expressed effectively using CG production technology in a cinematic expression suited to the situation rather than a live-action video to suit the main target, elementary school students. Using a real-time engine, a variety of scenes were created with a higher degree of freedom in expressing images compared to real-time images, and context information display and effect animation expressions were added to various types of models through an interaction event setting tool. User convenience was greatly considered through the development of Hold On Confirm-based UI optimization technology optimized for general VR devices. In other words, a natural interface was provided to control virtual and real objects on mobile VR devices. By applying gamification principles to English learning content, a reward and ranking system was provided to improve the sense of competition among learners and learning achievement. To provide interactive learning content using voice recognition, customized conversation recommendation technology was applied by determining the user's voice recognition and conversation level. Speech recognition was initially developed (Speech to Text) using the Google Cloud Speech API. Picture Analysis & Comprehension and Personal Experience & Expression analyzed the extent to which what the learner said was appropriate for the situation and created a database (see Table 2).

Table 2. Step-by-step implementation of tutoring service.

Step	Content description	Video in virtual reality
Starting	• Enter the name and manage history • Buttons placed to go directly to sections for each unit	
Control through head gesture	• Each section start point control with arrow • Start control by looking at the Start button	
Voice recognition interface	• Show the passage and instruct them • Check for using the response time indicator bar and voice recognition activation bar	
Correct/Incorrect Instructions	• Check for errors by displaying your answers	

3.2 Implementation of a Diagnostic System Through Data Analysis

We developed a conversation level discrimination algorithm for customized conversation recommendations through analysis of learners' conversation records. This technology uses artificial intelligence voice recognition technology and deep learning to quantify the conversation level of the learner and provide a service that induces a conversation suitable for the learner. In order to provide a learning service according to the situation, it is divided into stages such as daily life, preferences & hobbies, and experience to derive appropriate answers according to the user's situation, and presents questions tailored to the level of elementary school students and situational expressions according to the learner's answers.

In these interactive services, a conversation management method that allows conversations between humans and machines to proceed is a very important element. And specific domains to which the conversation system is applied are constructed based on learning materials. Therefore, this paper utilizes a knowledge-based approach in which conversation strategies are requested from experts or planners build them themselves to suit the specific situation and purpose used in the textbook. Among the learning structures of this artificial intelligence interactive learning content, Picture Analysis & Comprehension and Personal Experience & Expression clearly describe specific situations due to the structure of the textbook. Therefore, by creating a database of corpora (word combinations) that meet the purpose, it is possible to determine to what extent what the learner said is appropriate for the situation, and progressive conversation techniques are used to elicit the learner's response at each stage.

Table 3. Individual learning history and achievement status.

- (1) Learning situation				- (2) Pronunciation		
Division	Book1	Book2	...	Division	Book1	Poor pronunciation
Learning progress	■■■(100)	■■(30)		Word	85%	banana (80%), monkey (82%), ...
Study period	10.10.~.25 (6days 5hours)	10.26 (1 hour)		Sentence	82%	I like bananas (80%), ...

- (3) Comprehension			- (4) Self expression		
Division	Book1	Poor pronunciation	Division	Book1	Poor pronunciation
Situational	67%	1.Accurate representation (8/12) 2.Inaccurate expression (3/12) 3.No response (1/12)	Self-expression	67%	1.Expressions that fit the question (8/12) 2.Expressions that do not fit the context (3/12) 3.No response (1/12)

In this paper, we developed an individual diagnosis system through cumulative data analysis that can determine the learner's pronunciation accuracy and conversation level through quantitative indicators through pronunciation and conversation and present customized content. A LMS was developed so that teachers and parents can check and guide students' learning progress and performance, and can be accessed at any time through PC or mobile device environments. You can check learning achievement and encourage learning by presenting individual learning history and achievement status in the form of learning situation, pronunciation, comprehension, and self-expression. The table shows examples (result pages) of individual learning history and achievement status (See Tables 3, 4).

Table 4. Result pages of the learning history and achievement status.

Learning situation	Pronunciation	Comprehension	Self expression

4 Conclusion

In this paper, we proposed an interactive training service platform that provides situational English learning for elementary school students using artificial intelligence and VR technology. It provides customized tutoring services through AI answers based on voice recognition and an LMS (Learning Management System) that analyzes learner information through individual learning situations and statistical analysis. To this end, by applying gamification principles to English learning content, we created content that induces fun and interest in elementary school students. It is possible to respond to the absolute lack of learning content for elementary school students by providing a reward and ranking system within the content to improve competition among learners and learning achievement. Therefore, the artificial intelligence interactive learning tutor platform in a virtual reality environment will contribute to the dissemination of basic technology that creates new services in areas where non-face-to-face educational services are needed due to COVID-19. In addition, the developed technology can be used as a base technology for producing realistic content not only in education or games, but also in the film and broadcasting fields, and will also contribute to cultivating customized talent for industrial sites. In the future, we expect that it will be possible to apply it in the metaverse environment by applying it to various learning contents by applying developing artificial intelligence technologies in real time.

Acknowledgment. This work was supported by the Technology development Program (RS-2023-00224316) funded by the Ministry of SMEs and Startups (MSS, Korea).

References

1. Yoo, M., Kim, J., Koo, Y., Song, J.H.: A meta-analysis on effects of VR, AR, MR-based learning in Korea. J. Educ. Inf. Med. **24**(3), 459–488 (2018)
2. Jeon, H.B., Chung, H., Kang, B.O., Lee, Y.K.: Survey of recent research in education based on artificial intelligence. Electron. Telecommun. Trends **36**(1), 71–80 (2021)
3. Patel, S., Panchotiya, B., Patel, A., Budharani, A., Ribadiya, S.: A survey: virtual, augmented and mixed reality in education. Int. J. Eng. Res. Technol. (IJERT) **9**(05) (2020)
4. Qiu, X.B., Shan, C., Yao, J., Fu, Q.K.: The effects of virtual reality on EFL learning: a meta-analysis. Educ. Inf. Technol. 1–27 (2023)

5. Luckin, R., Cukurova, M.: Designing educational technologies in the age of AI: a learning sciences-driven approach. Br. J. Educ. Technol. **50**(6), 2824–2838 (2019)
6. Zhai, X., et al.: A review of artificial intelligence (AI) in education from 2010 to 2020. Complexity **2021**, 1–18 (2021)

Voice-Activated Pet Monitoring: An Integrated System Using Wit.ai and Jetbot for Effective Pet Management

Geon-U Kim[✉], Dong-Hee Lee, and Bong-Jun Choi

Department of Software, Dongseo University, 47 Jurye-ro, Sasang-gu, Busan 47011, Korea
rjsdn4403@gmail.com, bongjun.choi@dongseo.ac.kr

Abstract. In recent years, there has been a growing prevalence of single-person households, leading to an increased demand for pets as a source of emotional stability and stress relief. However, the social culture characterized by long working hours has resulted in a significant rise in the number of pets being left unattended at home. To address this issue, pet technology devices have emerged, however, research on more effective methods for pet management remains limited. In this paper, we propose an AI-based pet care robot system that leverages voice recognition technology for real-time pet management. The proposed pet care robot system incorporates natural language processing (NLP) through speech recognition to efficiently perform its functions. By integrating existing embedded-based systems such as pet tech devices into an AI robot environment and evaluating their performance, we demonstrate a mitigation of the pet management problem associated with neglected pets.

Keywords: Natural Language Processing · Embedded · Road Following

1 Introduction

The "Statistics on Single-Person Households" report, released by Statistics Korea in 2022, reveals that the total number of single-person households reached approximately 7.16 million in 2021, marking an increase of around 522,000 from the previous year. These single-person households now represent about 33.4% of all households, and projections indicate that this percentage will rise to an estimated 34.3% within the next four years. These societal changes have sparked an increased interest in pets as a means of attaining emotional stability and stress relief, subsequently driving up the demand for pets. The "2023 Korean Pet Report," published by KB Financial Group, indicates that the number of 'pet households' was approximately 5.52 million at the end of 2022, marking a 2.8% increase compared to the preceding year [1]. However, a significant issue arises as animals in approximately 75.3% of pet households frequently endure extended periods of solitude, resulting in various behavioral problems, including accidental elimination and separation anxiety, among others. To address these issues, the purpose of this research is to propose a system in which users can make real-time detection requests to a pet care robot through voice input. The overarching goal of this research is not only to mitigate the problems associated with pets being left unattended for extended periods but also to offer assistance for pet owners in managing their pets more effectively.

2 Related Research

In the study presented in this paper, the system was configured to employ JETBOT's road-following feature for navigating to areas of pet interest when receiving voice input from the user.

2.1 NVIDIA AI Embedded-System: Jetson Platform

The Jetson Nano Development Kit is an ultra-compact artificial intelligence (AI) computing platform developed by NVIDIA. Among the diverse Jetson platforms, the Jetson Nano platform is notable for being a favored option in the field of home robotics. This is primarily attributed to its capacity to effectively handle data from high-resolution sensors while maintaining the required performance and power efficiency [2].

2.2 Jetbot

Jetbot, an open-source AI robot project built upon the Jetson Nano platform, provides road-following capabilities using a built-in CSI camera. This system seamlessly integrates with deep learning frameworks like Tensorflow and Pytorch, encompassing software components that enable autonomous movement (road following) and object recognition [3].

2.2.1 Road Following Based-CNN

Jetbot's road tracking principles are rooted in the domains of machine vision and deep learning. When it comes to road following, the system predominantly utilizes Convolutional Neural Networks (CNNs), well-known for their proficiency in image processing and their capacity to identify vital features like edges, curves, and colors [4]. The CNN model structure consists of a feature extraction segment with a convolutional layer and a pooling layer [5] (Fig. 1).

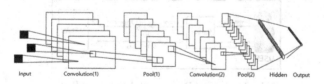

Input   Convolution(1)   Pool(1)   Convolution(2)   Pool(2)   Hidden   Output

Fig. 1. CNN (Convolutional Neural Networks) Architecture

As a result, Jetbot emerges as a suitable solution for pet care robots targeting the problems highlighted in this study. Expanding on this foundation provides the opportunity to develop a system that can transmit image information about your pet's areas of interest as it navigates the road and reaches its destination, providing effective management support.

2.3 Wit.ai

Wit.ai offers interfaces that understand user text input and generate appropriate responses. Furthermore, Wit.AI's Speech API enables the conversion of speech data into text [6]. In this study, the goal is to utilize these features to transform the user's voice into text, thus eliciting suitable responses. This process will establish the foundation for controlling the pet care robot through these responses.

3 Pet Care System Using Pet Care Robot

3.1 System Architecture

The AI pet care system proposed in this paper primarily consists of two components: the robot part and the application part.

3.1.1 Application Component

The application component of the system utilizes a Flask server, chosen for its user-friendly nature and seamless integration with AI and machine learning libraries. The process that captures a user's voice, converts it to text, and applies natural language processing (NLP) involves multiple steps, including real-time voice processing, signal processing, and voice recognition [7]. This study implemented this process using pyAudio. PyAudio records the user's voice in real-time and provides the option to save it as an audio file (.wav). The saved audio file is then transmitted to Wit.ai's Speech API for conversion into text. The converted text is subsequently passed to the natural language processing module for analysis. This module extracts 'intents' and 'entities' from the provided sentence. The information extracted in this manner is parsed from the API response, leading to the invocation of an appropriate function based on the parsed result. As a result, the system developed in this study can trigger specific functions within the server in response to the recognized purpose and object by accepting user voice input, converting it into text, and conducting natural language processing (Fig. 2).

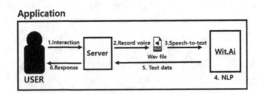

Fig. 2. Concept Diagram of Application Part

3.1.2 Robot Component

In this study, a Flask server is constructed for autonomous driving and image data transmission. The Flask server carries out self-guided navigation around predefined areas of interest in response to user requests. Image data generated during this process is saved in a specific file format within Jetson Nano. As a result, the system allows autonomous driving to user-specified areas of interest for pets and expeditiously stores image data at the destination, simplifying preparations for subsequent tasks [8, 9] (Fig. 3).

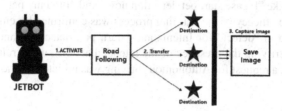

Fig. 3. Concept Diagram of Robot part

3.1.3 Websocket Protocol

In this study, the Websocket protocol, which facilitates two-way communication, was employed to exchange image data between two devices. This protocol enables real-time transmission of data, allowing user requests to be sent immediately to the Flask server of the robotic component. Based on the requests received in this manner, the pet care robot conducts autonomous driving, and the image data produced as a result of this process is sent to the Flask server via the Websocket protocol. As a result, the system developed in this study was designed to facilitate real-time data exchange between user requests and the robot's responses.

3.2 Model Training

The proposed system in this study consists of two models—a WIT.AI model for natural language processing (NLP) and a dedicated model for implementing the road following functionality.

3.2.1 WIT.AI Training and Results

Table 1. WIT.AI Response Sentence

Sentence	Intent	Entity
"Please start pet detection now"	detect	"start", "pet", "detection"
"Find my pet"	detect	"find", "pet"

In this study, learning for Natural Language Processing (NLP) was carried out using Wit.AI. In this context, the Intents were configured as 'detect,' and the Entities were composed of 'start,' 'detecting,' and 'find.' These entities primarily correspond to terms that indicate actions, such as 'start,' 'detect,' and 'find.' Sentences like "Start detection" were used as training data (Utterance), and a training dataset was constructed, taking into account various similar cases [10]. Additionally, the input sentence was sent to Wit.ai through the HTTP API, and the response was examined using the curl command. Input sentences like "Please start pet detection now" and "Find my pet" were employed in this study. The efficacy of the learning process was confirmed by ensuring that each sentence resulted in "detect" for Intents and [start, pet, detection] and [find, pet] for Entities. After verifying the presence of the intended actions, when relevant intentions were recognized, a request for Autonomous driving was initiated on the Pet Care Robot (Table 1).

3.2.2   Robot Training and Results

The process of data collection and model training takes place within Jupyter Notebook. Objects in images captured by a CSI camera are annotated to enable the robot to observe and track them. This process allows for the verification of recognized objects and predefined paths (Figs. 4 and 5).

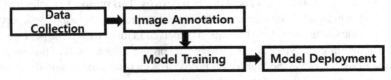

Fig. 4. Process of Road Following

Fig. 5. Image labeling and Object Detection (Color figure online)

In this procedure, the green circle represents the designated path, and the blue circle signifies what the robot detected for tracking. This process allowed the validation of data collection and real-time object detection results. The collected data was employed for model training using the following parameters: data count: 1000, epochs: 100. Subsequent to the training, a model file named 'road_following_model' was generated, and the loss function values were measured at 0.088 [11].

4 Conclusion

This study implemented a voice recognition-based pet care assistance system using Jetbot to address the issue of managing pets left unattended for extended periods in homes. The system allows users to issue voice commands and observe pets in the area of interest in real time. Jetbot's road following function, measured with a loss value of 0.088, was expected to perform well. However, during testing, it frequently deviated from the designated path. A lack of diversity in the training image data likely caused this phenomenon. If the system learns from various scenarios where it deviates left or right from the designated path, we expect it to demonstrate high performance in road navigation. Additionally, when running processes on Jetson Nano, there was an issue where the RAM allocation reached its maximum capacity of 4 GB. As a result, frame drops occurred in the CSI camera, causing some objects to not be detected during observation and moving away during autonomous driving. However, as confirmed in this study, by adjusting image resolution below 416 on Jetson Nano and setting batch size under 100, reduced RAM allocation was observed. This adjustment is expected to minimize the possibility of objects being ignored during autonomous driving.

Acknowledgement. Following are results of a study on the "University innovation" project, supported by the Ministry of Education and National Research Foundation of Korea.

References

1. Yoon, Y.K.: A study on mobile app interface design for companion animal care. J. Prod. Cult. Des. Stud. **65**, 293–302 (2021)
2. Mittal, S.: A survey on optimized implementation of deep learning models on the NVIDIA Jetson platform. J. Syst. Archit. **97**, 428–442 (2019)
3. Kawakura, S., Shibasaki, R.: Deep learning-based self-driving car: Jetbot with NVIDIA AI board to deliver items at agricultural workplace with object-finding and avoidance functions. Eur. J. Agric. Food Sci. **2**(3), 1–9 (2020)
4. Alzubaidi, L., et al.: Review of deep learning: concepts, CNN architectures, challenges, applications, future directions. J. Big Data **8**, 1–74 (2021)
5. Teja, K., et al.: Review on Convolutional Neural Networks (CNN) in vegetation remote sensing. ISPRS J. Photogramm. Remote Sens. **173**, 24–49 (2021)
6. Qaffas, A.A.: Improvement of Chatbots semantics using wit. ai and word sequence kernel: education Chatbot as a case study. Int. J. Mod. Educ. Comput. Sci. **11**(3), 16 (2019)
7. Chowdhary, K.R.: Natural language processing. In: Chowdhary, K.R. (eds.) Fundamentals of Artificial Intelligence, pp. 603–649. Springer, New Delhi (2020). https://doi.org/10.1007/978-81-322-3972-7_19
8. Sung, I., et al.: On the training of a neural network for online path planning with offline path planning algorithms. Int. J. Inf. Manag. **57**, 102–142 (2021). ISSN 0268-4012
9. Mitrevski, M.: Getting started with wit. ai. In: Mitrevski, M. (eds.) Developing Conversational Interfaces for iOS: Add Responsive Voice Control to Your Apps, pp. 143–164. Apress, Berkeley (2018). https://doi.org/10.1007/978-1-4842-3396-2_5
10. Lei, X., Pan, H., Huang, X.: A dilated CNN model for image classification. IEEE Access **7**, 124087–124095 (2019)

Fraudulent Practice Detection in Bullion Trade in Selling of Gold Jewellery Through AI Methods

Nagamani Molakatala[1,2,3,4,5,6](✉), Vimal Babu Undru[1,2,3,4,5,6],
Shalem Raju Tambala[1,2,3,4,5,6], M. Tejaswini[1], M. Teja Kiran[1,2,3,4,5,6],
M. Tejo Seshadri[1,2,3,4,5,6], and Venkateswara Sagar Juturi[1,2,3,4,5,6]

[1] University of Hyderabad, Hyderabad, India
nagamanics@uohyd.ac.in
[2] Legal Metrology Telangana and A.P., Hyderabad, India
[3] Osmania University, Hyderabad, India
[4] Gitam University, Hyderabad, India
[5] Asian Institute of Technology, Khlong Nueng, Thailand
[6] American International Group Inc., New York, USA

Abstract. The Industrial 5.0 movement aims to bring new trends in the detection of fraudulent practices in the bullion trade along with others, streamline the identification of human errors in making transactions, and improve transparency to safeguard the economy, human health, and wealth. This paper focuses on deceptive practices adopted in transactions of bullion (gold) ornaments. The implementation of AI-based techniques and cloud-based data storage to automate and standardize the integration of data from all sources within the industry, eliminating calculation discrepancies. Data collected from these transactions is stored in the cloud to assess customer losses using AI methods. Daily transactions are logged and stored in a metadata file, and computational techniques, including Machine Learning (ML) and Deep Learning (DL), are used to calculate costs and potential gains for business owners precisely. This integration of legacy methods with proposed AI and ML/DL automation procedures forms the crux of our work, aiming to bridge the gap between traditional practices and advanced technologies in the industry.

Keywords: Computational Techniques · value added tax · Weighing-malpractices in Gold transactions · F1 Weights · Billing Issues · Gross weight-Wastage-Making charges · AI and ML/DL

1 Introduction

In India, women regard gold as a form of investment and a financial security asset. Gold can be easily acquired or sold by traders at any time, distinguishing it from other commodities. It also serves as a viable option for obtaining loans through mortgage arrangements. The annual demand for gold in India is estimated to fall within the range of 600 to 800 tonnes. Furthermore, gold exhibits

B. J. Choi et al. (Eds.): HCI 2023, LNCS 14531, pp. 380–393, 2024.
https://doi.org/10.1007/978-3-031-53827-8_34

remarkable resistance to corrosion under various atmospheric conditions. Taking advantage of demand the traders are levying exorbitant charges on customers without any rationality. This paper presents the prevailing situation in the context.

Technical Characteristics of Gold and Gold Alloys: The ICRA Rating Services has developed a rating methodology for the Gold Jewelry Retail Industry. This methodology serves as a valuable reference tool for assessing the creditworthiness of companies operating within the jewelry retail sector.

Utilizing their advantageous position, individuals have the potential to create counterfeit ornaments using metals such as Tungsten, Copper, Silver, and Platinum to dilute the purity of Gold. Consequently, detecting the authenticity or assessing the purity of Gold becomes a challenging endeavor. Precise evaluation of Gold purity necessitates the use of destructive methods exclusively. Recognizing this need, the Government of India has introduced legislation for the facilitation of Gold purity determination through the HALLMARKING process, covering 14, 18, and 22 carats of gold jewelry and artifacts. Subsequently, a Hallmarking scheme for Silver articles was introduced in 2005, incorporating real-time online monitoring methods like weighment, XRF, Sampling, Fireassay, and laser marking. In the case of Silver, this scheme accounts for various fineness levels, including 990, 970, 925, 900, 835, and 800, with distinctive Assay center identification marks and numbers as well as Jeweler identification marks and numbers (as per IS 2112: 2014). Purity/Fineness Levels:

These standards align with Assaying and Hallmarking regulations (as per IS 1417:1999 and IS 1418:1999) [[]]

KDM gold refers to a specific gold alloy comprised of 92 percent gold and 8 percent cadmium, which gives rise to the term "KDM gold". This particular blend was utilized to achieve a superior level of gold purity. However, its use led to significant health concerns for artisans working with it, ultimately resulting in its prohibition by the Bureau of Indian Standards (BIS). Nowadays, cadmium has been replaced with more advanced solder metals, such as zinc and other alternatives.

1.1   Gold Alloys and Their Compositions

The following Table 1 consists the different compositions of alloys with their percentages show to categories the gold types.

The total cost of your jewelry is calculated as follows: Gold Price multiplied by Weight in grams, plus Making Charges, along with a 3% GST applied to the sum of the Price of Jewelry and Making Charges (Fig. 1).

Table 1. Purity or Fineness Levels of the Gold as per IS2112:2014 C in table indicate Carats [15]

Purity	C	Purity	C	Purity	C	Purity	C	Purity	C	Purity	C	Purity	C
0958	23	0916	22	0875	21	0750	18	0595	14	0375	14	0375	9

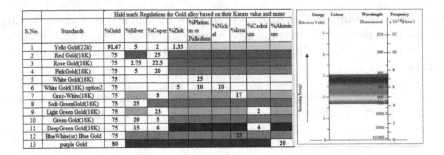

Fig. 1. Gold Alloys and Their Composition percentages []

The primary drivers that influence working women to buy gold jewelry include factors such as purity, quality, diversity of choices, recommendations from friends and family, pricing, brand recognition, and advertising. Research findings suggest that working women are discerning purchasers of gold jewelry, often acquiring it for investment purposes, special events like weddings and festivals, and to maintain a solid social presence during gatherings. They typically prefer shopping for gold jewelry at reputable branded stores. The importance of the research work on Gold and purchases and estimation of the burden charges on the customer to prevent as the Gold metal becomes day by day (Fig. 2).

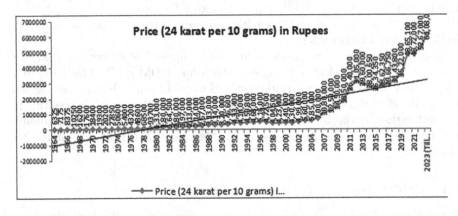

Fig. 2. Gold Price increase from 1964 to date []

2   Literature Survey

Bagging, also known as bootstrap aggregation, serves as a widely employed ensemble learning technique aimed at mitigating variance in noisy datasets. Its primary purpose is to prevent data over fitting and it is effectively applied in both regression and classification tasks, particularly when using decision tree algorithms. Ensemble learning aligns with the concept of the "wisdom of crowds", implying that decision-making within a larger group of individuals generally outperforms that of an individual expert [1] A non-destructive method employed is X-Ray Fluorescence (XRF) spectrometry. The reference materials encompass the entire spectrum of conventional colored and white carat gold jewelry alloys. Notably, the foils utilized in this process are of substantial thickness, exceeding the 'critical thickness', typically falling within the range of 100 to 200 microns.

This represents data-level techniques. Presently, the prevailing preprocessing approaches involve the Synthetic Minority Oversampling Technique (SMOTE) and its extended variations. SMOTE addresses class imbalance by augmenting the minority classes, achieved by generating data points strategically placed between a minority instance and its nearest neighbor [2].

In more recent studies, Generative Adversarial Networks (GANs) have been employed to synthesize data. The authors are examining the potential benefits of incorporating clustering into this process to enhance the quality of synthetic data. This exploration is conducted within the framework of widely utilized supervised classifiers and ensemble methods.

In a recent development, Swedbank, one of Sweden's largest financial institutions, employed Generative Adversarial Neural Networks (GANs) trained on NVIDIA GPUs as a crucial component of its strategy to combat fraud and money laundering. Swedbank has successfully devised innovative solutions to address these challenges by employing a combination of deep learning techniques on GPUs, resulting in cutting-edge methods for detecting suspicious activities.

In the process, Hopsworks leveraged its software platform and harnessed the power of NVIDIA V100 GPUs to work with a substantial dataset of up to 40 terabytes (TBs) in size. This enabled efficient feature engineering at scale and the simultaneous parallel training of GANs across multiple GPUs [3].

One of the domains significantly advanced by Deep Learning progress is Natural Language Processing. Deep Learning models, particularly those founded on attention mechanisms, are instrumental in processing sequences of symbols, generating another sequence while maintaining attention on the input throughout the generation process. Syndata has previously employed a recurrent model for this task; however, they are confident that a Transformer-based model could deliver superior performance. The objective is to compare the newly implemented model with the earlier recurrent model, with an expectation that the former will surpass the latter in performance.

Both models, the GPT architecture and LSTM, are autoregressive. On the other hand, the Bidirectional LSTM, structured with an Encoder-Decoder approach, and the Transformer are sequence-to-sequence models. These latter models with Encoder-Decoder architectures, such as the Transformer and Bidirectional LSTM, have demonstrated superior performance in synthetic data generation scenarios where the output, or the target field for prediction, depends on the other accompanying fields.

To forecast Gold Price returns, various Machine Learning Techniques have been employed. The project can be found on GitHub under the repository "Riazone/Gold-Return-Prediction". The approach encompasses supervised learning techniques, encompassing both regression and classification methods, followed by the utilization of Time Series techniques. Furthermore, there's an exploration into the integration of these methods to assess whether their predictive capabilities are enhanced through integration [4].

Gold exhibits exceptional malleability and ductility, with the ability to stretch one ounce into an astonishing 80 km of fine gold wire (diameter of 5 microns). It possesses Young's modulus of 79 GPa, a value remarkably close to that of silver. Gold's atomic properties include an atomic number of 79, an atomic mass of approximately 196.967, and an atomic radius measuring 0.1442 nm. The distinctive yellow color of gold is a result of the arrangement of its outer electrons around the gold nucleus.

The authors demonstrate how machine learning regression techniques can be employed to forecast the future direction of gold prices [5].

Machine learning for gold price prediction is a significant research domain that leverages historical data and algorithms to forecast future gold prices. This abstract underscores the pivotal role of machine learning in predicting gold prices and offers a concise overview of this field. Utilizing data science and machine learning methodologies, it anticipates fluctuations in gold prices [6]. Faulty rolling element bearings typically manifest vibration signals with non-linear and non-stationary attributes due to the intricate operational conditions. In this context, Convolutional Neural Network-based Hidden Markov Models (CNN-HMMs) are introduced as a methodology for classifying multiple faults in mechanical systems. HMMs serve as a robust tool for ensuring stability in fault classification. Both benchmark data and experimental data are utilized in the application of CNN-HMMs [7].

2.1 Methods Review

In the realm of Optical Character Recognition (OCR) and text detection, emerging trends include digital camera and mobile document image acquisition. To enhance OCR accuracy, novel approaches such as nonparametric and unsupervised methods are employed to address undesirable document image distortions effectively. Additionally, a highly efficient stack of document image enhancement techniques is utilized to rectify deformations across the entire document image [8]. Smartphones have evolved into a valuable agricultural tool due to their inherent mobility, affordability, and robust computing capabilities, enabling the development of a wide range of practical applications in farming [9].

In the realm of transportation networks, predicting travel times holds immense significance. Popular web mapping services such as Google Maps continually handle a substantial volume of travel time inquiries from both individual users and businesses. This paper introduces a Graph Neural Network (GNN) estimator for Estimated Time of Arrival (ETA), which has been implemented into the production system of Google Maps.

Upon deployment, the GNN has demonstrated its efficacy by notably decreasing instances of incorrect ETA predictions in multiple regions compared to the prior production baseline. In cities like Sydney, this reduction amounts to over

40 With the advancements in deep learning, particularly the utilization of deep generative models for image synthesis, there has been remarkable progress in the field of image synthesis. Among these generative models, Generative Adversarial Networks (GANs) stand out as one of the most impactful. GANs have found successful applications in various domains, including computer vision and natural language processing [11].

This work, provides a comprehensive and structured overview of the development and evolution of GANs-based text-to-image synthesis, offering a detailed and coherent account of its progression. For investors, putting money into gold is regarded as a secure investment due to its potential for high profitability and liquidity. Data analysis is conducted using the Chi-square test, revealing that there is no substantial association between factors such as age, marital status, education, occupation, and respondents' belief that gold investment is a superior choice likely to yield favorable returns [12].

Wolfram shares similarities with gold, as their densities are closely matched at 19.30 g/ml for gold and 19.25 g/ml for tungsten, making tungsten a potential chemical counterfeit for gold. Gold purity assessment through X-ray is an expensive method. However, a cost-effective approach is adopted by employing both sound and image processing techniques to distinguish between counterfeit and genuine gold.

3   Analysis of the Gold Purchage

In the image processing phase, a Convolutional Neural Network (CNN)-based toolbox is initially utilized to segment the gold material. Subsequently, deep CNNs are employed to differentiate between the colors of gold and copper materials. These combined sound and image processing techniques have yielded promising results in discerning genuine from counterfeit gold [13].

3.1   Multiple Machine Learning Techniques

– Part I: Establishing the Approach, Data Collection, and Preparation
– Part II: Regression Modeling with PyCaret
– Part III: Classification Modeling with PyCaret
– Part IV: Time Series Modeling with Prophet (Facebook)
– Part V: Assessing the Integration of These Approaches

Legal Metrology is the prime statutory authority that conducts inspections on Gold at frequent intervals of time with the Bureau of Indian Standards (BIS). The BIS will inspect for the Purity and Hall Marking of Jewellery.
The Department of Legal Metrology has no functional jurisdiction over the issues of

a) Value addition, b) Making Charges, c) Purity of Gold (as cartage machines are not included in Legal Metrology (General Rules), 2011 [], d) Gold Rate e)Cost of Precious Stone or Precious Metal.

Wight of the Gold issues in the context of Legal Metrology are classified as

i) Weighing Machine issues, ii) Short Weight issues ii) Billing Issues

3.2   Weighing Machine Issues:

The following are the weighing machine Issues in Gold cost deviation

1. Verify, whether Test weights are present or not with Non-Automatic Weighing Instrument (NAWI), if test weights are there, verify whether these test weights are stamped or not. It is Well and Good if F1 Weights are used as test weights.
2. Check NAWI to be positioned on a flat base or platform. Ensuring a bubble or watermark in an exact central location of the NAWI is a prerequisite. Some traders use this technique of positioning in the wrong manner to gain unlawfully.
3. Ensure that the glass windows of the weighing machine are not damaged, as this can also be exploited for unlawful gain.
4. The Glass windows of the weighing machine shall not be in broken condition. This technique can also be used to gain unlawfully. Calculate the error for that NAWI and ensure that the error shall be within tolerance.
5. Class-I NAWI with internal calibration will give better results

3.3   Precious Metals Usage

These metals and their alloys are finding growing utility in a variety of medical and biomedical applications, owing to their remarkable characteristics, which include bio-compatibility and resistance to corrosion. They are being employed in fields such as dentistry, therapy, tissue engineering, and bio-imaging, showcasing their versatility in healthcare contexts

3.4   Short Weighment Issues:

1. Generally, we can observe short weighment in milligrams compared to bill/invoice, mainly because of negligent weighed motto to gain unlawfully and wrong positioning of weighing machine. Short weighment cases can be established by adopting the procedure LMO's apply at Fair Price shops and Mandal Level Stock (MLS) Points i.e., waiting for the completion of transaction and re-weighing the items mentioned in invoice and establishing the facts.
2. Ornamented jewelry, particularly those adorned with stones or stud should be considered in terms of their gross weight. The gold rate should be calculated based on the weight of the ornaments minus the studs or stones. Studded ornaments or ornaments with stones are treated as Gross weight. Gold rate has to be calculated on ornaments minus studs or stones. But, some of the traders issue invoice/bill with same Gross weight and Net weight. In studded ornament Gross weight and Net weight never be equal. So some of the traders are collecting Gold rate for weight of stones and also collecting stone amount.

4   Preliminary Experiments and Results

Fig. 3. Gold ornament types for price calculation.

Billing Issues:

– Instead of using Bullion terminology like Gross weight, Net Weight they confuse with Jewel Weight etc., Gold purity has to be mentioned in cartage eg. 22 kt, 24 kt, 18 kt. They confuse with the word BIS (Fig. 3).

4.1   Empirical Data Set1 and Data Set2

The gold price calculation for studs and tops.

4.2   Price Deviations of Ornaments of Gold

Ornaments and their Gold price actual to customer paid charges detail

– Not mentioning of cartage or wrong quoting of Karatage in the invoice or cash bill in order to deceive the customer. 18 ct Gold can be shown and issue invoice as 22 carat. Eg. Generally Diamond, Platinum ornaments are made with 18 ct Gold only. But wrongly they mention as 22 carat for Diamond and Platinum ornaments and collects the difference of amount.
– Sometimes even calculation errors plays great role in deceiving customer. Given below is the example for wrong calculation observed in one of the corporate jewellery shop (Figs. 4 and 5).

Old Gold issues: Majority of the customers are cheated in numerous ways in selling of old Gold. Old Gold = (Gold + Impurities). Depreciation of Gold shall be up to the extent of weight of impurities only Table 1 measures. Table 1 gives the details of the sample1 output of the NAWI data in the below table.

S. NO	DESCRIPTION	PCS	Gross Weight	Net Weight	STONE VALUE	VALUE ADDED	AMOUNT
1	DIAMOND PENDANTS DIAMOND : 0.11 C 72000/- 7290.00 DIAMOND : 0.37 C 60000/- 2200.00 CERTIFICATE : 1 P 1500/- 1500.00	1	3.140	3.140	31620.00	4516.00	44928.00
2	DIAMOND TOPS DIAMOND : 0.70 C 60000/- 2000.00 CERTIFICATE : 1 P 1500/- 1500.00	2	6.710	6.710	43500.00	7835.65	69470.00

Fig. 4. Data Analysis Dimond set.

S.no	Description	Pcs	Jewel weight	Purity	Stone /Mc	Amount
1	GOLD STUD(VA:0.926)	2	6.170	BIS	0	19684.00
2	GOLD STUD (VA:0.625)	2	3.290	BIS	1158	12018.00

Fig. 5. Data Analysis with Gold Studs.

5 Results and Analysis

Possible Modifications in Rules/Act:

1. Specifications for Gold Purity Testing Machine may be incorporated in Legal Metrology (General Rules), 2011 to attain sanctity.
2. Accuracy wise e-value - 1 mg, Class-I NAWI with auto calibration is vital for bullion trade.
3. Guidelines to be issued in buying /selling of old Gold.
4. F1 Weight set as test weights shall be used in jewellery shops for testing. Frequent and timely inspections and proper calibration of NAWI in bullion trade makes Department of Legal Metrology more visible in the vicinity of Traders and Public (Fig. 6).

Gold Particulars	Gold Price (₹)	Gold Weights of Jewel (g)	Wastage/Value Added (₹)	Labour charges for making jewel articl e (₹)	Stone Weight (g)	Other Tax levied/ VAT /other (₹)	Total Gold cost leveid to the custo mer (₹)	Actual Cost of Gold purchased by the customer (₹)	Excess amount collected from the customer (₹)	Excess weight of Gold levied on customer for making ornament (g)
Gold Stud1	2774	6.17	2372	0	0	197	19684	17115.58	2568.42	0.925890411
Gold Stud2	2774	3.29	1853	1158	0	120	12018	9126.46	2891.54	1.042372026
Top	2795	15.1	2173	8052	0.61	552	55203	42204.5	12998.5	4.650626118
Necklace	2795	61.41	8610	2672	3.97	2239	222970	171640.95	51329.05	18.3645975
Necklace Mix1	2695	17.73	10018	2800	0	606	60600	47782.35	12817.65	4.756085343
Back Chain1	2695	3.78	2386	0	0	126	12574	10187.1	2386.9	0.88567718
Necklace Mix2	2819	16.56	8795	6800	0	623	62278	46682.64	15595.36	5.532231288
Back Chain2	2819	4.05	2435	0	0	139	13850	11416.95	2433.05	0.863089748
Back Chain3	2695	3.78	18.98	0	0	126	12574	10187.1	2386.9	0.88567718
ST Kondoli	2833	95.648	74775	0	0	3112	311250	270970.784	40279.216	14.21786657
Bangles	2596	50.85	17065	19360	0	1574	157448	132006.6	25441.4	9.800231125
Necklace Srt	2819	48.01	33995	0	0	1400	140000	135340.19	4659.81	1.653001064

Fig. 6. Gold Price with Estimated cost and wastage and actual cost paid by cutomer

5.1 Old Gold Issues

Majority of the customers are cheated in numerous ways in selling of old Gold. Old Gold = (Gold + Impurities). Depreciation of Gold shall be up to the extent of weight of impurities only (Figs. 7 and 8).

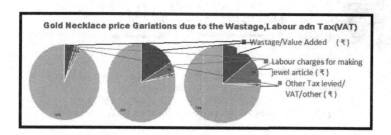

Fig. 7. Gold price for various type Necklace with Mix.

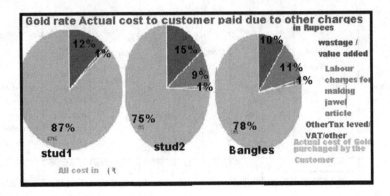

Fig. 8. Gold Stud1, Stud2, and Bangle Cost of Actual to Customer paid charges in Rupees

Frequent and timely inspections and proper calibration of NAWI in bullion trade makes the Department of Legal Metrology more visible in the vicinity of Traders and the Public.

6 Summarizing and Discussion

6.1 Outcomes of the Research

Observations from the filed work of Gold merchants and customer choice of ornament and purchases and based on the three possible ways of merchant advantageous and standards by the legal metrology and BIS regulation the following are the summarized observations (Fig. 9)

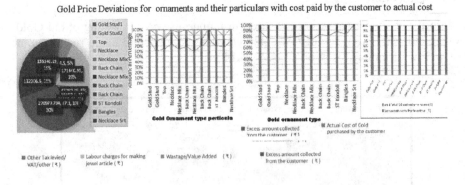

Fig. 9. Data Sets -Gold price Deviations to Actual cost

1. There is no control of statutory authorities over the value-added services put forth by the Jewelers. The statute prescribes no standards. The jewelers are exploiting the consumers by adding their own Value-added services. The value-added services are ranging from 4% to 30%. The jewelers are claiming that the intricacy of the article-making is the root cause of levying the value-added service to the customer. The result is the customer is paying more than the expected cost of the jewel article.

2. There is no labour charge prescribed to make the particular jewelry article for the artisan. At present the traders are charging the customer of their choice to pay the artisan. The artisan charges may vary from trader to trader. It is noted that the traders are paying the artisan to make the ornament on daily wages, but not on the intricacy of the ornament, and the artisan charges are levied from the customer on the piece rate/intricacy of the ornament. So both the artisan and the customer are at a loss.

3. The best practice adopted by the trader to attract the customer is giving a discount on the ornament in question for sale. The trader is decreasing the value-added service charge (for about 2%-4%) from the already billed amount.

4. On practical and survey it is noticed that there will not be more than 5% (in practice the traders are levying the value added charge up to 30%) of loss of Gold metal as wastage while making the ornament. The wastage may vary from article to article. Eg: the wastage to make a coin may be 0.25% (source: Customs duty manual). But the traders are charging the customer ranging from 4% to 6%.

5. With the adoption of mandatory hallmarking to the ornaments the wrong declaration of "Karatage" (purity) is controlled. The customers are recommended to purchase the ornaments which are Hall Marked to get the exact price for the ornament.

6. The traders are collecting the old ornaments that are in use from the customers on the pretext of OLD GOLD and intent to make new ornaments. Here it is informed that there is no such 'old gold' or 'new gold'. If the proposal of the trader is conceded by the customer it shall benefit the trader and the customer is at a loss. Immediately the trader adopts a destructive method (melting the existing ornament) to ascertain the Karatage (purity) and as usual adopts regular practice for gain in the name of Value added charges ie, levies, making charges, value-added charges (Wastage charges), Taxes, adding additional metal to the ornament, etc. in addition from the customer.

7. The adoption of the AI system to the billing with Machine learning for gold price prediction and algorithms to forecast the cheating of customers by the trading community. The pivotal role of machine learning in prediction and concise overview.

6.2 Future Scope

The Government shall take measures to curb the prevailing practices in selling gold ornaments by levying the incorrect Value-added services, making charges, and Wastage charges in the interest of the customer and the public at large. Deep Learning models, particularly those in billing, levying the charges to customers, practices adopted, in processing, generating another sequence while maintaining attention on the input throughout the generation process.

Acknowledgement. The authors acknowledge the Legal Metrology departments of Telangana and Andra Pradesh state for the observation and also the Bureau of the Indian Standards department for valuable information and motivation for the research directions.

References

1. Krishnan, A., Nandhini, M.: A study on the factors which leading customers to purchase gold jewellery with special reference to working women. Int. J. Mech. Eng. Technol. (IJMET) **8**(12), 1020–1029 (2017)
2. Stankiewicz, W., Bolibrzuch, B., Marczak, M.: Gold and gold alloy reference materials for XRF analysis. Gold Bull. **31**, 119–125 (1998)
3. Mamaghani, M., Ghorbani, N., Dowling, J.: Detecting financial fraud using GANs at swedbank with Hopsworks and NVIDIA GPUs (2021)
4. Ekstedt, E., Torres, D., Gustafsson, J.: Synthetic data generation using transformer networks. Comput. Sci. (2021). School of Electrical Engineering and Computer Science Host company: Text generering med transformatornätverk Swedish subtitle: Skapa text från ett syntetiskt dataset i tabellform @ 2021 Pedro Campos
5. Gold price prediction: step by step guide using python machine learning. Quantitative Finance & Algo Trading Blog by QuantInsti (2023)
6. Bhajantri, R.M., Puttappanavar, R.F., Shashank, Vishal, Rao, B.: Gold price prediction using machine learning. Int. J. Adv. Res. Innov. Ideas Educ. **9**(3), 1664–1668 (2023)
7. Wang, S., Xiang, J., Zhong, Y., Zhou Y.: Convolutional neural network-based hidden Markov models for rolling element bearing fault identification (2018). College of Mechanical and Electrical Engineering, Wenzhou University, Wenzhou, PR China
8. Harraj, A., Raissoun, N.: OCR accuracy improvement on document images through a novel pre-processing approaches. Journal (SIPIJ) **6**(4) (2015). https://doi.org/10.5121/SIPIJ.2016.6401
9. Pongnumkul, S., Chaovalit, P., Surasvadi, N.: A systematic review of research (2015). National Electronics and Computer Technology Center (NECTEC), 112 Thailand Science Park, Phahonyothin Road, Khlong Nueng, Khlong Luang, Pathum Thani 12120, Thailand
10. She, J., et al.: ETA prediction with graph neural networks in Google maps. ArXiv (2021). https://doi.org/10.1145/3459637.3481916
11. Zhou, R., Jiang, C., Xu, Q.: A survey on generative adversarial network-based text-to-image synthesis. Neurocomputing **451** (2021). https://doi.org/10.1016/j.neucom.2021.04.069
12. Rupa, R., Sailini, P.: A study on consumer's preferences towards Goad as an investment with reference to Coimbatore city Peripex. Indian J. Res. **V**(ix) (2016)
13. Can, Y.S.: Classification of original and counterfeit gold matters by applying deep neural networks and support vector machines.]ursa Uludağ Univ. J. Fac. Eng. **27**(1) (2022)

14. Indulia, B.: Protection of jewellery under law and standardisations—SCC Blog. SCC Blog (2022. https://www.scconline.com/blog/post/2022/07/22/protection-of-jewellery-under-law-and-standardisations/. In-Text Citation: (Indulia, 2022)

15. Das, S., Biswas, A.: Deployment of information diffusion for community detection in online social networks: a comprehensive review. IEEE Trans. Comput. Soc. Syst. 8(5), 1083–1107 (2021). https://doi.org/10.1109/TCSS.2021.3076930

Machine Learning Strategies for Analyzing Road Traffic Accident

Sumit Gupta[1][✉] and Awadhesh Kumar[2]

[1] Department of Computer Science, Institute of Science, BHU, Varanasi, India
Sumitkugp123@gmail.com
[2] Department of Computer Science, MMV, BHU, Varanasi, India
akmcsmmv@bhu.ac.in

Abstract. Road safety and accidents have been an important concern for the entire world and everyone is putting effort into resolving the long-standing problem of road safety and accidents. In every country on earth, there is traffic and reckless driving. This has a negative impact on a lot of pedestrians. They become victims, although having done nothing wrong. The number of traffic accidents is rising quickly due to the enormous increase in road cars. Accidents like these result in harm, impairment, and occasionally even fatalities. Numerous things like weather changes, sharp curves, and human error all contribute to the high number of traffic accidents. In this research paper various machine learning techniques such as, K Nearest Neighbors, Random Forest, Logistic Regression, Decision Tree, and XGBoost etc., are used to investigate why road traffic accidents occur in various nations throughout the world. For evaluating and analyzing these algorithm several metrics, including precision, recall, accuracy and F1-Score are used to improve the performance of the dataset and predicts accuracy by approximately more than 85%.

Keywords: Machine learning algorithm · Supervised Learning Feature Analysis · Road Accident · metric parameter

1 Introduction

Road traffic accidents cause a substantial amount of fatalities and injuries each year, making it a major global concern. For successful efforts to prevent these accidents and enhance road safety, it is essential to under-stand the elements causing them. Road traffic accident analysis is essential for determining the reasons behind these accidents as well as their patterns and trends. The goals of this study on road traffic accident analysis aim to examine various factors such as over speeding, drunken driving, red light jumping, environmental factors, human behaviors, vehicles, and road conditions etc. which leads to severity of injuries. By conducting thorough accident analysis, researchers can uncover valuable insights that inform policymakers, traffic engineers, law enforcement agencies, and other stakeholders in developing targeted interventions. These interventions may include traffic regulations, road design improvements, driver education programs, and enforcement strategies to reduce accident rates and minimize their impact. This research

B. J. Choi et al. (Eds.): IHCI 2023, LNCS 14531, pp. 394–405, 2024.
https://doi.org/10.1007/978-3-031-53827-8_35

paper aims to contribute to the existing body of knowledge on road traffic accidents by conducting a comprehensive analysis of a specific time period over population. It also identifies the key factors contributing to accidents and propose evidence-based recommendations for improving road safety and reducing the frequency of road traffic accidents by examining the available data sources using appropriate statistical methods, and utilizing machine learning (ML) techniques. Decision trees, K Neighbours, Logistic Regression, XGB Classifier, and Random Forest are some machine learning algorithms. etc. are used to improve the performance of evaluation metrics that are used to predict the accidental cause of severity. The primary objective of this research is to contribute a range of machine learning techniques to improve the precision of accident cause prediction models. ML strategies have ability to capturing complex connections and managing large datasets. We want to determine the most effective methods for precise accident cause prediction through a thorough review of these algorithms. To assess the performance of our models, we employ several metric measurements commonly used in machine learning evaluation such as Precision, Recall, and F1-score etc. We also evaluate the models' predictive power, sensitivity, and overall effectiveness by looking at these metrics.

2 Related Work

There are several machine learning strategies are used to predict the causes responsible for the severity of injuries caused by road traffic. In paper [1], the authors employ data mining techniques to pinpoint the locations where the number of accidents is a significant contributing factor, analyses these locations using k-means clustering to divide them into k groups, and then uses association rule mining algorithm to uncover relationships based on the characteristics of the locations. In paper [2], the author employed three data mining techniques: decision trees, Naive Bayes, and KNN, to investigate the correlations between recorded road features and accident severity in Ethiopia. They have created a number of suggestions to enhance road safety in Ethiopia Considering the guiding principles. To determine how road flaws affect accident severity, the authors of the research [3] employed logistic regression based on the post-accident status of the vehicle. The authors of paper [4] suggested a method for calculating the probability of accidents occurring on Bangladeshi roadways using vision-based techniques. They learned from roadside camera data, and in specific circumstances they had an accuracy rate of 85%. The author studied 892 traffic events on National Highway N-5 in Bangladesh, using a variety of decision tree induction methods in an effort to identify patterns to make the highway safer and prevent accidents, they also set rules for the trees [5]. On the other hand, many cutting-edge and useful research projects in this subject are being conducted in other industrialized nations also. The K-means clustering method and mining association rules were used by the authors to find the key elements associated to traffic accidents [6]. The authors in their study [7] proposes a framework for storing data on traffic accidents in a big data platform that employs deep learning methods to create prediction and classification models from the data on traffic accidents. The framework [7] focuses on predicting and identifying accident hotspots on the roads in an effort to reduce the number of accidents. The paper [8], authors analyzing the road accident using k- means and machine learning algorithm on a different hotspot region in a metropolitan cities where a variety of things happen, such as atmospheric fluctuations, abrupt bends, and human errors.

3   Methodology

Machine learning (ML) is a subset of an artificial intelligence focuses on improving accuracy and categorized into three categories as supervised, unsupervised, and reinforcement learning [9].

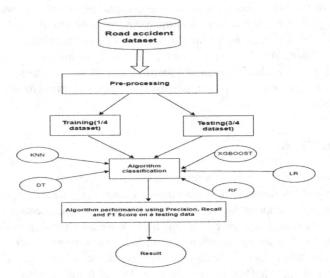

Fig. 1. Methodology using various Classification Algorithms

In this study, supervised machine learning techniques are used to analyze dataset and make prediction about the cause of road accident. Every machine learning algorithm has three parts: representation, evaluation, and optimization. The dataset is preprocessed by using a variety of methods and after that, low-correlated attributes, null values, duplicate data, and outlier representations are eliminated. A dataset's object data type is also changed to a numerical array. Once all of the attribute values have been converted to numbers, the dataset is divided into sections for testing and training. In the context of traffic accidents, the dataset is classified into three severity levels: Slight, Serious and fatal. The paper uses Extreme Gradient Boosting (XGBoost) classifier, Random Forest (RF), Logistic Regression (LR), K-Nearest Neighbours (KNN), and Decision Tree (DT) algorithms to investigate the causes of increased traffic incidents and severity levels [10, 11]. Furthermore, the analysis is performed on seven parameters, which are used for predicting the severity level of accidents. Figure 1 depicts the architecture of the proposed framework. In this case, nominal, numeric, and mixed data types are used for seven different attributes. The chosen data is pre-processed, and then various machine learning algorithms (RF, KNN, DT, LR, and XGB) are used to compare and compute the prediction depending on different severity levels (i.e., Slight, Serious, and Fatal). [9–11]. Decision tree has a hierarchal tree like structure consisting of root node, internal nodes, branches, leaf nodes and are used to classification as well as regression task [10, 11]. The performance of a decision tree depends on the amount of records and attributes in the data and is not relied on probability distribution. It also more accurately handles high-dimensional data. [12]. Random Forest (RF) is a method of ensemble learning that combines different decision trees to get accurate predictions [9, 13]. The K Nearest neighbors (KNN) is a non-parametric for categorizing new items based on learning data that is closest to the new entity. KNN is unsuccessful for data with large dimensions. In certain situations, reducing the dimension will improve performance [10, 12]. The Logistic Regression algorithm goal is to obtain the greatest fit for describing the connection between our target variable and the predictor variable that is diagnostically plausible [9, 11]. XGBoost [9–11], a powerful machine learning technique is a scalable and reliable variant of gradient boosting. XGBoost focuses on increasing goals and achieving peak performance. Its achievements are in fast tree building and distributed search methods, making it a cornerstone in machine learning [14]. In the proposed framework, first preprocess the data set, then partition the dataset into training and testing, and after that apply several machine learning algorithms to evaluate the different evaluation metrics measure their performance.

4   Implementation Platform and Empirical Result

The result has been put into practice in the "COLAB" Google Research laboratory. The platform enables the execution of Python programs using a Web browser and an internet connection and is well suited for applications in machine learning, deep learning, and data science. COLAB is an online Jupyter Notebook application that requires no setup and provides free access to computing resources, including GPUs. We employed a Jupyter notebook for our analysis. Through a web browser any one can write, update and execute code using google COLAB. Along with the some packages with jupyter Notebook, Pandas was essential to the analysis. Pandas is a tool for analyzing and manipulating data. It provides specific data structures and working methods for handling time series and mathematical tables. It offers high-performance, user-friendly data analysis structures and tools. In this paper, we explored different machine learning techniques on a dataset that contains information about traffic accidents. These techniques include Decision Tree, Random Forest, KNN, Logistic Regression, and XGBoost [8–10].

The dataset used in this investigation is taken from the KAGGLE which comprises total 048575 records and 7 attributes with specific accident-related features listed in Table 1. We used dictionary method for pre-processing technique to classify data into separate groups. The dataset is then divided into two parts: one for training and the other for testing [16]. This allowed us to use machine learning techniques to evaluate its performance. In this study, two separate experiments were carried out based on the accident severity condition to assess the performance of the suggested methodologies. The effectiveness of each method in terms of various parameters was assessed in our initial experiment (Accuracy and mean absolute error) and results are shown in Table 2.

Table 1. Dataset features and its type

S.NO	Features	Data type
1	Date	datetime64
2	Day_of_Week	Categorical
3	Junction_Control	Categorical
4	Year	Numeric
5	Road_Type	Categorical
6	Accident_Severity	Categorical
7	Light_Conditions	Categorical

Table 2. Accuracy and MAE score of various machine learning algorithms

Algorithm	Accuracy		MAE
	Training	Testing	
Random Forest Classifier	85.44%	85.40%	0.160
K Neighbors Classifier	84.37%	84.34%	0.170
Logistic Regression	85.39%	85.45%	0.159
Decision Tree Classifier	85.44%	85.37%	0.160
XGB Classifier	85.39%	85.45%	0.159

Table 2, compares the performance of five machine learning algorithms namely Random Forest Classifier, K Neighbors Classifier, Logistic Regression, Decision Tree Classifier, and XGB Classifier in context of predicting road traffic accidents. Training Accuracy, Testing Accuracy, and Mean Absolute Error (MAE) are the performance metrics examined. Based on the outcomes, every algorithm achieves accuracy rates higher than 84%, showing strong prediction performance overall. In the training phase, Random Forest Classifier, Decision Tree Classifier, and XGB Classifier have the maximum accuracy of 85.44%. During testing, however, Random Forest Classifier and Decision Tree Classifier maintain similar high accuracy of 85.40% and 85.37%, respectively, although XGB Classifier slightly increase to 85.45%. When compared to the other algorithms, K-Neighbors Classifier and Logistic Regression have slightly lower accuracy rates. In terms of MAE, all the algorithms have relatively low values of MAE and Logistic Regression having the lowest MAE of 0.15931 among rest of tested machine learning algorithms. The MAE scores for Random Forest Classifier, Decision Tree Classifier, and XGB Classifier are practically comparable at 0.160, 0.160 and 0.159, respectively. As a result, Random Forest, Decision Tree and XGB perform better in our scenario since their MAE is lower in every algorithm case. We also determined the other metrics like accuracy score, Precision, Recall value, and F1-Score for categorizing our machine learning model [15–17]. Because of our dataset is imbalanced, to address this issue and enhance the performance, we can utilize metric techniques having certain parameter that are effective for handling imbalanced classes. To achieve better results, we should calculate Precision score, Recall value, and F1-Score using Eqs. 1, 2, 3, and 4 respectively [9–12]. The ratio of true positives to total positive (i.e. sum of true positive (TP) and false positive (FP)) predictions is what we might refer to as precision whereas accuracy determines the proportion of actual positives among expected positives. Recall is the ratio of true positive (TP) and sum of true positive (TP) and false negative (FN). Recall also measures how well a model predicts favorable results. F1-Score combine the precision and recall score to access model accuracy and its effectiveness. The F1 score is calculated using the harmonic mean of the precision and recall scores and in the range of 0 to 100%, higher value indicating a higher quality classifier. The Eqs. 1 to 4 are used to compute Precision, Recall, Accuracy and F1-Score respectively.

$$\text{Precision} = \frac{TP}{TP + FP} \tag{1}$$

$$\text{Recall} = \frac{TP}{TP + FN} \tag{2}$$

$$\text{Accuracy} = \frac{\text{True Positive} + \text{True Negative}}{\text{Total number of Predictions}} \tag{3}$$

$$\text{F1SCORE} = \frac{2}{\frac{1}{\text{Precision}} + \frac{1}{\text{Recall}}} = \frac{2 \times \text{precision} \times \text{recall}}{\text{precision} + \text{recall}} \tag{4}$$

We also compute some other evaluation metrics such as macro-averaged F1-score, micro-average F1-score, and sample weighted F1-score. The results of class-wise F1 scores are simply averaged to provide the macro-averaged F1 score of a model while the micro-averaged F1 score is a useful metric for multi-class data distribution. The macro-averaged F1 score is only useful if the dataset being utilized has an equal amount of data points for each of its classes. Micro-averaged F1 score metric is computed using "net" TP, FP, and FN values. The net TP is the total of the class-specific TP scores for a dataset, which are deter-mined by breaking down the confusion matrix into one-vs-all matrices for each class. The sample-weighted F1-score is the most effective method for calculating the net F1-score for a class-imbalanced distribution of data. It is an average of the F1 scores obtained for each class, weighted according to the number of samples for each class. These evaluation metrics are formulated mathematically as follows for a dataset with "n" classes.

$$Macro\,F1\,Score = \frac{\sum_{i=1}^{n} F1\,Score}{n} \tag{5}$$

$$Micro\,F1\,Score = \frac{Net\,TP}{Net\,TP + 0.5(Net\,FP + Net\,FN)}$$

$$= \frac{\sum_{i=1}^{n} M_{ii}}{\sum_{i=1}^{n} M_{ii} + 0.5[\sum_{i=1}^{n} \sum_{j=1;i\neq j}^{n} M_{ij} + \sum_{i=1}^{n} \sum_{j=1;i\neq j}^{n} M_{ji}]} \tag{6}$$

where M is the confusion matrix and M_{ij} denotes the element for the ith row and ij^{th} column. A micro F1 score for a binary class dataset is simply the accuracy score and is computed as follows.

$$Micro\ F1\ Score(binary\ Class) = \frac{Net\ TP}{Net\ TP + 0.5(Net\ FP + Net\ FN)} \quad (7)$$

$$= \frac{M_{11} + M_{12}}{M_{11} + M_{12} + 0.5[(M_{12} + M_{21}) + (M_{21} + M_{12})]}$$

$$= \frac{M_{11} + M_{22}}{M_{11} + M_{12} + M_{21} + M_{22}}$$

$$= \frac{TP + TN}{TP + FP + FN + TN} = Accuracy$$

$$Weighted\ F1\ Score = \sum_{i=1}^{N} w_i \times F1\ Score, \quad (8)$$

$$where\ w_i = \frac{Number\ of\ samples\ in\ class\ i}{Total\ number\ of\ samples}$$

To effectively handle the complexity of our target variable with multiple classes, we incorporated the 'average' parameter, which includes choices such as 'micro', 'macro', and 'weighted'. Additionally, we took into account the 'zero_division' parameter to manage potential instances of division by zero. By applying the Eqs. 1 to 4, we compute Precision, Recall, F1 score and then use Eqs. 5, 6, 7 and 8 to compute 'macro', 'micro', and 'weighted' calculations. These computations were instrumental in providing a comprehensive evaluation of the machine learning algorithms under consideration. These approaches allowed us to gain valuable insights into their performance across various attributes. Our findings and results are summarized in Table 3 & 4, where we display the values that we have calculated for different measurement parameters. On the basis of using several parameters to assess ML algorithm's performance, we have created Fig. 2 and 3 which compares the algorithm over various metrics (Precision, recall value, and F1-Score) and computes the overall sum value of given parameter. By applying the Eqs. 1 to 4, we compute Precision, Recall, F1 score and then use Eqs. 5, 6, 7 and 8 to compute 'macro', 'micro', and 'weighted' calculations. These computations were instrumental in providing a comprehensive evaluation of the machine learning algorithms under consideration. These approaches allowed us to gain valuable insights into their performance across various attributes.

Our findings and results are summarized in Table 3 & 4, where we display the values that we have calculated for different measurement parameters. On the basis of using several parameters to assess ML algorithm's performance, we have created Fig. 2 and 3 which compares the algorithm over various metrics (Precision, recall value, and F1-Score) and computes the overall sum value of given parameter. The algorithms consistently show good performance across micro averages, indicating their ability to handle different types of data effectively while the Decision Tree Classifier perform well in training using macro and weighted averages. On the other hand, the K Neighbors Classifier,

Table 3. Performance Analysis of Machine Learning Algorithms with Different Average Parameters value

ALGORITHM	Zero Division	Weighted Attribute						Macro attribute						Micro Attribute					
		F1-SCORE		RECALL VALUE		PRECISION SCORE		F1-SCORE		RECALL VALUE		PRECISION SCORE		F1-SCORE		Recall Value		PRECISION SCORE	
		0	1	0	1	0	1	0	1	0	1	0	1	0	1	0	1	0	1
RF	Training	0.7881	0.788	0.8544	0.8544	0.8479	0.8479	0.3119	0.3119	0.3357	0.3357	0.8317	0.8317	0.8544	0.8544	0.8544	0.8544	0.8544	0.8544
	Testing	0.8541	0.8541	0.8541	0.8541	0.7483	0.7483	0.3333	0.3333	0.3333	0.3333	0.3354	0.3354	0.8541	0.8541	0.8541	0.8541	0.854	0.8541
KNN	Training	0.7869	0.7869	0.8437	0.8437	0.7553	0.7553	0.8437	0.8437	0.3364	0.3364	0.3641	0.3641	0.3198	0.8437	0.8437	0.8437	0.8437	0.8437
	Testing	0.8434	0.8434	0.8434	0.8434	0.7533	0.7533	0.3355	0.3355	0.3355	0.3355	0.3553	0.355	0.8434	0.8434	0.8434	0.8434	0.8434	0.8434
LR	Training	0.7866	0.7866	0.8539	0.8539	0.7291	0.8752	0.3071	0.3071	0.3333	0.3333	0.2846	0.9513	0.8539	0.8539	0.8539	0.8539	0.8539	0.8539
	Testing	0.8545	0.8545	0.8545	0.8545	0.7303	0.8757	0.3333	0.3333	0.3333	0.3333	0.2848	0.9515	0.8545	0.8545	0.8545	0.85455	0.8545	0.8545
DT	Training	0.7878	0.7878	0.8544	0.8544	0.8667	0.8667	0.3112	0.3112	0.3354	0.3354	0.9291	0.9291	0.8544	0.8544	0.8544	0.8544	0.8544	0.8544
	Testing	0.8538	0.8538	0.8538	0.8538	0.7462	0.7462	0.3333	0.3333	0.3333	0.3333	0.3299	0.3299	0.8538	0.8538	0.8538	0.8538	0.8538	0.8538
XGB	Training	0.7866	0.7866	0.8539	0.8539	0.7291	0.8752	0.3071	0.3071	0.3333	0.3333	0.2846	0.9513	0.8539	0.8539	0.8539	0.8539	0.8539	0.8539
	Testing	0.8545	0.8545	0.8545	0.8545	0.7303	0.8757	0.3333	0.3333	0.3333	0.3333	0.2848	0.9515	0.8545	0.8545	0.8545	0.8545	0.8545	0.8545

Table 4. Aggregate performance of Algorithmic using Different metric using Average parameter value

Algorithm		Micro	Macro	Weighted
Random Forest Classifier	Training	5.124	2.96	4.9807
	Testing	5.124	2.002	4.913
K Neighbours Classifier	Training	4.54	2.565	4.7718
	Testing	5.058	2.05	4.8802
Logistic Regression	Training	5.124	2.516	4.8853
	Testing	5.129	2.569	5.024
Decision Tree Classifier	Training	5.124	3.15	5.0178
	Testing	5.124	1.994	4.9076
XGB Classifier	Training	5.124	2.516	4.8853
	Testing	5.129	2.568	5.024

Logistic Regression, and XGB Classifier perform particularly well in specific situations on the testing dataset. From a comparative perspective, it's evident that the algorithms perform consistently well across micro averages, both in training and testing datasets. This indicates their ability to maintain a balanced performance, treating all instances with equal importance.

From Table 2, 3 and Fig. 3, 4, we found that on the testing dataset, the K Neighbors Classifier demonstrates the highest macro average, indicating its capability to excel in class-wise performance. This suggested that the Decision Tree classifier perform well in capturing inherent class characteristics during training while the K-Neighbors Classifier dynamically adapts well to the testing dataset classes. It is also obtain from the above said tables and figures that weighted averages provide valuable insights into class distribution. On the training dataset, the Decision Tree classifier performs well whereas Logistic Regression and XGB Classifier exhibit robust weighted average scores on the testing dataset, signifying their adaptability to diverse class distributions. In addition, we experimented with the dataset feature to see how its parameters effect on traffic accidents and making the right decisions which lowering the accident rate.

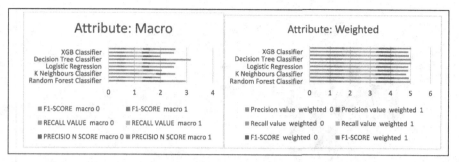

Fig. 2. Performance of various machine learning algorithm in Macro and Weighted attribute

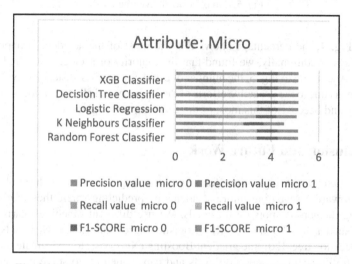

Fig. 3. Performance of various machine learning algorithm in Micro

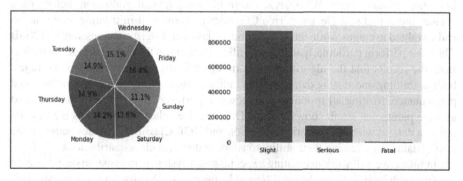

Fig. 4. No of casualities and cases of accident

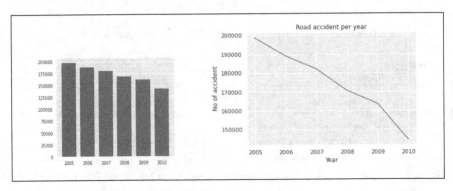

Fig. 5. No of accident over the year

From Fig. 4, the percentage shows that the most of the accident is happened on the workdays. Additionally, we found that the majority of accidents resulted in slight outcomes. According to Fig. 5, the number of accidents decreased steadily over the year. This decline is due to the proactive efforts of the Ministry and all stakeholders to improve road safety and because of this most of the accidents are slight.

5   Conclusion and Future Work

Traffic accidents are a big reason for deaths in many countries of the globe but we can work towards preventing future accidents by predicting where they might happen and finding dangerous spots on the roads. We use different Machine Learning (ML) algorithm such as Random Forest (RF), Logistic Regression (LR), K Nearest Neighbors (KNN), Decision tree, Extreme gradient Boosting (XGBoost) to investigate the reasons of higher traffic accidents/severity levels and can be helpful to solve problems using data. Because the number of traffic accidents has been going up in recent years, it's important to study them. We analyze the result using several evaluation metrics on a dataset and found that Decision Tree Classifier perform well in training using macro and weighted averages while the K-Neighbors Classifier, Logistic Regression, and XGB Classifier perform particularly well in specific situations on the testing dataset. It is clear from the results that the algorithms perform consistently well across micro averages, both in training and testing datasets. This indicates their ability to maintain a balanced performance, treating all instances with equal importance. In the case of the weighted average parameter, result shows that the Decision Tree classifier performs well on the training dataset whereas Logistic Regression and XGB Classifier provides better on the testing dataset, indicating their ability to adapt to different class distributions.

In future we will work in creating a special smart system using custom deep learning model which include Convolutional Neural Networks (CNNs) for Image Analysis. This will help us predict traffic accidents even more accurately.

Acknowledgement. This research is supported by research seed grant under IoE, BHU [grant No. R/Dev/D/IoE/SEED GRANT/2020-21/Scheme No.6031/Dr. Awadesh kumar].

References

1. Kumar, S., Toshniwal, D.: A data mining approach to characterize road accident locations. J. Mod. Transp. (2016)
2. Beshah, T., Hill, S.: Mining road traffic accident data to improve safety: role of road-related factors on accident severity in Ethiopia. In: AAAI Spring Symposium: Artificial Intelligence for Development (2010)
3. Esmaeili, A., Khalili, M., Pakgohar, A.: Determining the road defects impact on accident severity; based on vehicle situation after accident, an approach of logistic regression. In: 2012 International Conference on Statistics in Science, Business and Engineering (ICSSBE), Langkawi (2012)
4. Elahi, M.M.L., Yasir, R., Syrus, M.A., Nine, M.S.Q.Z., Hossain, I., Ahmed, N.: Computer vision based road traffic accident and anomaly detection in the context of Bangladesh. In: 2014 International Conference on Informatics, Electronics & Vision (ICIEV), Dhaka (2014)
5. Satu, M.S., Ahamed, S., Hossain, F., Akter, T., Farid, D.M.: Mining traffic accident data of N5 national highway in Bangladesh employing decision trees. In: 2017 IEEE Region 10 Humanitarian Technology Conference (R10-HTC), Dhaka, pp. 722–725 (2017)
6. Nandurge, P.A., Dharwadkar, N.V.: Analyzing road accident data using machine learning paradigms. In: 2017 International Conference on I-SMAC (IoT in Social, Mobile, Analytics, and Cloud) (I-SMAC), Palladam (2017)
7. Naseer, A., Nour, M.K., Alkazemi, B.Y.: Towards deep learning based traffic accident analysis. IEEE (2020)
8. Patil, J., Prabhu, M., Walavalkar, D.: Road accident analysis using machine learning. In: 2020 IEEE Pune Section International Conference (PuneCon). Vishwakarma Institute of Technology, Pune (2020)
9. Kumar, A., Mishra, M.K., Kumar, A., Gupta, S.: Machine learning approaches for cardiac disease prediction. In: Intelligent Systems and Smart Infrastructure: Proceedings of ICISSI 2022, vol. 391 (2023)
10. Kumar, A., Kumar, A.: Human sentiment analysis on social media through naïve bayes classifier. J. Sci. Res. 66(1) (2022)
11. Singh, M., Singh, D., Jara, A.: Secure cloud networks for connected & automated vehicles. In: 2015 International Conference on Connected Vehicles and Expo (ICCVE), Shenzhen, China, pp. 330–335 (2015). https://doi.org/10.1109/ICCVE.2015.94
12. Kumar, A., Kumar, A.: Analysis of machine learning algorithms for facial expression recognition. In: Woungang, I., Dhurandher, S.K., Pattanaik, K.K., Verma, A., Verma, P. (eds.) ANTIC 2021. CCIS, vol. 1534, pp. 730–750. Springer, Cham (2021). https://doi.org/10.1007/978-3-030-96040-7_55
13. Clarke, D.D., Ward, P.J., Jones, J.: Processes and countermeasures in overtaking road accidents. Ergonomics 42(6), 846–867 (1999)
14. Singh, M.: Vehicle mobile data analysis for driving safety and security. In: Singh, M. (ed.) Information Security of Intelligent Vehicles Communication. SCI, vol. 978, pp. 141–153. Springer, Singapore (2021). https://doi.org/10.1007/978-981-16-2217-5_10
15. Kumar, A., Kumar, A.: DEEPHER: human emotion recognition using an EEG-based deep learning network model. Eng. Proc. 10(1), 32 (2021)
16. Shen, X., Wei, S.: Application of XGBoost for hazardous material road transport accident severity analysis (2020)
17. Bokaba, T., Doorsamy, W., Paul, B.S.: Comparative study of machine learning classifiers for modelling road traffic accidents. Appl. Sci. 12(2), 828 (2022). https://doi.org/10.3390/app12020828

Attendance Monitoring System Using Facial and Geo-Location Verification

Garima Singh[✉], Monika kumari, Vikas Tripathi, and Manoj Diwakar

Graphic Era (Deemed to Be) University, Dehradun, Uttarakhand, India
garimasinghgryffindor@gmail.com, {vikastripathi.cse,
dr.manojdiwakar}@geu.ac.in

Abstract. Face verification is the most distinctive method used in the most effective image processing software, and it is essential in the technical world. Face verification is a method that can be used instead of face recognition. Face recognition is the act of recognising a person from a given image, whereas face verification is the process of confirming that a given face belongs to a particular person. As a result, after the face is validated, the attendance is automatically recorded, and an update is sent. In addition, an anti-spoofing measure of liveness detection is considered to ensure that the system is foolproof. Eye-blink has been used to detect liveness. Additionally, the geographical coordinates of the user are also verified before the attendance is marked. The creation of this system aims to digitalize the outdated method of collecting attendance by calling names and keeping pen-and-paper records. Because current methods of taking attendance are cumbersome and time-consuming.

Keywords: Attendance Monitoring System · Facial Verification · Geo-location Verification

1 Introduction

Attendance management's main aim is to track the working or office hours of employees, faculty as well as students in college. The old method of attendance marking system is a very lengthy task in many schools and colleges. It does create a big work-load on faculty as they have to manually keep a record or documentation of students being present or absent which might take about 5 to 10 min of the entire session. This is time-consuming for students as well as for the faculties too. There are also some chances of proxy attendance. To reduce this many universities and colleges have started deploying many other techniques for recording attendance like Radio Frequency Identification(RFID), iris recognition, face recognition, face recognition with GSM Notification [1].

Biometric-based techniques have emerged as the most promising option for recognizing individuals in recent years since, instead of authenticating people and granting them access to physical and virtual domains based on passwords, PINs, smart cards, plastic cards, tokens, keys and so forth, these methods examine an individual's physiological and/or behavioral characteristics in order to determine and/or ascertain his

© The Author(s), under exclusive license to Springer Nature Switzerland AG 2024
B. J. Choi et al. (Eds.): IHCI 2023, LNCS 14531, pp. 406–416, 2024.
https://doi.org/10.1007/978-3-031-53827-8_36

identity. Passwords and PINs are hard to remember and can be stolen or guessed; cards, tokens, keys and the like can be misplaced, forgotten, purloined or duplicated; magnetic cards can become corrupted and unreadable. However, an individual's biological traits cannot be misplaced, forgotten, stolen or forged [2]. Biometric characteristics, such as iris recognition, which are employed for security purposes for limited areas in some organisations, can be utilised to track attendance [3]. The initiative by M. Khari et al. to create a software product that enables facial recognition for admission and leave from metro platform gates. A smart attendance monitoring system can employ a similar strategy [4].

The primary goal of developing an attendance system was to use the verification technique and involve the general public in order to assess its potential benefits for the educational system. Here, the face of a person and their location are taken into account while recording attendance [5]. Since the population is growing by 0.81% year [6], the attendance system is becoming monotonous and wearying. To enable comparisons across organisations and jurisdictions, educational entities are being asked more frequently to report attendance data in a uniform manner. The correlation between student attendance and academic achievement is the main justification for high-quality attendance data [7].

2 Literature Review

According to the research paper [8], the author spoke about how in Covid-19 the biggest problem with inability in normal life was going to in-person classes, attending programs and going to jobs have been impossible without a large amount of risk being involved. In order to make sure educational institutes can function properly and efficiently over an internet connection, they devised a system which was given the title "Attendance management system based on face recognition". The proposed system aims to improve the already existing systems for video conferences that have poor attendance management systems, as also it tries to cover the gap between attendees and hosts by providing features other than just detection of faces and analysis of attendance like keeping track of students. The algorithm used here was converting an RGB image to greyscale and then focusing on one grid of 8 * 8 pixels for each pixel, horizontal and vertical gradients are calculated and then calculated gradient magnitude and gradient angle for each of 64 pixels are compressed and for which histogram of magnitudes and angles are plotted. Now with that 8 * 8 grid will slide along the whole image and after interpreting the full histogram and plotting the HOG features find that the structure of the object or face is well maintained, losing all the insignificant features.

Kowsalya et al. [9] developed a face recognition algorithm-based automatic attendance management system. The camera at the entrance records each person who enters the classroom. After that, the face region is removed and prepared for further processing. Face detection requires less labour because more than one person can be in the entire classroom at once. Face recognition has advantages over other systems that have been discussed for analysing facial recognition and recording attendance. Issues like occlusion persist in this system. There is a high possibility of a person being present and still not being marked present.

Bah et al. [10] study project had two major sections: While the second half concentrated on the attendance management system based on the identified human faces, the first section primarily focused on strengthening the face recognition algorithm. To capture images of employees entering an office or building, a digital live camera will be used in the first section. To improve the quality of the images, advanced image processing techniques, such as contrast adjustment, noise reduction using bilateral filter, and image histogram equalisation, will be applied to the captured images. Next, the Haar Algorithm will be applied to the images to detect specific faces, which will then be used as an input to the Facial Recognition System. The system fails to provide anti-spoofing capabilities.

Some techniques require minimal user involvement and rely exclusively on face recognition. These techniques are also integrated with CCTV systems. When there is a gathering of people in the CCTV feed, this system automatically records attendance [11]. Face Recognition, not Face Verification, is employed in this instance; the two are distinct. Face Verification includes verifying a person's identity by comparing their face with a template that has been stored. It is a one-to-one comparison that is utilised for authentication. In face recognition, a person is recognised by their face being compared to a database of reference pictures. It is a one-to-many comparison that is used to identify someone. Therefore, it is reasonable to say that Face Recognition takes longer because there are more comparisons. Additionally, this technique is not very reliable because there is a chance that some areas may not be covered by the cameras, and occlusion may prevent some people from being marked as present. Systems that rely on facial recognition and verification exist, however, they are easily deceived by spoofing attempts.

The proposed method requires each user to mark their attendance and is not an automatic system like the CCTV-based attendance marking system. The user must log in and engage with the system. Facial Verification would function far better to match the face than Face Recognition, which has a higher Time-Complexity since the user must be logged in before marking the attendance. Additionally, this approach is easily spoofable, necessitating the use of reliable anti-spoofing. It is necessary to use a liveness detection system. Eye-blink has been used to identify liveness. Additionally, we included a function that checks the user's geographic location to ensure that they are indeed at the institution or office. This feature makes our system even more secure.

3 Proposed System

The proposed system improves the existing attendance monitoring system by using Biometric Facial data (i.e.; Inherence factors) (see Fig. 1) for the authentication purpose, which is the highest level of authentication in a 3-Factor Authentication. A 3-factor authentication (3FA) uses three different types of authentication elements to confirm a user's identity. The information or traits that a user must submit as proof of their identity in order to utilise a system, application, or resource are known as authentication factors. In 3-factor authentication, the following three factors are frequently used: Something You Know: This is the standard knowledge-based security measure, such as a password,

PIN, or responses to security questions. Only the authorised user should be aware of it, Something You Have: Possession of a tangible item that is specific to the user, such as a smartphone, a security token, or a smart card, constitutes this factor. One-time codes, which the user must supply during the authentication procedure, are generated or received by this object, Something you are: This element uses biometric data, such as voice, facial, or retinal scans, fingerprint, or retinal scans. Biometric information is a reliable authentication component since it is challenging to fake or duplicate.

Here, in our system, we are also using geographical location checks (see Fig. 1) using the coordinates of the user to make the system even more strict and precise. Thereby adding another factor to our 3FA, which makes the system even more secure and difficult to fool. All the steps are shown in Fig. 1. The following steps explain how the proposed system works:

1. While registering the user will provide a photo of his/her face to be saved in the database for matching in the later phase.
2. The user login their account with their ID and Password. This is the first factor of a 3FA which is "what you know". We can also use OTP here after this step to encompass the second factor of "what you have" as well.
3. Marking the attendance of a particular class will only be enabled in its time slot.
4. Also, the button that will allow the marking of attendance will only be enabled when the geographical coordinates of the user match the geographical coordinates of the Institution (which will be set during the deployment).
5. After the Time Slot check and Geographical Check the button to mark the attendance will be enabled. As and when the user will click on this button, the webcam will be activated.
6. The user will be required to capture their photos. After the spoofing analysis is performed. Then the picture is compared to the original data in the database linked with that account.
7. If matched, the user is required to capture the photo of the faculty whose lecture is being conducted. If the face is matched, the user will be finally verified for attendance and will be marked present in the database. Thereby accomplishing the third factor of the 3FA.

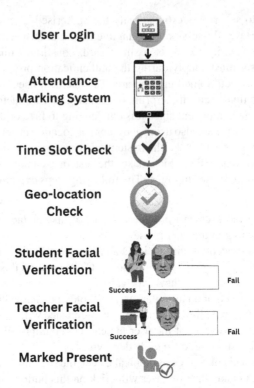

User Login

Attendance
Marking System

Time Slot Check

Geo-location
Check

Student Facial
Verification

Success   Fail

Teacher Facial
Verification

Success   Fail

Marked Present

Fig. 1. The overall design of the suggested method to allow attendance marking without any spoofing efforts

3.1 Facial Verification

The facial verification part of the proposed system is divided into the following four phases (see Fig. 3):

Face Detection. A computer vision technology called face detection is used to recognise and locate human faces in digital photos or video frames (see Fig. 3). It is an essential stage in many applications, such as security systems, facial identification, and emotion analysis. The main objective of face detection is to identify any faces in a frame of an image or video and, if any are present, to return the coordinates of the bounding boxes around those faces.

Face Embeddings Extraction. Before we can apply the algorithms that will compare the faces for verification, the face detected in the above step (which is still in the image form) needs to be converted into information or a numerical vector (see Fig. 3) that the algorithm can understand and operate on. The system computes measures based on landmarks and face characteristics like the eyes, nose, and mouth to obtain these embeddings. 68 facial landmarks (see Fig. 2), often called facial keypoints [12].

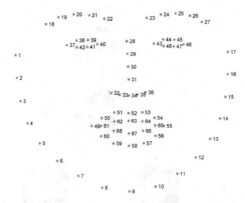

Fig. 2. Sample Facial Landmarks. Picture from PyImageSearch, Adrian Rosebrock [13]

Face Matching. The technique of verifying whether or not two provided face photos belong to the same individual is known as face verification (see Fig. 3). It entails determining whether a person is who they say they are, in other words. Either the faces match (belong to the same person) or they don't, which is typically a binary choice. The more comprehensive problem of face recognition is determining a person's identification from a set of known people. Face recognition, unlike verification, does not involve a binary choice of "same" or "different". Instead, it seeks to narrow down the candidate identities in order to identify the person. For matching faces, we are using verification instead of recognition. Face verification is frequently used for tasks including employing facial recognition to unlock devices, gaining access to protected locations, or verifying people for online transactions [14].

Anti-spoofing. Systems for facial recognition and verification must include anti-spoofing (see Fig. 3) in order to guarantee the process' security and reliability. It is intended to stop attackers from getting around these defences by employing fictitious representations of faces, such as 2D images, films, masks, or other manufactured tools [15]. Anti-spoofing methods try to spot irregularities or discrepancies in the given face that point to a spoofing attempt. These techniques can involve analyzing various factors such as texture, depth, motion, liveness, and even physiological responses like blood flow or blinking (see Fig. 3) [16].

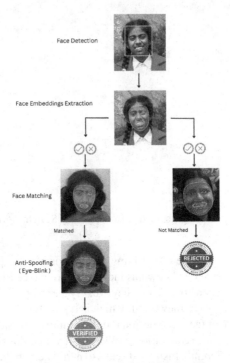

Fig. 3. Step-by-step Facial Verification Process

3.2 Face Detection

We are using MTCNN (Multi-task Cascaded Convolutional Networks) for Face Detection. It is a modern tool to detect faces, consisting of three stages of convolutional networks. The representation of the effective working of all three stages is given in Fig. 4.

Stage 1: P-Network (Proposal-Network). To recognise faces of various sizes, the image is resized numerous times. Here, the threshold for detection is low and hence, many false positives (see Fig. 4) but it is all intentional. It is supposed to work like that. The low detection threshold used by the P-network means that it is sensitive to even subtle face-like features in the image. This sensitivity helps in capturing potential faces that might be missed if a higher threshold were used. While this sensitivity leads to more false positives, it ensures that actual faces are not overlooked [17].

Stage 2: R-Network (Refine-Network). The output from the P-Net is the input to this network. Here, many false positives received from the above network will be rectified (see Fig. 4). And we will obtain precise bounding boxes [17].

Stage 3: O-Network (Output-Network). This is the final stage. Final filtering happens here and also we obtain very precise bounding boxes (see Fig. 4) [17].

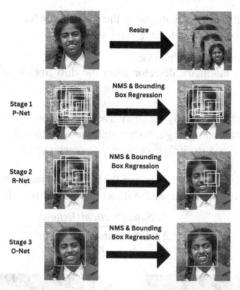

Fig. 4. Face detection with the MTCNN method.

3.3 Facial Embeddings Extraction

For Facial Embeddings extraction, we are using Keras-VGGFace Model. We are using only feature extraction layers of the model with VGGFace(include_top = False) initiation. Keras-VGGFace is a popular deep-learning model implementation in Keras that is specifically designed for face recognition tasks. It is based on the VGG16 architecture, which is a deep convolutional neural network (CNN) model originally developed for image classification.

3.4 Face Matching

The process of detecting whether two face photographs belong to the same person or not is known as facial verification. Therefore, the task effectively consists of calculating the separation between two face vectors or embeddings acquired from the previous stage [18]. Therefore, this step of appropriate distance metrics is critical for facial verification accuracy.

For facial verification, we employ a metric called the cosine similarity. This technique has the best accuracy recorded in the literature when tested on the cutting-edge dataset Labelled Faces in the Wild (LFW) [18].

3.5 Anti-spoofing

Eye-Blink Detection. In order to detect any spoofing attempts we need some form of liveness detection. For that, we are proposing Eye-blinking based liveness detection. For the purpose of detecting eye blinks, OpenCV, Python, and dlib are used. In addition to the imutils package, which contains several functions that will enable us to convert the landmarks to a NumPy array, we are using the dlib library for facial recognition [19]. The following are the main steps for eye-blink detection:

1. Using dlib to detect facial landmarks:
   Using the face landmark detector from the dlib library, one can detect facial landmarks. [19].
2. Eye Landmarks:
   Because we can extract any facial characteristic from the 68 Facial Landmarks discovered in the preceding stage. As a result, we will extract the ocular landmarks, which are 6 (x, y) (see Fig. 5(a)) coordinates for each eye [19].
3. Eye Aspect Ratio (EAR)
   This ratio establishes a link between the horizontal and vertical measures of the eye. [17]. The formula to calculate EAR can be referred from Eq. (1).

$$EAR = \frac{SumofVerticalDistance}{2 * HorizontalDistanceofEye} \tag{1}$$

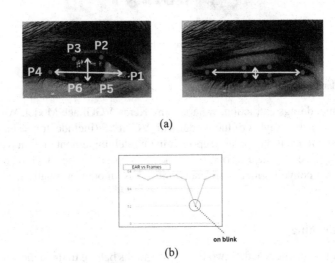

(a)

(b)

Fig. 5. (a) Opened Eye & Eye when blinked (b) EAR graph when the eye is blinking [19]

When the eye is opened (see Fig. 5(a)), the EAR remains constant, but it abruptly drops (see Fig. 5(b)) when the eye is blinked (see Fig. 5(b)). The average of the EAR from both eyes will be retained. Then, we'll see if it falls below a predetermined threshold (see Fig. 5(b)).

When the EAR falls below the predetermined threshold, we shall keep track of those frames. And if the count matches the predetermined count (as determined by the FPS), it might be 4 or 5, etc. A blink is only regarded as a blink after that.

3.6 Geographic Location Verification

Along with facial verification, the students will also be required to verify their locations. Because of this, the system so build is even stronger, and the system bypassing attempts possibilities would be narrowed. All the conditions to be satisfied in order for the attendance to be marked are illustrastrated in Fig. 6.

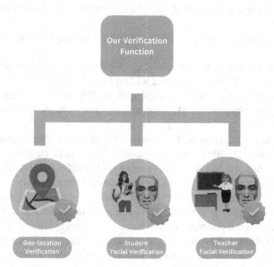

Fig. 6. Verification aspects we are considering - (1) Geo-location, (2) Student's Face, and (3) Teacher's Face.

4 Conclusion

From this system, we found out that Attendance Monitoring System is required in every field whether it comes to universities, schools or offices. Face verification will reduce the time of marking attendance. Along with facial verification, we also added functionality to capture the geolocation of the user and verify it as well. Most notably, the system does not rely on a widely used system as in [20] to automatically take pictures of everyone there and mark their attendance; in that scenario, there is a great likelihood that attendance won't be marked owing to problems like occlusion. By requiring user interaction to register their own attendance, our approach lowers the likelihood of fake absences. We added an anti-spoofing mechanism for liveness detection which involves eye-blink detection, but it can also be spoofed. Therefore, the system can be further improved by adding a number of other liveness detection strategies as well like texture analysis, 3D depth analysis, reflection analysis, behavioural analysis etc. Overall an idea has been presented in the paper which although has scope for further improvement but carries the potential to be incorporated as a real-life application.

The author has used her own image to showcase the result and have no issue on publishing her image in the paper.

References

1. Patel, U.A., Priya, S.: Development of a student attendance management system using RFID and face recognition: a review. Int. J. Adv. Res. Comput. Sci. Manage. Stud. **2**(8), 109–119 (2014)
2. Jafri, R., Arabnia, H.R.: A survey of face recognition techniques. J. Inf. Process. Syst. **5**(2), 41–68 (2009)
3. Dua, M., Gupta, R., Khari, M., Crespo, R.G.: Biometric iris recognition using radial basis function neural network. Soft. Comput. **23**(22), 11801–11815 (2019)
4. Dalal, R., Khari, M., Arbab, M.N., Maheshwari, H., Barnwal, A.: Smart metro ticket management by using biometric. In: Multimodal Biometric Systems, pp. 101–110. CRC Press (2021)
5. Puthea, K., Hartanto, R., Hidayat, R.: A review paper on attendance marking system based on face recognition. In: 2017 2nd International Conferences on Information Technology, Information Systems and Electrical Engineering (ICITISEE), pp. 304–309. IEEE (2017)
6. macrotrends. https://www.macrotrends.net/countries/IND/india/population-growth-rate
7. Aden, A.A., Yahye, Z.A., Dahir, A.M.: The effect of student's attendance on academic performance: a case study at Simad university Mogadishu. Acad. Res. Int. **4**(6), 409 (2013)
8. Nalini, N.: Attendance monitoring system based on face recognition (2021)
9. Kowsalya, P., Pavithra, J., Sowmiya, G., Shankar, C.K.: Attendance monitoring system using face detection & face recognition. Int. Res. J. Eng. Technol. (IRJET) **6**(03), 6629–6632 (2019)
10. Bah, S.M., Ming, F.: An improved face recognition algorithm and its application in attendance management system. Array **5**, 100014 (2020)
11. Rakshitha, S.: Face based CCTV attendance monitoring system using deep face recognition (2021)
12. Hangaragi, S., Singh, T., Neelima, N.: Face detection and recognition using face mesh and deep neural network. Procedia Comput. Sci. **218**, 741–749 (2023)
13. pyimagesearch. https://pyimagesearch.com/2017/04/03/facial-landmarks-dlib-opencv-python/
14. Beltrán, M., Calvo, M.: A privacy threat model for identity verification based on facial recognition. Comput. Secur. 103324 (2023)
15. Saraswat, D., Bhattacharya, P., Shah, T., Satani, R., Tanwar, S.: Anti-spoofing-enabled contactless attendance monitoring system in the COVID-19 pandemic. Procedia Comput. Sci. **218**, 1506–1515 (2023)
16. Parveen, S., Ahmad, S.M.S., Hanafi, M., Adnan, W.A.W.: Face anti-spoofing methods. Curr. Sci. 1491–1500 (2015)
17. Ku, H., Dong, W.: Face recognition based on mtcnn and convolutional neural network. Front. Signal Process. **4**(1), 37–42 (2020)
18. Nguyen, H.V., Bai, L.: Cosine similarity metric learning for face verification. In: Kimmel, R., Klette, R., Sugimoto, A. (eds.) ACCV 2010. LNCS, vol. 6493, pp. 709–720. Springer, Heidelberg (2010). https://doi.org/10.1007/978-3-642-19309-5_55
19. geeksforgeeks. https://www.geeksforgeeks.org/eye-blink-detection-with-opencv-python-and-dlib/
20. da Rosa Righi, R., et al.: Designing Cloud-Friendly HPC Applications. In: Borin, E., Drummond, L.M.A., Gaudiot, J.L., Melo, A., Melo Alves, M., Navaux, P.O.A. (eds.) High Performance Computing in Clouds, pp. 99–126. Springer, Cham (2023). https://doi.org/10.1007/978-3-031-29769-4_6
21. Tomar, A., Kumar, S., Pant, B.: Crowd analysis in video surveillance: a review. In: 2022 International Conference on Decision Aid Sciences and Applications (DASA), pp. 162–168. IEEE (2022)

Blockchain Approach to Non-invasive Gastro-Intestinal Diagnosis System

Aman Singh[1], Madhusudan Singh[2] (ID), and Dhananjay Singh[3](✉) (ID)

[1] COIKOSITY Pvt. Ltd., Prayagraj 211012, UP, India
[2] Oregon Institute of Technology, Klamath Falls, OR 97601, USA
[3] ReSENSE Lab, School of Professional Studies, Sain Louis University, St. Louis, MO 63108, USA
dhananjay.singh@slu.edu

Abstract. In this paper, we have examined the clinical importance of monitoring stomach dysrhythmias and the fundamental role of Interstitial Cells of Cajal (ICCs) in generating slow waves. Gastroparesis and functional nausea and vomiting syndrome (FNVS) are associated with a decrease in interstitial cells of Cajal (ICCs) and abnormalities in the onset and spread of slow waves. The research presents a novel blockchain-based infrastructure for storing EGG data in order to improve the security and accessibility of healthcare data. Blockchain technology guarantees the safe and reliable storage and transfer of EGG data, thereby addressing problems related to patient confidentiality and compliance with data protection requirements such as HIPAA. The decentralized nature of blockchain allows for secure and transparent storage of EGG readings and patient records, facilitating efficient data sharing among healthcare participants.

Keywords: Electrogastrography · Decentralized · Diagnosis IoT · Medical Device · Digital Healthcare

1 Introduction

Electrogastrography (EGG) is a non-invasive diagnostic technique employed to record electrogastrograms, which serve as valuable indicators of the slow waves occurring within the stomach [1]. These slow waves, generated by the depolarization of smooth muscle cell plasma membranes, have long been recognized as essential components of gastric muscle activity. They manifest as rhythmic electrical oscillations, known as electrical slow waves, playing a pivotal role in regulating the timing and speed of gastric peristaltic contractions, which are responsible for mixing and emptying ingested food. In healthy humans, the gastric pacesetter potential typically exhibits a frequency of approximately 3 cycles per minute (cpm), making EGG signals at 3-cpm a noninvasive means of monitoring the stomach's electrical activity. However, disruptions in this rhythmic pattern, referred to as gastric dysrhythmias, can lead to irregularities in gastric motility. Tachygastria, characterized by excessively fast electrical rhythms, and Bradygastria, marked by abnormally slow rhythms, are commonly observed in patients with functional dyspepsia, unexplained nausea, diabetic or idiopathic gastroparesis, and post-surgical gastroparesis. The process of recording electrogastrograms bears similarities to electrocardiography, attracting growing interest in electrogastrography for diagnostic purposes. Healthcare professionals, particularly internists, family medicine physicians, and diabetologists, managing patients with dyspepsia, gastric dysfunction, or diabetes, find EGG recordings intriguing due to their potential diagnostic significance. Research has revealed that electrical slow waves are generated by specialized pacemaker cells known as interstitial cells of Cajal (ICCs). These ICCs form an interconnected network between the circular and longitudinal muscle layers of the stomach, actively propagating slow waves throughout the distal stomach musculature. While ICCs are responsible for generating these slow waves, smooth muscle cells lack this capability but possess voltage-dependent ion channels, such as $Ca2+$ channels. These channels are activated during slow wave depolarization, leading to the influx of calcium ions into gastric smooth muscle cells, initiating the contraction process. Electrogastrography (EGG) has been employed for registering the gastric slow wave activity in these situations as a diagnostic tool. Despite its potential, it has not gained broad acceptance in clinical practice [2]. Both gastroparesis and FNVS have been linked to a reduction in interstitial cells of Cajal (ICC) and irregularities in the initiation and propagation of gastric slow waves [3–5]. Slow wave generation primarily occurs within the tunica muscularis of the proximal corpus along the greater curvature of the stomach, with subsequent circumferential and downward propagation towards the pylorus, forming an excitation ring that underpins gastric peristaltic contractions.

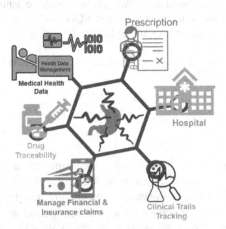

Fig. 1. Digital Healthcare System

Disruptions in the gastric frequency gradient, caused by factors like changes in pacemaker frequencies or abnormalities in signal propagation pathways, can give rise to motility disorders. Such disorders hinder the synchronization of pacemakers along the stomach, impeding the normal propagation of peristaltic contractions. Consequently, impaired gastric motility and motility disorders may occur, underscoring the clinical significance of monitoring and understanding gastric dysrhythmias. While changes in electrophysiological processes could potentially play a mechanistic role in functional nausea and vomiting syndrome and gastroparesis, the extent to which slow wave irregularities contribute to the underlying pathophysiology of these conditions remains uncertain [6]. In this context, the implementation of a blockchain-based EGG data storage platform holds immense promise, particularly in the healthcare sector. Blockchain's cryptographic techniques provide secure storage and transmission of sensitive EGG data, ensuring patient privacy and compliance with data protection regulations like HIPAA. EGG readings and patient records stored on the blockchain become tamper-resistant and transparent, guaranteeing. The accuracy and integrity of historical data. Furthermore, blockchain facilitates seamless data exchange among various healthcare stakeholders, empowering doctors, researchers, and patients to access and share EGG data efficiently. Patients gain control over their medical information, allowing them to grant access to specific doctors, researchers, or industries as needed. This approach not only enhances the safety and reliability of non-invasive gastrointestinal disorder detection but also addresses the critical concern of securing sensitive medical data in an era of digital healthcare. Figure 1 has represents the digital healthcare System.

The remainder of this paper is organized as follows: Sect. 2 of this paper discusses related work. Section 3 describes GIT diagnosis healthcare monitoring system. Section 4 describes Electrogastrography in presence of diseases. Section 5 contains considerations for Blockchain-based system. Section 6 discusses further challenges in this domain.

2 Related Works

The field of electrogastrography (EGG) has seen significant research efforts aimed at understanding its clinical applications and addressing concerns sur- rounding its reliability and correlation with gastric motility. In this section, we review several key studies and contributions in the field. Paper [7] focuses on utilizing electrogastrography in a heterogeneous group of patients experiencing unexplained nausea and vomiting. It identifies a subgroup with abnormal myoelectrical activity. While gastroenterologists are keen on exploring the clinical applications of EGG, concerns linger regarding its reliability and the analysis of EGG data, as well as its correlation with gastric motility. The primary objective of paper [8] is to address the aforementioned concerns surrounding EGG. It aims to comprehensively review the potential clinical applications of EGG technology, shedding light on its diagnostic and therapeutic implications. Paper [9] employs multi-channel electrogastrography to investigate whether patients suffering from functional dyspepsia exhibit impaired propagation or coordination of gastric slow waves in the fasting state when compared to healthy controls. The study delves into the physiological aspects of gastric motility in relation to EGG data. The objective of paper [10] is to present the consensus opinion of the American Motility Society Clinical GI Motility Testing Task Force regarding the performance and clinical utility of electrogastrography. This authoritative perspective provides valuable insights into the practical applications of EGG in clinical settings. Paper [11] introduces a model that describes the electrogastrogram as the result of field potentials generated by depolarization and repolarization dipoles. Such a model offers a theoretical framework for understanding the underlying electrical phenomena in the stomach that EGG captures (Fig. 2).

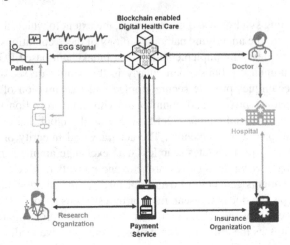

Fig. 2. Overview of EEG signal data management

The purpose of paper [12] is to discuss various aspects of EGG, including methodologies, validation techniques, and its broad range of applications. It provides a comprehensive overview of the current state of EGG research.

3   GIT Diagnosis System

In recent years, the advancement of IoT (Internet of Things) technology has led to the development of more compact and portable devices, making them increasingly accessible for use in clinical settings. These devices serve various purposes, including the diagnosis and monitoring of critical health parameters such as skin temperature, electrogastrograms (EGG), electrocardiograms (ECG), heart rate, respiratory rate, blood oxygen levels (SpO2), stress levels, and mental health. This section delves into the design of an IoT-based EGG system, exploring its potential in gastroenterological signal analysis and clinical applications. Figure 3 illustrates the basic architecture of a Gastrointestinal Tract (GIT) diagnosis de- vice, emphasizing the relevance of EGG in contemporary medical practice.

3.1   Clinical Role of Gastroenterological Signals

In contrast to electrocardiograms (ECGs), which physicians can interpret visually by inspecting the waveform, the electrogastrogram (EGG) necessitates computerized spectral analysis for evaluation. This requirement arises because the EGG signal represents a composite measure of gastric slow waves within the stomach. The waveform's shape is influenced by various uncontrollable factors, and there are no established diagnostic criteria based solely on the EGG wave- form. Additionally, the EGG signal may contain respiratory artifacts, typically ranging from 12 to 25 cycles per minute (cpm), and occasionally even electro- cardiogram (ECG) artifacts (less than 60 cpm). In certain instances, the EGG signal might also capture the slow wave activity of the small intestine

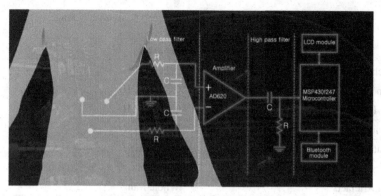

Fig. 3. Capturing of Electrogastrogram signal and inside of an EGG device.

(9–12 cpm). While these external influences can distort the gastric slow waves within the EGG signal, their frequencies do not align with those of the gastric slow waves. Figure 4 has shown the phases of electrogastrogram from stomach to smartphone.

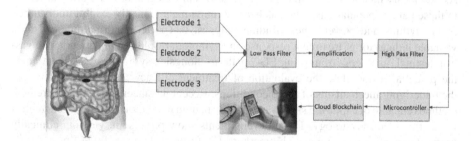

Fig. 4. Phases of electrogastrogram from stomach to smartphone.

Consequently, spectral analysis becomes a viable method for distinguishing genuine gastric slow waves from these interferences. The non-recursive band-pass filter plays a pivotal role in processing the discrete EGG signal to extract meaningful information. Mathematically, this operation involves the convolution of the signal (xj) with the digital filter operator (hj) to generate the filtered output (yi), as described by Eq. (1).

$$y_i = \sum_{j=0}^{m-1} h_j x_{i-j} \tag{1}$$

The smoothing process further utilizes the Goodman-Enoxon-Otnes (GEO) window, following the 7-point algorithm, to enhance the signal, as indicated by Eq. (2).

$$\bar{x}_i = \sum_{l=-3}^{3} a_i x_{i+1} \tag{2}$$

Table 1. Categorization of different signals to frequency ranges.

Signal	Frequency
Stomach	0.03–0.07 Hz
Ilenium	0.08–0.12 Hz
Skinny Intestine	0.13–0.17 Hz
Duodenum	0.18–0.22 Hz
Respiration	0.2–0.4 Hz
ECG	1–1.33 Hz
Motion Artefacts	Whole Range

While a detailed mathematical analysis goes beyond the scope of this paper, interested readers can refer to the literature [13] for in-depth insights. Electrogastrography serves as one of the valuable tests for assessing gastrointestinal function, particularly in patients with unexplained symptoms such as nausea, vomiting, and dyspeptic complaints. As previously mentioned, significant abnormalities in EGG patterns are often observed in these patient populations, whereas healthy individuals typically exhibit undisturbed EGG rhythms and power. These findings suggest that EGG abnormalities can be a valuable indica- tor of treatment response and symptom resolution in patients with gastrointestinal disorders. Monitoring EGG patterns alongside symptom assessment holds the potential to facilitate the evaluation of treatment efficacy and offer insights into the underlying mechanisms of these conditions. Recent advancements in measurement methods capable of capturing slow wave activity in high resolution have renewed interest in gastric electrophysiology. These developments show promise in yielding clinically useful diagnostic and therapeutic biomarkers. For further reference, Table 1 provides an overview of the frequency ranges associated with different parts of the body, aiding in the understanding of the diverse applications of EGG technology.

3.2 Considerations for Blockchain Technology

As the quantity of Health IoT devices used by consumers increases, the associated organizational, ethical, and economic challenges also expand accordingly. In this section, we offer our insights into different aspects of the blockchain-based system. Figure 5 explains how a GIT signal is captured from stomach through electrodes and then sent to a blockchain platform, where it is securely stored and made available for the use of different entities like doctors, patients, the Research and Development Division, etc.

3.3 High-Level Requirements for the Blockchain-Based Platform

The following points are the high-level requirements for a blockchain-based plat- form:

- Support for personal health devices offered by various manufacturers, enabling the measurement of diverse bio-signals.

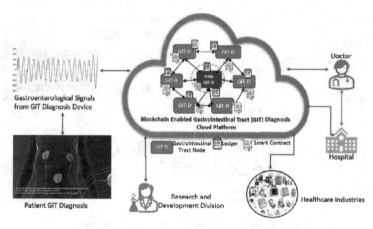

Fig. 5. A description of blockchain platform.

- Capability to easily add or remove personal health devices as needed.
- Functionality to operate even when one or more personal health devices are offline.
- Capacity to create a consistent personalized health data record by gathering data from multiple devices.
- Automated validation for each new addition to a data record.
- Maintenance of an enduring chronological history of data record updates.
- Provision of a user-friendly interface/dashboard for an individual's health data.
- The ability to share data with third-party systems (like clinicians or care-givers) based on user consent, allowing health data to be used for analysis and diagnosis.
- Protection of the database from unauthorized access.
- Validation and confirmation of each data record without relying on central regulatory authorities or a central database server.

3.4 BchainGIT- Blockchain Gastro-Intestinal Track Architecture

There are a lot of medical IoT gadgets and apps on the market right now that can help customers. There are usually specialized mobile apps that come with these devices that connect to them, collect health data, and then send that data to the backend to be stored.

The design of BchainGIT includes a key adapter part that makes this process easier. The adapter acts as a go-between, getting data from the device's backend store and making it safe to send to the blockchain for long-term keeping. The adapter part is also responsible for getting information from the blockchain and sending it smoothly to the mobile app that goes with it. In turn, this app is a user interface that shows the saved health data in a way that is clear and easy to understand.

Using adapter components in the BchainGIT architecture is encouraged by the adapter software architecture pattern. This pattern encourages the reuse of existing functions, which makes the system more efficient and modular as a whole. BchainGIT's design is made up of four main layers, which can be seen in Fig. 6.

- **Blockchain Layer:** Blockchain layer has all blockchain database setup and maintenance features. In the BchainGIT design, medical IoT devices store health data in blockchain. The health data owner must allow viewing.

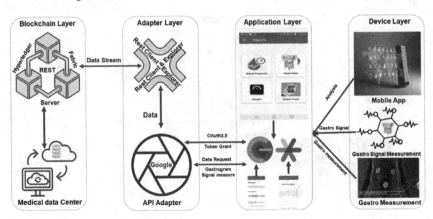

Fig. 6. BchainGIT layered architecture system.

Public and private blockchains exist. The main distinction is blockchain membership. A public blockchain, like bitcoin, lets anybody join, contribute, and observe its log, blocks, and data. In "Permissioned blockchains", network members can accomplish some things but not others. Users with permission can join private blockchain networks. Public blockchains are unsuitable for PHIMS since anybody can join but only approved entities can stay.

- **Adapter Layer:** Integration between the application and blockchain layers is handled via the adaptor layer. As a client of blockchain REST APIs and application layer services, the adapter layer connects these two components.

  The adaptor layer uses OAuth 2.0 to verify user consent and secure health data. After acquiring user health data, the adapter layer writes it securely to the blockchain via blockchain layer REST APIs. As part of the blockchain business network, the adaptor layer retrieves health data using REST APIs. This data can be sent to an application layer dashboard application so consumers can evaluate their consolidated health information.

- **Application Layer:** The application layer includes mobile apps that collect data from consumer health devices, health data sources, and apps that display blockchain-sourced consolidated health data. Health devices' mobile apps save health data in the application backend, commonly in a cloud database. User credentials are needed to access this info.

  The gastro measurement store stores data from gastro apps on user activity, exercise, food, and weight. The Google health software records user activity using the mobile device accelerometer and data from supported devices like Strava, Polar Balance, and Wear OS smartwatches in the Google store. These apps use OAuth 2.0 to authorize third-party apps to access user accounts with user consent and share device data.

- **Device Layer:** The device layer comprises medical IoT devices, including health watches, activity trackers, weighing scales, and smart pill bottles. These IoT-enabled devices connect to handheld mobile devices using short-range wireless technologies like Bluetooth. They serve as data sources, collecting person-specific or entity-specific data related to health. Typically, consumers operate these devices through corresponding mobile apps. Handheld mobiles can also function as health data devices if relevant applications are installed.

In the BChainGIT system, the patient holds the responsibility for updating personal health information and can access this information through direct authorization. The patient possesses the authority to verify other stakeholders, including clinicians, nurses, and caregivers, who have access to the patient's data, and vice versa. Utilizing blockchain technology in the Personal Health Information Management System (PHIMS), mutual verification is employed for accessing stored data.

As depicted in Fig. 6, the patient grants consent to the application layer for obtaining health data from the backend of IoT devices. Additionally, the patient provides data sharing consent to the doctor.

3.5 Blockchain Implementation Considerations

The code runs on the peer nodes within a blockchain network, and the ledger gets modified whenever the code is triggered. This code is written in the Go language and adheres to a predefined pattern designed to meet the needs of Hyperledger Fabric. In this part, our emphasis is solely on significant components, including the data structure, the Init () and Invoke () functions, and other functions that the Invoke () function will utilize. The data structures comprise three key elements. The Patient structure contains fundamental details regarding the patient.

Algorithm 1: Definition of Data Structures

Data: Patient, PatientRecord, DataShareRequest

1 data structure (*Patient*) field PatientName: string;
2 field Gender: string.
3 field Address: string;
4 field Phone: string;
5 data structure (*PatientRecord*) field UserId: string;
6 field RecordDate: string;
7 field PatientRecordData: string.
8 data structure (*DataShareRequest*) field From: string;
9 field to: string;
10 field EndDate: string;

The PatientRecord structure stores data pertaining to the patient's health record. The JSON format employed is adaptable and can handle a range of health data types. The DataShareRequest structure symbolizes the request for sharing health record information. The Init() function is activated when the code is first instantiated within the Hyperledger Fabric network. It serves as a means to establish numerous initial states. Typically, it is considered a best practice to keep this function empty, and instead, the initLedger () function is called from outside. The initLedger function loads two sets of mock patient data into the ledger using the PutState () API. This operation is performed just once. The Invoke () function specifies the actions carried out when the client application triggers the code. The arguments provided during code invocation include the name of the function to execute and its respective argument list.

Algorithm 2: Initialization of Smart Contract

Data: SmartContract, APIstub
Result: sc.Response
1   **function** *Init(APIstub: shim.ChaincodeStubInterface) : sc.Response*:
2   ⌊ **return** *shim.Success(nil)*

Algorithm 3: Initialization of Ledger in Smart Contract

Data: SmartContract, APIstub, patients, i
Result: sc.Response
1   **function** *initLedger(APIstub: shim.ChaincodeStubInterface) : sc.Response*:
2   patients ← [Patient{PatientName: "user1", Gender: "male", Address: "address1", Phone: "1234567890"}, Patient{PatientName: "user2", Gender: "female", Address: "address2", Phone: "1234567890"}];
3    $i \leftarrow 0$;
4   ⌊ **while** $i < len(patients)$ **do**
5 | patientsBytes, ← json.Marshal(patients[i]);
6 | APIstub.PutState(" PATIENT" + strconv.Itoa(i), patientBytes);
7   ⌊ $i \leftarrow i + 1$;
8   **return** *shim.Success(nil)*;

The functions to be executed include: initLedger (), createPatient (), createPatientRecord (), queryRecord (), queryRecordByDate(), createDataShareRequest(), and queryDataShareByTo. These functions are briefly explained in this section. The createPatient () function is responsible for adding a patient to the ledger, with the necessary data being provided as part of the Patient structure. The createPatientRecord () function is used to create a patient's health data record in the ledger, with the required data supplied through the PatientRecord structure. The queryRecord () function enables the querying of an individual patient's record based on their UserId. The queryRecordByDate () function allows querying an individual patient's record based on their UserId and a specific date. The createDataShareRequest () function generates a request from a patient to a doctor, requesting access to the patient's record. This request includes an expiration date. The queryDataShareByTo () function shares patient data with a doctor according to received specifications [14].

Algorithm 4: Invoke Smart Contract Method

Data: SmartContract, APIstub, functionn, args
Result: sc.Response
1 **function** *Invoke (APIstub: shim.ChaincodeStubInterface): sc.Response*:
2   functionn, args ← APIstub.GetFunctionAndParameters();
3   logger.Infof ("' Function name is: logger.Infof("' Args length is:
4   **switch** *functionn* **do**
5   **case** *"initLedger"* **do**
6   **return** *s.initLedger(APIstub)*;
7   **case** *"createPatient"* **do**
8   **return** *s.createPatient(APIstub, args)*;
9   **case** *"createPatientRecord"* **do**
10   **return** *s.createPatientRecord(APIstub, args)*;
11   **case** *"queryRecord"* **do**
12   **return** *s.queryRecord(APIstub, args)*;
13   **case** *"queryRecordByDate"* **do**
14   **return** *s.queryRecordByDate(APIstub, args)*;
15   **case** *"createDataShareRequest"* **do**
16   **return** *s.createDataShareRequest(APIstub, args)*;
17   **case** *"queryDataShareByTo"* **do**
18   **return** *s.queryDataShareByTo(APIstub, args)*;
19   **otherwise do**
20   **return** *shim.Error("Invalid Smart Contract function name.")*;

---

4 Electrogastrography in Presence of Diseases

In below, Fig. 7 demonstrates a sample disease in the stomach. Now, we will explore the role of electrogastrography (EGG) in the presence of various diseases, shedding light on the connections between gastric motility disorders, EGG abnormalities, and specific medical conditions. Understanding these relationships is crucial for both diagnosis and treatment strategies.

4.1 Abnormalities in Gastric Electrical Activity

It is well-established that numerous clinical conditions associated with delayed gastric emptying often coincide with abnormalities in EGG frequency and amplitude responses following a meal. Two common abnormalities observed in patients with diabetic gastroparesis are tachygastria (excessively fast gastric rhythms) and bradygastria (abnormally slow gastric rhythms). Additionally, some individuals with diabetes experience a simultaneous decrease in amplitude response after meal ingestion. Elevated blood sugar levels, or hyperglycemia, have also been linked to increased dysrhythmic activity in the stomach. These observations underscore the close relationship between gastric motility disorders, EGG abnormalities, and the presence of diabetes and its associated metabolic changes [15]. Gastric dysrhythmias are not exclusive to diabetic patients; they are also frequently observed in patients with conditions such as gastroesophageal reflux disease (GERD) and delayed gastric emptying. Additionally, disorders like motion sickness, gastroparesis resulting from ischemia, chronic renal failure, and specific paraneoplastic

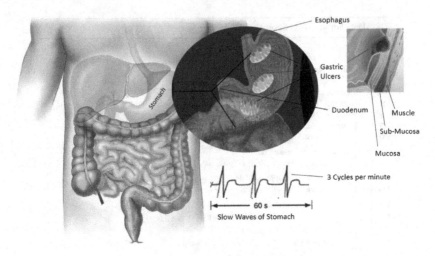

Fig. 7. An ulcer in stomach

syndromes have been associated with gastric dysrhythmias. This diverse range of medical conditions underscores the significance of understanding and addressing disruptions in gastric electrical activity across various clinical contexts [16, 17].

4.2 EGG in Functional Gastrointestinal Disorders

Apart from the well-established clinical conditions mentioned earlier, a significant proportion of patients experiencing unexplained nausea and vomiting also exhibit disturbances in slow wave activity. Studies have shown that gastric dysrhythmias are present in 69% of patients diagnosed with functional dyspepsia [18, 19]. Among individuals with dyspepsia, EGG rhythm disturbances are found in 75% of those with delayed gastric emptying, compared to only 25% of dyspeptic individuals with normal emptying [19]. In cases of idiopathic gastroparesis, where delays in gastric emptying are observed, patients may also display EGG dysrhythmias, suggesting a potential role for these primary disruptions in slow wave rhythm in the pathogenesis of these conditions. Furthermore, investigations have identified patients with unexplained nausea and vomiting who do not exhibit abnormalities in gastric emptying but do display EGG dysrhythmias or abnormal amplitude responses to meal ingestion. These findings emphasize the relevance of gastric dysrhythmias in understanding and managing symptoms associated with gastrointestinal disorders characterized by nausea and vomiting [7, 20].

4.3 Clinical Application and Diagnosis

Several vendors offer EGG equipment for clinical use, including 3CPM Company, Sandhill Scientific, Inc., Medtronic/Synectics, RedTech, and MMS. During EGG recording, it is imperative to create a quiet environment to minimize interference from extraneous electrical signals that could be detected by cutaneous electrodes. Additionally, any distractions that may induce patient movement should be minimized. Patients can be positioned in various conformations, from a supine position to a 45-degree inclination, as long as their comfort and stability can be maintained throughout the testing period. Changes in position can impact EGG parameters, particularly signal amplitude, so it is crucial for patients to remain still during recording. Data from functional magnetic resonance imaging (fMRI) demonstrate that the resting brain is synchronized with the gastric rhythm through an extensive cortical network. This network encompasses well-known visceral-sensitive areas, including the primary and secondary somatosensory cortices, along with the parieto-occipital region where the coupling with gastric alpha waves was detected [21]. Before the recording, patients are advised to use the restroom to minimize interruptions during the test and eliminate potential motion artifacts that could arise due to urinary or fecal urgency. Cutaneous electrogastrography, a technique used to record and analyze gastric electrical activity, plays a pivotal role in the detection of gastric arrhythmias. A study conducted to determine the specificity of electrogastrography, as well as to assess the prevalence and patterns of abnormalities in individuals diagnosed with functional dyspepsia and irritable bowel syndrome, aimed to provide insights into the diagnostic value and frequency of electrogastrography abnormalities in these particular conditions [22].

4.4 EGG as an Assessment Tool

EGG is capable of detecting gastric arrhythmias, including tachygastrias (fast frequency waves) and bradygastrias (slow frequency waves). Studies using serosal transducers and manometry have revealed a correlation between tachygastrias and the absence of antral contractions. Bradygastrias have been associated with both strong and absent antral contractions. Research has also explored the relationship between EGG and disorders affecting gastric emptying. Notably, studies have proposed that a significant postprandial arrhythmia detected by EGG can predict gastroparesis with an accuracy of 78%, underlining the potential usefulness of EGG in evaluating gastric motility disorders and their associated symptoms [23]. Several studies, although conducted with small sample sizes, have suggested a correlation between symptom improvement with drug therapy and the resolution of EGG abnormalities in patients with dyspepsia. For instance, a study involving diabetic patients with gastroparesis found that reductions in nausea symptoms after six months of treatment coincided with the correction of tachygastrias and bradygastrias [24]. This body of evidence underscores the valuable role of EGG in diagnosing, monitoring, and understanding the complexities of gastrointestinal disorders, providing clinicians with a non-invasive tool to assess gastric electrical activity and its implications for patient care. Figure 3 illustrates pathway of GIT signal from stomach to mobile phone. While Fig. 4 shows the sample EGG signal on a mobile phone. Figure 5 displays the filtered gastrogram and enterogram signals.

5   BchainGIT Test Results

Electrogastrography serves as one of the valuable tests for assessing gastrointestinal function, particularly in patients with unexplained symptoms such as nausea, vomiting, and dyspeptic complaints. As previously mentioned, significant abnormalities in EGG patterns are often observed in these patient populations, whereas healthy individuals typically exhibit undisturbed EGG rhythms and power. These findings suggest that EGG abnormalities can be a valuable indica- tor of treatment response and symptom resolution in patients with gastrointestinal disorders. Monitoring EGG patterns alongside symptom assessment holds the potential to facilitate the evaluation of treatment efficacy and offer insights into the underlying mechanisms of these conditions. Recent advancements in measurement methods capable of capturing slow wave activity in high resolution have renewed interest in gastric electrophysiology. These developments show promise in yielding clinically useful diagnostic and therapeutic biomarkers. For further reference, Table 1 provides an overview of the frequency ranges associated with different parts of the body, aiding in the understanding of the diverse applications of EGG technology as shown in Fig. 8.

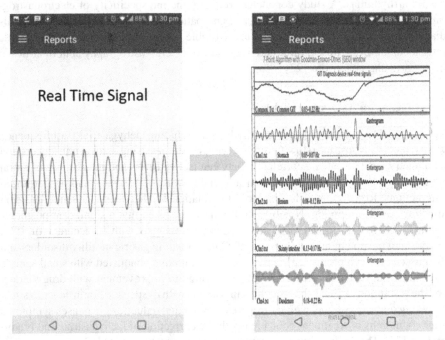

Fig. 8. Sample EGG signal with filtered Gastrogram and enterograms on a smartphone of a healthy patient

5.1 BchainGIT Client Application

In Hyperledger Fabric, the external world interacts with the fabric network and chaincode using a client application. This interaction is facilitated through Hyperledger SDK using languages Java and Node. Being a permissioned blockchain platform, every participant including a client application must be authorized to interact with the fabric network. A Certificate Authority in the BchainGIT Chain network provides appropriate certificate for a client application to participate. The client application consists of the logic to interact with the fabric network and the chaincode. The client application interacts with the peer nodes for record approval and with the orderer nodes for block generation respectively. For this purpose, the access point of peers and orderers, as well as the channel name and chaincode name, are to be specified by the client application. Similarly, the client application shouldsupply the correct function name and required arguments while performing query or invoking the chaincode. The BchainGIT client application is developed onAndroid platform using Java language. As shown in the Fig. 8, the client application supports various functionality such as registration of doctor and patients, performing OAuth2.0 flow with the health-IoT device backends and showing health record on the mobile device. The client application also allows a patient to authorize viewing of health records to a doctor based on patient consent (Fig. 9).

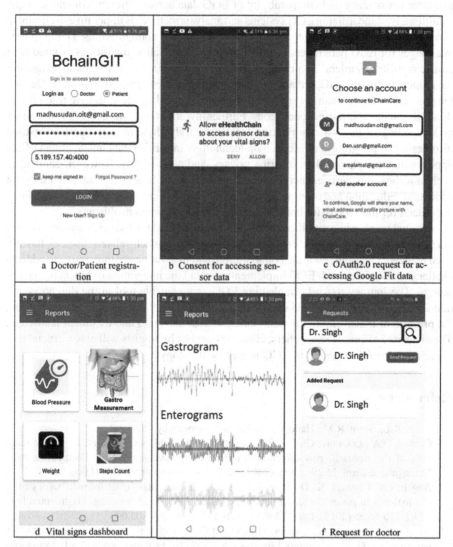

Fig. 9. BchainGIT client application screenshots. (**a**) Doctor/patient registration. (**b**) Consent for accessing sensor data. (**c**) OAuth2.0 request for accessing Google health data. (**d**) Vital signs dashboard. (**e**) Gastrogram and Enterogram overview. (**f**) Request for doctor to review the signals.

6 Final Discussion and Further Challenges

Electrogastrography (EGG) has emerged as a promising diagnostic tool for assessing stomach motor dysfunction in various gastrointestinal disorders, including gastroparesis and functional dyspepsia. However, several challenges must be addressed to establish EGG as a reliable and widely accepted diagnostic solution. One critical aspect is the standardization of EGG implementation, involving optimal electrode placement, recording duration, and selection of test meals. Standardized protocols in these areas are essential to ensure consistency and comparability of EGG data across different clinical settings. Additionally, standardizing EGG systems and analysis methods is imperative to eliminate variations introduced by different hardware and software configurations. Defining normal ranges for EGG parameters is crucial for identifying abnormalities associated with gastric motility disorders. Achieving consensus on these factors is pivotal in advancing the clinical utility of EGG, enhancing the reliability and comparability of data. Despite the promise of EGG, unanswered questions persist regarding the causative role of dysrhythmias in gastrointestinal conditions. Further investigation is needed to gain a deeper understanding of the intricate relationship between dysrhythmias and gastrointestinal disorders. The integration of blockchain technology into EGG data management offers opportunities to enhance data security, privacy, and accessibility. However, challenges such as technical expertise, energy consumption, scalability, and user adoption must be addressed. Education and training for users, including patients, doctors, and researchers, are crucial for the successful implementation of blockchain in healthcare.

In conclusion, while significant progress has been made in EGG technology, its full potential as a comprehensive diagnostic solution for stomach-related diseases has not been realized. Ongoing research, collaboration, and innovation are imperative to address challenges associated with EGG implementation and the integration of blockchain technology. The journey toward establishing EGG as a valuable tool in the diagnosis and management of gastrointestinal conditions is ongoing, with future innovations holding the promise of transforming the gastroenterological diagnosis and treatment landscape. The concerted efforts of researchers, clinicians, and technologists will play a crucial role in realizing the full potential of EGG in gastroenterology.

References

1. Koch, K.L., Stern, R.M.: Handbook of Electrogastrography (2003)
2. Carson, D.A., O'Grady, G., Du, P., Gharibans, A.A., Andrews, C.N.: Body surface mapping of the stomach: new directions for clinically evaluating gastric electrical activity. Neurogastroenterol. Motil. 33(3), e14048 (2021). https://doi.org/10.1111/nmo.14048
3. Angeli, T.R., Cheng, L.K., Du, P., et al.: Loss of interstitial cells of cajal and patterns of gastric dysrhythmia in patients with chronic unexplained nausea and vomiting. Gastroenterology 149(1), 56-66.e5 (2015). https://doi.org/10.1053/j.gastro.2015.04.003
4. Grover, M., Farrugia, G., Lurken, M.S., et al.: Cellular changes in diabetic and idiopathic gastroparesis. Gastroenterology 140(5), 1575–85.e8 (2011). https://doi.org/10.1053/j.gastro.2011.01.046

5. O'Grady, G., Angeli, T.R., Du, P., et al.: Abnormal initiation and conduction of slow-wave activity in gastroparesis, defined by high-resolution electrical mapping. Gastroenterology **143**(3), 589-598.e3 (2012). https://doi.org/10.1053/j.gastro.2012.05.036

6. O'Grady, G., Wang, T.H., Du, P., Angeli, T., Lammers, W.J., Cheng, L.K.: Recent progress in gastric arrhythmia: pathophysiology, clinical significance and future horizons. Clin. Exp. Pharmacol. Physiol. **41**(10), 854–862 (2014). https://doi.org/10.1111/1440-1681.12288

7. Geldof, H., van der Schee, E.J., van Blankenstein, M., Grashuis, J.L.: Electrogastrographic study of gastric myoelectrical activity in patients with unexplained nausea and vomiting. Gut **27**(7), 799–808 (1986). https://doi.org/10.1136/gut.27.7.799

8. Chen, J.D., McCallum, R.W.: Clinical applications of electrogastrography. Am. J. Gastroenterol. **88**(9), 1324–1336 (1993)

9. Lin, X., Chen, J.Z.: Abnormal gastric slow waves in patients with functional dyspep- sia assessed by multichannel electrogastrography. Am. J. Physiol. Gastrointest. Liver Physiol. **280**(6), G1370–G1375 (2001). https://doi.org/10.1152/ajpgi.2001.280.6.G1370

10. Parkman, H.P., Hasler, W.L., Barnett, J.L., Eaker, E.Y.: American motility society clinical GI motility testing task force. Electrogastrography: a document prepared by the gastric section of the american motility society clinical GI motility testing task force. Neurogastroenterol. Motil. **15**(2), 89–102 (2003). https://doi.org/10.1046/j.1365-2982.2003.00396.x

11. Smout, A.J.P.M., Van Der Schee, E.J., Grashuis, J.L.: What is measured in electrogastrography? Digest. Dis. Sci. **25**, 179–187 (1980). https://doi.org/10.1007/BF01308136

12. Yin, J., Chen, J.D.: Electrogastrography: methodology, validation and applications. J. Neurogastroenterol. Motil. **19**(1), 5–17 (2013). https://doi.org/10.5056/jnm.2013.19.1.5

13. Zaynidinov, H., Makhmudjanov, S., Rajabov, F., Singh, D.: IoT-enabled mobile device for electrogastrography signal processing. In: Singh, M., Kang, D.K., Lee, J.H., Tiwary, U.S., Singh, D., Chung, W.Y. (eds.) IHCI 2020. LNCS, vol. 12616, pp. 346–356. Springer, Cham (2021). https://doi.org/10.1007/978-3-030-68452-5_36

14. Pawar, P., Parolia, N., Shinde, S., Edoh, T.O., Singh, M.: EHealthChain-a blockchain-based personal health information management system. Ann. Telecommun. **77**(1–2), 33–45 (2022). https://doi.org/10.1007/s12243-021-00868-6

15. Jebbink, R.J., Samsom, M., Bruijs, P.P., et al.: Hyperglycemia induces abnormalities of gastric myoelectrical activity in patients with type I diabetes mellitus. Gastroenterology **107**(5), 1390–1397 (1994). https://doi.org/10.1016/0016-5085(94)90541-x

16. Liberski, S.M., Koch, K.L., Atnip, R.G., Stern, R.M.: Ischemic gastroparesis: resolution after revascularization. Gastroenterology **99**(1), 252–257 (1990). https://doi.org/10.1016/0016-5085(90)91255-5

17. Lin, X., Mellow, M.H., Southmayd, L., 3rd., Pan, J., Chen, J.D.: Impaired gastric myoelectrical activity in patients with chronic renal failure. Dig. Dis. Sci. **42**(5), 898–906 (1997). https://doi.org/10.1023/a:1018856112765

18. Koch, K.L., Medina, M., Bingaman, S., Stern, R.M.: Gastric dysrhythmias and visceral sensations in patients with functional dyspepsia (abstract). Gastroenterology **102**, A469 (1992)

19. Parkman, H.P., Fisher, R.S.: Disorders of gastric emptying. In: Yamada T (ed.) Yamada Textbook of Gastroenterology 4th edn., Lippincott, Williams & Wilkins, Philadelphia (2003)

20. Camilleri, M., Hasler, W.L., Parkman, H.P., Quigley, E.M., Soffer, E.: Measurement of gastrointestinal motility in the GI laboratory. Gastroenterology **115**(3), 747–762 (1998). https://doi.org/10.1016/s0016-5085(98)70155-6

21. Wolpert, N., Rebollo, I., Tallon-Baudry, C.: Electrogastrography for psychophysiological research: practical considerations, analysis pipeline, and normative data in a large sample. Psychophysiology **57**(9), e13599 (2020). https://doi.org/10.1111/psyp.13599

22. Leahy, A., Besherdas, K., Clayman, C., Mason, I., Epstein, O.: Abnormalities of the electrogastrogram in functional gastrointestinal disorders. Am. J. Gastroenterol. **94**(4), 1023–1028 (1999). https://doi.org/10.1111/j.1572-0241.1999.01007.x

23. Hasler, W.L., Soudah, H.C., Dulai, G., Owyang, C.: Mediation of hyperglycemia-evoked gastric slow-wave dysrhythmias by endogenous prostaglandins. Gastroenterology **108**(3), 727–736 (1995). https://doi.org/10.1016/0016-5085(95)90445-x

24. Koch, K.L., Stern, R.M., Stewart, W.R., Vasey, M.W.: Gastric emptying and gastric myo-electrical activity in patients with diabetic gastroparesis: effect of long-term dom- peridone treatment. Am. J. Gastroenterol. **4**(9), 1069–1075 (1989)

Author Index

B. J. Choi et al. (Eds.): IHCI 2023, LNCS 14531, pp. 435–437, 2024.
https://doi.org/10.1007/978-3-031-53827-8

Printed in the United States
by Baker & Taylor Publisher Services